Lecture Notes in Mathematics

Edited by A. Dold and B. Eckmann

726

T0211532

Yau-Chuen Wong

Schwartz Spaces, Nuclear Spaces and Tensor Products

Springer-Verlag
Berlin Heidelberg New York 1979

Author

Yau-Chuen Wong
Department of Mathematics
United College
The Chinese University of Hong Kong
Shatin, N.T./Hong Kong

AMS Subject Classifications (1970): 46A05, 46A15, 46A45, 46M05, 47B10, 47D15

ISBN 3-540-09513-6 Springer-Verlag Berlin Heidelberg New York
ISBN 0-387-09513-6 Springer-Verlag New York Heidelberg Berlin

Library of Congress Cataloging in Publication Data
Wong, Yau-chuen. Schwartz spaces, nuclear spaces, and tensor products. (Lecture notes in mathematics; 726) Bibliography: p. Includes index. 1. Schwartz spaces. 2. Nuclear spaces (Functional analysis) 3. Tensor products. I. Title. II. Series: Lecture notes in mathematics (Berlin); 726.
QA3.L28 no. 726 [QA322] 510'.8s [515'.73] 79-16330

Printing and binding: Beltz Offsetdruck, Hemsbach/Bergstr.
2141/3140-543210

TO MY WIFE

Judy, Dick-Ha Nei

INTRODUCTION

Nuclear spaces were discovered by A. Grothendieck, and the most important part of the theory of nuclear spaces was developed in his famous article (see Grothendieck [2]). Unfortunately, the machinery of tensor products used there is very cumbersome; this makes the theory some unnecessary difficulties and complicacy. Pietsch [1] was the first to simplify the theory of nuclear spaces by using locally convex spaces of summable and absolutely summable families, instead of locally convex tensor products. It is my impression that the idea of using seminorms satisfying some expected properties to investigate some classes of locally convex spaces, would make the development of the whole theory easier. This has been done by the author [1] for studying the topology σ_S of uniform convergence on order-bounded sets. Thus, one of the purposes of these lecture notes is to embark upon a unifying treatment of Schwartz spaces, nuclear spaces, and λ-nuclear spaces by using this idea. Therefore, this monograph is actually a continuation of that written by the author [1].

These lecture notes contain six chapters, starting with Chapter 0 in which terminology and some elementary results in topological vector spaces, that will be used through these notes, are given. Chapter 1 deals with Schwartz spaces by means of precompactness of seminorms. Chapter 3 is concerned with nuclear spaces in terms of prenuclearity of seminorms; in order to show that our definition of nuclearity coincides with the usual one, we use the concept of summability which is studied in the first two sections of Chapter 2.

In Chapter 4, the s-norm on a tensor product of normed spaces, studied by Cohen [1], is constructed in a natural way by asking when its Banach dual is isometrically isomorphic to the well-known Banach space of all absolutely summing mappings equipped with the absolutely summing norm. Using the same idea we construct the ℓ-norm on a tensor product of ordered normed spaces from the well-known Banach space of all cone-absolutely summing mappings equipped with a suitable norm. The ℓ-norm has been studied by Schaefer [2] for the tensor product of Banach lattices. Also in Chapter 5, we study two locally solid topologies on the tensor product, equipped with the projective and biprojective cone, of two locally solid spaces.

A few words should be said about the reference system in these notes, (i, j, k) will denote the k-th proposition or theorem in Chapter i, Section j.

Parts of these notes were delivered at the University of Western Australia, during the period from November to December, 1976. The author would like to thank both the University of Western Australia and The Chinese University of Hong Kong for financial and moral support during his stay in Perth.

I express my sincere thanks to my colleagues and friends in two Universities who give me many encouragements, in particular, to Dr K.F. Ng and Dr P.K. Tam for many helpful commends. Finally, I would like to thank Mr. Billy P.M. Lam for his truly outstanding job of typing the manuscript.

Y.C.W.

CONTENTS

CHAPTER 0. GENERAL NOTATIONS

The purpose of this preliminary chapter is to clarify terminology and notation in topological vector spaces, which we shall need in what follows. The following notation is used: $\mathbb{K}$ will denote either the field $\mathbb{R}$ of real numbers or the field $\mathbb{C}$ of complex numbers, $\mathbb{N}$ will denote the set of all natural numbers, and Λ will be a non-empty indexed set.

Topological vector spaces considered in these notes are assumed to be Hausdorff, unless a statement is made to the contrary. We follow without reference the most usual terminology and results of Köthe [1] and Schaefer [1]; the background material concerning summability of families and absolutely summing mappings can be found in Pietsch [1].

Throughout these notes, $(E, \mathcal{P})$ and $(F, \mathcal{J})$ will denote locally convex spaces over the same field, $\mathcal{U}_E$ and $\mathcal{U}_F$ will denote neighbourhood bases at 0 for $\mathcal{P}$ and $\mathcal{J}$ respectively which consist of closed absolutely convex and absorbing sets, the gauge of any $V \in \mathcal{U}_E$ will be denoted by p_V, and the gauge of any $U \in \mathcal{U}_F$ will be denoted by q_U. We also denote by $L^*(E, F)$ the vector space of all linear maps from E into F, and by $L(E, F)$ the vector subspace of $L^*(E, F)$ consisting of continuous linear maps. In particular, we write $L(E)$ for $L(E, E)$, E^* for $L^*(E, \mathbb{K})$ (the algebraic dual of E), and E' for $L(E, \mathbb{K})$ (the topological dual of $(E, \mathcal{P})$). The composition of two maps T and S will be denoted by either TS or $T \circ S$. If $\mathfrak{S}$ is a family of bounded subsets of E which covers E, then we denote by $L_{\mathfrak{S}}(E, F)$ the space $L(E, F)$ equipped with the topology of $\mathfrak{S}$-convergence; when $\mathfrak{S}$ is respectively the family of all finite subsets of E, of all $\sigma(E, E')$-compact absolutely convex subsets of E, and of all bounded subsets of E, we take the notation $L_\sigma(E, F)$, $L_\tau(E, F)$ and $L_\beta(E, F)$.

For any $T \in L^*(E, F)$, we denote by $Q_T : E \to E/_{\text{Ker } T}$ the quotient map, and by $j_T : T(E) \to F$ the canonical embedding map. Then there is a unique bijective linear map $\check{T}$ from $E/_{\text{Ker } T}$ onto $T(E)$ such that

$$T = j_T \check{T} Q_T .$$

$\check{T}$ is referred to as the bijection associated with T, and the injective linear map $j_T \check{T}$ from $E/_{\text{Ker } T}$ into F is called the injection associated with T, and denoted by $\hat{T}$. Clearly, T is continuous if and only if $\check{T}$ is continuous, and this is the case if and only if $\hat{T}$ is continuous.

The weak topology on E (or E') with respect to the dual pair $\langle E, E' \rangle$ is denoted by $\sigma(E, E')$ (resp. $\sigma(E', E)$), the Mackey topology on E (or E') with respect to $\langle E, E' \rangle$ is denoted by $\tau(E, E')$ (resp. $\tau(E', E)$), and the strong topology on E (or E') with respect to $\langle E, E' \rangle$ is denoted by $\beta(E, E')$ (resp. $\beta(E', E)$). E_σ, E_τ and E_β will denote respectively E equipped with $\sigma(E, E')$, $\tau(E, E')$ and $\beta(E, E')$. Dually, E'_σ, E'_τ and E'_β will denote respectively E' equipped with $\sigma(E', E)$, $\tau(E', E)$ and $\beta(E', E)$. More generally, if $\mathfrak{S}$ is any family of bounded subsets of E which covers E, then $E'_\mathfrak{S}$ (or $(E', \mathcal{L}(\mathfrak{S}))$) will denote E' equipped with the topology of $\mathfrak{S}$-convergence. When E' is considered as a locally convex space, it will be assumed to be equipped with the strong topology $\beta(E', E)$, unless the contrary is stated.

Denote by $L(E_\sigma, F_\tau)$ the vector space of all continuous linear maps from E_σ into F_τ ; and similarly we might define the space $L(E, F_\sigma)$ and $L(E'_\tau, F)$, etc.. As the continuity depends on topologies, it can be shown, for instance, that

$$L(E_\sigma, F_\tau) \subset L(E, F_\tau) \subset L(E_\tau, F_\tau) \text{ , and}$$

$$L(E_\sigma, F_\tau) \subset L(E_\sigma, F) \subset L(E, F) \subset L(E_\sigma, F_\sigma) \text{ ;}$$

here the inclusions may be strict. Moreover, we have

$$L(E_\sigma, F_\sigma) = L(E, F_\sigma) = L(E_\tau, F_\sigma) = L(E_\tau, F) = L(E_\tau, F_\tau) \text{ , and}$$

$$L(E'_\sigma, F_\sigma) = L(E'_\tau, F) = L(E'_\tau, F_\tau) \subset L(E'_\beta, F) \text{ .}$$

For any $V \in \mathcal{U}_E$, let $N(V)$ be the null space, i.e., $N(V) = p_V^{-1}(0)$, let $E_V = E/_{N(V)}$, let $Q_V : E \to E_V$ be the quotient map, and let $x(V)$ (or $\hat{x}$) the equivalent class $x + N(V)$ modulo $N(V)$. Then the quotient seminorm of p_V , denoted by $\hat{p}_V$, is actually a norm on E_V such that

$$\hat{p}_V(x(V)) = \hat{p}_V(Q_V(x)) = p_V(x) \quad \text{for all} \quad x \in E \text{ .}$$

For simplicity of notation, we write E_V for the normed vector space $(E_V, \hat{p}_V)$. It should be noted that E_V is, in general, not complete, even if E is; its completion will be denoted by $\tilde{E}_V$. If p is a continuous seminorm on E and if $V = \{x \in E : p(x) \leq 1\}$, then we let $E_p = E_V$ and $Q_p = Q_V$. If V and W , in $\mathcal{U}_E$, are such that $V \subseteq W$, then the canonical map from E_V onto E_W , denoted by $Q_{W,V}$, is clearly continuous, hence it has a unique continuous extension on $\tilde{E}_V$ into $\tilde{E}_W$, which is again said to be canonical, and also denoted by $Q_{W,V}$. Clearly, Q_W is the composition of $Q_{W,V}$ and

Q_V , that is,

$$Q_W = Q_{W,V} Q_V .$$

Dually, if A is an absolutely convex bounded subset of E , then we denote by $E(A)$ the normed vector space obtained by furnishing the vector subspace $\bigcup_{n\geq 1} nA$ of E generated by A equipped with the norm r_A induced by A . If A is closed, then the unit ball of $E(A)$ is identical with A , and if A is complete in E , then $E(A)$ is a Banach space. Clearly the relative topology on $E(A)$ induced by $\mathscr{P}$ is coarser than the norm-topology, thus the (natural) embedding map from $E(A)$ into $(E, \mathscr{P})$, denoted by j_A , is a continuous linear map. Likewise, if A and C are absolutely convex bounded subsets of E and if $A \subsetneq C$, then $E(A) \subseteq E(C)$ and $r_C \leq r_A$ on $E(A)$, thus the canonical embedding map from $E(A)$ into $E(C)$, denoted by $j_{C,A}$, is a continuous linear map. Clearly, j_A is the composition of j_C and $j_{C,A}$, that is

$$j_A = j_C \, j_{C,A} .$$

By an <u>infracomplete subset</u> of E we mean an absolutely convex subset A of E such that the normed space $E(A)$ is complete. It can be shown without difficulty that every absolutely convex bounded subset of E which is sequentially complete in itself (in particular, $\sigma(E, E')$-compact) is infracomplete, hence if $V \in \mathcal{U}_E$, then $E'(V^o)$ is a Banach space, where V^o is the polar of V taken in E' .

For any $V \in \mathcal{U}_E$, the adjoint map Q'_V of Q_V is an isometric isomorphism from the Banach dual $(E_V)'$ of E_V <u>onto</u> the Banach space $E'(V^o)$. If A is an absolutely convex bounded subset of E, then the injection $\hat{j}'_A$ associated with the adjoint map j'_A of j_A is an isometric isomorphism from the normed space $(E')_{A^o}$ <u>into</u> the Banach dual $E(A)'$ of $E(A)$.

An $T \in L^*(E, F)$ is said to be <u>bounded</u> (resp. <u>precompact</u>, <u>compact</u> and <u>weakly compact</u>) if it sends some o-neighbourhood in E into a subset of F which is bounded (resp. $\mathcal{T}$-precompact, $\mathcal{T}$-compact and $\sigma(F, F')$-compact).

The set consisting of all bounded (resp. precompact, compact) linear maps, denoted by $L^{\ell b}(E, F)$ (resp. $L^p(E, F)$ and $L^c(E, F)$), is a vector subspace of $L(E, F)$; therefore $L^c(E, F_\sigma)$ is the space of all weakly compact linear maps from E into $F_\sigma = (F, \sigma(F, F'))$. There are continuous linear maps which are not bounded; for instance, the identity map on $(c_o, \sigma(c_o, \ell'))$ is continuous but not bounded, where c_o is the Banach space of all null-sequences of numbers equipped with the sup-norm and ℓ^1 is the Banach space of all summable sequences of numbers equipped with the usual ℓ^1-norm. But we have

$$L^{\ell b}(E, F) = L(E, F) \quad \text{whenever } F \text{ is } \underline{\text{normable}}.$$

As bounded subsets of F are $\sigma(F, F')$-precompact, it follows that

$$L^{\ell b}(E, F) = L^p(E, F'_\sigma) .$$

Let $V \in \mathcal{U}_E$ and let B be an absolutely convex bounded subset of F. Then the set, defined by

$$P(V, B) = \{T \in L^{\ell b}(E, F) : T(V) \subseteq B\} ,$$

is clearly an absolutely convex subset of $L^{\ell b}(E, F)$ which is bounded for the topology of simple convergence, hence the gauge of $P(V, B)$, denoted by $r_{(V,B)}$, is a norm defined on the vector subspace $\bigcup_n n P(V, B)$ of $L^{\ell b}(E, F)$. Denoting this normed space by $L^{\ell b}(P(V, B))$ (or $(L^{\ell b}(P(V, B)), r_{(V,B)})$ when there is need to specify the norm $r_{(V,B)}$). Clearly the map

$$T_V \mapsto j_B T_V Q_V \quad \text{for all} \quad T_V \in L(E_V, F(B))$$

is an algebraic isomorphism from $L(E_V, F(B))$ onto $L^{\ell b}(P(V, B))$. Furthermore, if B is closed, then this map is also an isometry. Therefore, for any $T \in L^{\ell b}(P(V, B))$ there is a unique $T_{(V,B)} \in L(E_V, F(B))$, called the <u>induced map</u> of T, such that

$$T = j_B T_{(V,B)} Q_V .$$

Denote by $B^*(E, F)$ the vector space consisting of all bilinear forms on $E \times F$, by $S(E, F)$ the vector subspace of $B^*(E, F)$ of all separately continuous bilinear forms, and by $B(E, F)$ the vector subspace of $S(E, F)$ of all continuous bilinear forms. For each $T \in L^*(E, F^*)$, the map b_T, defined by

$$b_T(x, y) = \langle y, Tx\rangle \quad \text{for all} \quad (x, y) \in E \times F ,$$

is clearly a bilinear form on $E \times F$ which is called the <u>associated bilinear form</u> of T. It is not hard to show that the map $T \mapsto b_T$ is an algebraic isomorphism from $L^*(E, F^*)$ onto $B^*(E, F)$ (see Robertson-Robertson [1]), and that this map also carries an algebraic isomorphism from $L(E, F'_\sigma)$ onto $S(E, F)$. As $B(E, F) \subset S(E, F)$, we are going to seeking some vector subspace of $L(E, F'_\sigma)$ which is algebraically isomorphic to $B(E, F)$. To this end, let us define

$$L^{eq}(E, F(\mathcal{J})') = \{T \in L^*(E, F') : \exists V \in \mathcal{U}_E \text{ and } U \in \mathcal{U}_F \text{ s.t. } T(V) \subset U^\circ\} .$$

By using the bipolar theorem it is easy to show that $L^{eq}(E, F(\mathcal{J})')$ is a vector subspace of $L^*(E, F')$. On the other hand, it is also easy to check that the map $T \mapsto b_T$ is also an algebraic isomorphism from $L^{eq}(E, F(\mathcal{J})')$ onto $B(E, F)$, hence we identify $L^{eq}(E, F(\mathcal{J})')$ with $B(E, F)$.

Clearly, $L^{eq}(E, F(\mathcal{J})')$ depends on the topologies on E and F . Since $\sigma(F', F)$-compact convex subsets of F' are $\beta(F', F)$-bounded, it follows that

$$L^{eq}(E, F(\mathcal{J})') \subset L^c(E, F'_\sigma) \subset L^{\ell b}(E, F'_\beta) \subset L(E, F'_\beta) .$$

Moreover, we have

$$L^{eq}(E, F(\mathcal{J})') = L^c(E, F'_\sigma) \quad \text{whenever} \quad \mathcal{J} = \tau(F, F') ,$$

and

$$L^{eq}(E, F(\mathcal{J})') = L(E, F') \quad \text{whenever } F \text{ is a normed space.}$$

CHAPTER 1. SCHWARTZ SPACES

1.1 Precompact linear mappings

It is easily seen that the composition of two continuous linear maps, in which one of them is precompact (resp. compact), is precompact (resp. compact). On the other hand, as every totally bounded set in a semimetric space is separable, and the unit ball in a normed space is absorbing, it follows that every precompact (in particular, compact) linear map from one normed space into another has a separable range.

For normed spaces X and Y, precompact linear maps from X into Y can be characterized in terms of the compactness of their adjoint maps as shown by the following important result.

(1.1.1) Theorem (Schauder). Let X, Y be normed spaces and $T \in L(X, Y)$. Then $T \in L^p(X, Y)$ if and only if $T' \in L^c(Y', X')$.

Proof. Assume that $T \in L^p(X, Y)$. Then T' is continuous for the topology $p(Y', Y)$ on Y' of precompact convergence and the norm topology on X'. On the closed unit ball Σ^* in Y', the topologies $\sigma(Y', Y)$ and $p(Y', Y)$ coincide, it then follows that $T'(\Sigma^*)$ is compact in Y', and hence that $T' \in L^c(Y', X')$.

Conversely, we assume that $T' \in L^c(Y', X')$. Then $T'' \in L^c(X'', Y'')$

by the necessity of this theorem. Let $e_X : X \to X''$ and $e_Y : Y \to Y''$ be the evaluation maps. Then e_X and e_Y are isometries (into) and $e_Y T = T'' e_X$, thus $T \in L^p(X, Y)$ since compact linear maps are precompact.

Precompact linear maps whose range is a metrizable locally convex space can be factorized through normed spaces as shown by the following

(1.1.2) Lemma. *Let F be metrizable. Then every $T \in L^p(E, F)$ is the composition of the following continuous linear maps:*

$$E \xrightarrow{\;Q\;} X \xrightarrow{\;\tilde{T}\;} Y \xrightarrow{\;J\;} F ,$$

where X and Y are normed spaces and $\tilde{T} \in L^p(X, Y)$.

Proof. Let V, in $\mathcal{U}_E$, be such that $T(V)$ is a precompact subset of F. Since F is metrizable, there exists an absolutely convex bounded subset B of F such that $T(V)$ is a precompact subset of the normed space $F(B)$. As $\operatorname{Ker} p_V \subset \operatorname{Ker} T$, it follows from the precompactness of $T(V)$ in $F(B)$ that the induced map $T_{(V,B)}$ of T belongs to $L^p(E_V, F(B))$ because of $T_{(V,B)}(Q_V(V)) = T(V)$. Therefore the lemma is proved since $T = j_B \circ T_{(V,B)} \circ Q_V$.

Before giving another characterization of precompact linear maps, we require the following terminology and result: Let V, W be in $\mathcal{U}_E$, and let $n \geq 0$ be any integer. If V is absorbed by W, then the *n-th diameter of V with respect to W* is defined by

$\alpha_n(V, W) = \inf\{\mu > 0 : V \subset \mu W + G,\ G$ is a subspace of E with $\dim G \leq n\}$.

In particular, if $(X, \|\cdot\|)$ is a normed space and if Σ is the closed unit ball in X, then we have for any bounded subset B of X that

$$B \subset \alpha_o(B, \Sigma)\Sigma \quad \text{and} \quad \alpha_o(B, \Sigma) \geq \alpha_1(B, \Sigma) \geq \ldots \geq 0 .$$

Moreover, B is precompact if and only if $\lim_n \alpha_n(B, \Sigma) = 0$ as shown by the following result.

(1.1.3) Lemma. <u>A bounded set</u> B <u>in a normed space</u> $(X, \|\cdot\|)$ <u>is precompact if and only if</u> $\lim_n \alpha_n(B, \Sigma) = 0$, <u>where</u> Σ <u>is the closed unit ball in</u> X.

<u>Proof</u>. <u>Necessity</u>. For any $\delta > 0$ there exists a finite set $\{x_1, \ldots, x_n\}$ in B such that $B \subset \bigcup_{i=1}^{n} (x_i + \delta\Sigma)$. Let G be the vector subspace of X generated by $\{x_1, \ldots, x_n\}$. Then $\dim G \leq n$ and $B \subset \delta\Sigma + G$, thus

$$\alpha_k(B, \Sigma) \leq \alpha_n(B, \Sigma) \leq \delta \quad \text{for all} \quad k \geq n .$$

<u>Sufficiency</u>. For any $\delta > 0$ there is a positive integer n such that $\alpha_n(B, \Sigma) < \delta$, hence there is a vector subspace G of X with $\dim G \leq n$ such that

$$B \subset \delta\Sigma + G . \tag{1.1}$$

$\Sigma_o = \Sigma \cap G$ is a bounded subset of the finite-dimensional normed space G, hence Σ_o is a precompact subset of G, and surely precompact in X, thus there exists a finite subset $\{x_1, \ldots, x_m\}$ of Σ_o such that

$$(\delta + \alpha_o(B, \Sigma))\, \Sigma_o \subset \bigcup_{j=1}^{m} (x_j + \delta \Sigma) \ .$$

In view of (1.1), any $x \in B$ is of the form

$$x = \delta y + z \quad \text{with} \quad y \in \Sigma \quad \text{and} \quad z \in G \ ,$$

hence we obtain from $B \subset \alpha_o(B, \Sigma)\Sigma$ that

$$\|z\| \leq \|\delta y\| + \|x\| \leq \delta + \alpha_o(B, \Sigma)$$

which shows that $z \in (\delta + \alpha_o(B, \Sigma))\Sigma$. Consequently,

$$x = \delta y + z \in \delta \Sigma + (\delta + \alpha_o(B, \Sigma))\Sigma \subset \bigcup_{j=1}^{m} (x_j + 2\delta \Sigma) \ .$$

Therefore B is precompact.

The following result, due to Terzioğlu [2], gives a characterization of a precompact map from one normed space into another.

(1.1.4) Lemma. <u>Let</u> $(X, \|\cdot\|)$ <u>and</u> (Y, q) <u>be normed spaces and</u> $T \in L(X, Y)$. <u>Then</u> $T \in L^p(X, Y)$ <u>if and only if there exist an</u> $(\zeta_n) \in c_o$ <u>and an equicontinuous sequence</u> $\{x'_n\}$ <u>in</u> X' <u>such that</u>

$$q(Tu) \leq \sup\{|\zeta_n \langle u, x'_n\rangle| : n \geq 1\} \ \underline{\text{for all}} \ u \in X \ . \tag{1.2}$$

<u>Proof</u>. <u>Necessity</u>. Let Σ_Y be the closed unit ball in Y . By Schauder's theorem, $T'(\Sigma_Y^o)$ is a compact subset of the Banach space X' , hence there exists a null sequence $\{u'_n\}$ in X' such that each element in $T'(\Sigma_Y^o)$ is of the form

$$\Sigma_{n=1}^{\infty} \mu_n u'_n \quad \text{with} \quad \Sigma_{n=1}^{\infty} |\mu_n| \leq 1 .$$

For $\|u'_n\| \neq 0$ we set

$$\zeta_n = \|u'_n\| \quad \text{and} \quad x'_n = u'_n / \|u'_n\| .$$

Then $(\zeta_n) \in c_o$ and $\{x'_n\}$ is an equicontinuous sequence in X' such that

$$q(Tu) = \sup\{|\langle Tu, y'\rangle| : y' \in \Sigma_Y^o\} \leq (\Sigma_{n=1}^{\infty} |\mu_n|) \sup\{|\langle u, u'_n\rangle| : n \geq 1\}$$

$$\leq \sup\{|\zeta_n \langle u, x'_n\rangle| : n \geq 1\} \quad \text{for all} \quad u \in X .$$

<u>Sufficiency</u>. Let $u'_n = \zeta_n x'_n$ $(n \geq 1)$. Then $\{u'_n\}$ is a null sequence in X' , hence for any $\delta > 0$ there is an $k > 0$ such that

$$\|u'_n\| < \delta \quad \text{for all} \quad n \geq k . \tag{1.3}$$

Let Σ_X and Σ_Y be the closed unit balls in X and Y respectively, and let

$$M = \{x \in X : \langle x, u'_i\rangle = 0, i = 1, 2, \ldots, k\} .$$

Then M is a vector subspace of X and the polar M^o of M is a finite-dimensional vector subspace of X' . Furthermore, (1.2) and (1.3) show that $T(\Sigma_X \cap M) \subset \delta \Sigma_Y$, hence

$$T'(\Sigma_Y^o) \subset \delta \overline{co}(\Sigma_X^o \cup M^o) \subset \delta \overline{(\Sigma_X^o + M^o)} = \delta\Sigma_X^o + M^o .$$

Therefore $T'(\Sigma_Y^o)$ is compact by (1.1.3), thus $T \in L^p(X, Y)$ by Schauder's theorem.

(1.1.5) Corollary. <u>Let</u> X <u>and</u> Y <u>be normed spaces</u>, <u>let</u> $\tilde{Y}$ <u>be</u>

the completion of Y and $T \in L(X, Y)$. Then $T \in L^p(X, Y)$ if and only if $T \in L^c(X, \tilde{Y})$.

By combining the results of (1.1.2) and (1.1.4), we obtain the following

(1.1.6) Theorem. Let F be metrizable and $T \in L(E, F)$. Then $T \in L^p(E, F)$ if and only if there exist an $(\zeta_n) \in c_o$, an equicontinuous sequence $\{x_n'\}$ in E' and an absolutely convex bounded subset B of F such that

$$r_B(Tu) \leq \sup\{|\zeta_n \langle u, x_n' \rangle| : n \geq 1\} \text{ for all } u \in E,$$

where r_B is the gauge of B defined on $F(B)$.

As a consequence we obtain the following result concerning with a compact factorization of a compact map.

(1.1.7) Corollary. Let F be metrizable and $T \in L(E, F)$. Then $T \in L^c(E, F)$ if and only if T admits a compact factorization through a closed subspace M of c_o, namely, there exists a closed subspace M of c_o such that T is the composition of the following two compact maps:

$$E \xrightarrow{T_1} M \xrightarrow{\tilde{T}_2} F .$$

Proof. The sufficiency is obvious. To prove the necessity, we know from (1.1.6) that there exist an $(\zeta_n) \in c_o$, an equicontinuous sequence

$\{x'_n\}$ in E' and an absolutely convex bounded subset B of F such that

$$r_B(Tu) \leq \sup\{|\zeta_n^2 \langle u, x'_n\rangle| : n \geq 1\} \quad \text{for all} \quad u \in E .$$

Let $T_1 : E \to c_o$ be defined by

$$T_1(u) = (\zeta_n \langle u, x'_n\rangle)_{n\geq 1} \quad \text{for all} \quad u \in E ,$$

and let M be the closure of $T_1(E)$ in c_o. Then $T_1 \in L^c(E, M)$ by (1.1.6) on account of the completeness of M. Let $D : c_o \to c_o$ be defined by

$$D((\eta_n)) = (\zeta_n \eta_n)_{n\geq 1} \quad \text{for all} \quad (\eta_n) \in c_o .$$

Then $D \in L^c(c_o)$ by (1.1.6). As r_B is a norm, it follows that $\text{Ker}\,(DT_1) \subset \text{Ker}\,T$, and hence that there is an $S \in L(DT_1(E), F(B))$ such that $T = SDT_1$. Therefore $T_2 = j_B \circ S \circ D \in L^c(T(E), F)$, and thus there exist a o-neighbourhood W in $T_1(E)$ and a compact subset K of F such that $T_2(W) \subset K$. Since T_2 is uniformly continuous, its restriction to W has a unique continuous extension to the closure $\overline{W}$ of W in M with values in K because K is complete. It is clear that this extension is the restriction to $\overline{W}$ of a linear map $\tilde{T}_2 : M \to F$ which is compact, and that $T = \tilde{T}_2 T_1$.

1.2 Precompact seminorms and Schwartz spaces

A seminorm p on E is said to be precompact if there exist an $(\zeta_n) \in c_o$ and an equicontinuous sequence $\{x'_n\}$ in E' such that

$$p(x) \leq \sup\{|\zeta_n \langle x, x'_n\rangle| : n \geq 1\} \quad \text{for all} \quad x \in E .$$

Clearly precompact seminorms on E are continuous and every $\sigma(E, E')$-continuous seminorm on E is precompact.

In terms of precompact linear maps we present some alternative characterizations of precompact seminorms, due to Randtke [1], as follows.

(1.2.1) Lemma. Let p be a continuous seminorm on E and $V_p = \{x \in E : p(x) \leq 1\}$. Then the following statements are equivalent.

(a) p is a precompact seminorm.

(b) The quotient map $Q_p : E \to E_p$ is a precompact linear map.

(c) There exists a continuous seminorm r on E with $p \leq r$ such that the canonical map $Q_{p,r} : E_r \to E_p$ is precompact.

(d) There exists an absolutely convex o-neighbourhood W in E with $W \subset V_p$ such that the canonical embedding map from $E'(V_p^o)$ into $E'(W^o)$ is compact.

Proof. The equivalence of (a) and (b) is a consequence of (1.1.6), the equivalence of (c) and (d) follows from Schauder's theorem (1.1.1), and the implication (c) $\Rightarrow$ (b) is obvious on account of $Q_p = Q_{p,r} \circ Q_r$. To prove the implication (b) $\Rightarrow$ (c), let $U \in \mathcal{U}_E$ be such that $Q_p(U)$ is a precompact subset of E_p. Then the injection $\hat{Q}_p$ associated with Q_p belongs to $L^p(E_U, E_p)$ and $Q_p = \hat{Q}_p \circ Q_U$. Let r be the gauge of $W = V_p \cap U$. Then r is a continuous seminorm on E with $p \leq r$. As $W \subset U$, we have $Q_{U,r} \in L(E_r, E_U)$. The precompactness of $Q_{p,r}$ then follows from the precompactness of $\hat{Q}_p$ and $Q_{p,r} = \hat{Q}_p \circ Q_{U,r}$.

(1.2.2) Corollary. Let $T \in L(E, F)$ and let q be a continuous seminorm on F. If one of T and q is precompact, then $q \circ T$ is a precompact seminorm on E.

Proof. It is clear that $Q_q \circ T \in L^p(E, F_q)$. Let $G = T(E)$ and $r = qT$. Then

$$\hat{r}(Q_r(x)) = r(x) = q(Tx) = \hat{q}(Q_q(Tx)) \quad \text{for all } x \in E ;$$

it follows that the map S, defined by

$$S(Q_r(x)) = Q_q(Tx) \quad \text{for all } x \in E ,$$

is an isometry from E_r onto G_q, hence $Q_r \in L^p(E, E_r)$ and thus qT is precompact by (1.2.1).

The following result is due to Randtke [1].

(1.2.3) Lemma. If p and q are precompact seminorms on E, then so do $p + q$ and ζp for any $\zeta \geq 0$.

Proof. Clearly ζp is precompact. To see the precompactness of $p + q$, let us define $T : E \to E_p \times E_q$ by setting

$$T(x) = (Q_p(x), Q_q(x)) \quad \text{for all } x \in E .$$

Then T is linear and $\operatorname{Ker} T = \operatorname{Ker}(p + q)$, hence the injection $\hat{T}$ associated with T is an isometric isomorphism from the normed space E_{p+q} onto the subspace $T(E)$ of the normed space $E_p \times E_q$ because of

$$\|\hat{T}(Q_{p+q}(x))\| = \|Tx\| = \|(Q_p(x), Q_q(x))\| = p(x) + q(x) = (\widehat{p + q})(Q_{p+q}(x)).$$

Since p and q are precompact seminorms, it is easily seen that T is precompact. It then follows from $T = \hat{T}\, Q_{p+q}$ that $Q_{p+q} \in L^p(E, E_{p+q})$, and hence from (1.2.1) that $p + q$ is a precompact seminorm.

A locally convex space $(E, \mathcal{P})$ is called a <u>Schwartz space</u> if every continuous seminorm on E is precompact.

In terms of precompact linear maps we present some alternative characterizations of Schwartz spaces as follows.

(1.2.4) Theorem. <u>The following statements are equivalent.</u>

(a) $(E, \mathcal{P})$ <u>is a Schwartz space</u>.

(b) $Q_p \in L^p(E, E_p)$ <u>for any continuous seminorm</u> p on E.

(c) <u>For any continuous seminorm</u> p <u>on</u> E <u>there is a continuous seminorm</u> r <u>on</u> E <u>with</u> $p \leq r$ <u>such that</u> $Q_{p,r} \in L^p(E_r, E_p)$.

(d) <u>For any continuous seminorm</u> p <u>on</u> E <u>there is a continuous seminorm</u> r <u>on</u> E <u>with</u> $p \leq r$ <u>such that the canonical embedding from</u> $E'(V_p^o)$ <u>into</u> $E'(V_r^o)$ <u>is compact</u>.

(e) $L(E, Y) = L^p(E, Y)$ <u>for any normed (or Banach) space</u> Y.

(f) <u>For any</u> $W \in \mathcal{U}_E$ <u>there is an</u> $V \in \mathcal{U}_E$ <u>such that for any</u> $\zeta > 0$ <u>it is possible to find a finite subset</u> $\{x_1, \ldots, x_m\}$ <u>of</u> E <u>such that</u>

$$V \subset \bigcup_{j=1}^{m} (x_j + \zeta W).$$

(g) <u>For any</u> $W \in \mathcal{U}_E$ <u>there is an</u> $V \in \mathcal{U}_E$ <u>which is absorbed by</u> W <u>such that</u> $\lim_n \alpha_n(V, W) = 0$, <u>where</u> $\alpha_n(V, W)$ <u>is the</u> n-th <u>diameter of</u> V <u>with respect to</u> W.

(h) (i) <u>Bounded subsets of</u> E <u>are precompact, and</u>

(ii) <u>for any</u> $W \in \mathcal{U}_E$ <u>there is an</u> $V \in \mathcal{U}_E$ <u>such that for any</u> $\zeta > 0$ <u>it is possible to find a bounded subset</u> B <u>of</u> E <u>with</u> $V \subset B + \zeta W$.

<u>Proof</u>. In view of (1.2.1), the statements (a) through (d) are mutually equivalent. The implication (e) ⇒ (b) and (f) ⇒ (b) are obvious. On the other hand, it is clear that $\alpha_n(Q_W(V)) \leq \alpha_n(V, W)$, where $\alpha_n(Q_W(V))$ is the n-th diameter of the bounded set $Q_W(V)$ in E_W. It then follows from (1.1.3) that (g) implies (b). Therefore we complete the proof by showing the indicated implications (b) ⇒ (e), (b) ⇒ (f), (c) ⇒ (g) and (b) ⇒ (h) ⇒ (f).

(b) ⇒ (e): Let $T \in L(E, Y)$ and let p be the gauge of $T(\Sigma)$, where Σ is the closed unit ball in Y. Then $\operatorname{Ker} p \subset \operatorname{Ker} T$ and the normability of Y ensures that $L^{\ell b}(E, Y) = L(E, Y)$, thus there exists an $\tilde{T} \in L(E_p, Y)$ such that $T = \tilde{T} Q_p$; consequently, $T \in L^p(E, Y)$.

(b) ⇒ (f): For any $W \in \mathcal{U}_E$, the assumption that $Q_W \in L^p(E, E_W)$ implies that there is an $V \in \mathcal{U}_E$ such that $Q_W(V)$ is a precompact subset of the normed space E_W. As $Q_W(W)$ is the unit ball in E_W, for any $\zeta > 0$ there is a finite set $\{x_1, \ldots, x_m\}$ in V such that $Q_W(V) \subset \bigcup_{j=1}^{m} ((\zeta/2) Q_W(W) + Q_W(x_j))$, hence

$$V \subset \bigcup_{j=1}^{m} ((\zeta/2) W + x_j) + p_W^{-1}(0) \subset \bigcup_{j=1}^{m} (\zeta W + x_j) \tag{2.1}$$

because of $p_W^{-1}(0) \subset \delta W$ for any $\delta > 0$.

(c) ⇒ (g): For any $W \in \mathcal{U}_E$ there is an $V \in \mathcal{U}_E$ with $V \subset W$ such that $Q_{W,V} \in L^p(E_V, E_W)$, hence $Q_W(V) = Q_{W,V}(Q_V(V))$ is a precompact subset of E_W. For any $\mu > 0$ there is a finite set $\{x_1, \ldots, x_m\}$ in E such that (2.1) holds. Denote by M the vector subspace of E generated by $\{x_1, \ldots, x_m\}$. Then (2.1) implies that $V \subset \mu W + M$, hence $\alpha_m(V, W) \leq \mu$. It is obvious that $\alpha_n(V, W) \geq \alpha_{n+1}(V, W)$ for all $n \geq 1$, it then follows that $\lim_n \alpha_n(V, W) = 0$.

(b) ⇒ (h): For any absolutely convex bounded subset A of E and any $V \in \mathcal{U}_E$, the canonical map $K_{V,A} : E(A) \to E_V$ (i.e., $K_{V,A} = Q_V \circ j_A$) is precompact, hence A must be precompact in E, therefore the condition (i) holds.

To prove the condition (ii), let $W \in \mathcal{U}_E$ and let $V \in \mathcal{U}_E$ be such that $Q_W(V)$ is a precompact subset of E_W. For any $\zeta > 0$, by the same argument given in the proof of (b) ⇒ (f) of this theorem, there is a finite set $\{x_1, \ldots, x_m\}$ in V such that (2.1) holds. Let $B = \{x_1, \ldots, x_m\}$. Then B is bounded in E and $V \subset \zeta W + B$, which obtains (ii).

(h) ⇒ (f): By (ii), for any $W \in \mathcal{U}_E$ there is an $V \in \mathcal{U}_E$ such that for any $\zeta > 0$ it is possible to find a bounded subset B of E with $V \subset (\zeta/_2)W + B$. By (i), B is precompact, there is a finite set $\{x_1, \ldots, x_m\}$ in E such that $B \subset \bigcup_{j=1}^m ((\zeta/_2)W + x_j)$, thus

$$V \subset \bigcup_{j=1}^m (\zeta W + x_j) .$$

The equivalence of (a) and (b) in (1.2.4) is due to Randtke [1], while the equivalence between (a), (b), (f) and (g) are due to Terzioğlu [3].

It is well-known that a topological vector space E is finite-dimensional if there is a precompact o-neighbourhood. Therefore we obtain an immediate consequence of (1.2.4)(h).

(1.2.5) Corollary. *Every normed Schwartz space is finite-dimensional*.

Grothendieck [1] calls a locally convex space $(E, \mathcal{P})$ *quasi-normable* if it satisfies the condition (ii) of (1.2.4)(h). It is not hard to show that $(E, \mathcal{P})$ is quasi-normable if and only if for any $W \in \mathcal{U}_E$ there exists an $V \in \mathcal{U}_E$ with $V \subset W$ such that on W^o the topology induced by $\beta(E', E)$ coincides with the norm topology induced by $E'(V^o)$. In view of (1.2.4), $(E, \mathcal{P})$ is a Schwartz space if and only if it is quasi-normable and bounded subsets of E are precompact.

We conclude this section with results concerning with some important properties of Schwartz spaces.

(1.2.6) Proposition. *For a Schwartz space* $(E, \mathcal{P})$, *the following assertions hold*:

(a) *Bounded subsets of* E *are precompact*, *hence* $\beta(E', E)$ *coincides with the topology of precompact convergence*, *and equicontinuous subsets of* E' *are* $\beta(E', E)$-*relatively compact*.

(b) E <u>is quasi-complete if and only if it is a semi-Montel space</u>.

(c) <u>The normed space</u> E_V <u>is separable for any</u> $V \in \mathcal{U}_E$, <u>hence every metrizable Schwartz space is separable</u>.

(d) <u>For any</u> $V \in \mathcal{U}_E$ <u>and any absolutely convex bounded subset</u> A <u>of</u> E , <u>the canonical map</u> $K_{V,A} : E(A) \to E_V$ <u>defined by</u>

$$K_{V,A}(x) = Q_V(x) \quad \text{for all} \quad x \in E(A) ,$$

<u>is a precompact linear map, hence</u> $K_{V,A}(A)$ <u>is precompact and separable in</u> E_V .

(e) <u>For any equicontinuous subset</u> M <u>of</u> E' <u>there exists an</u> $V \in \mathcal{U}_E$ <u>such that</u> M <u>is relatively compact and separable in</u> $E'(V^o)$, <u>hence</u> M <u>is separable for</u> $\beta(E', E)$.

(f) <u>If, in addition,</u> E <u>is a dual metric space in the sense of</u> Pietsch [1, p.11] (<u>namely,</u> E <u>is</u> σ<u>-infrabarrelled and possesses a fundamental sequence of bounded sets</u>), <u>then</u> $(E', \beta(E', E))$ <u>is a Fréchet-Montel space</u>, <u>and</u> $\mathcal{P}$ <u>is the topology of uniform convergence on</u> $\beta(E', E)$-<u>relatively compact subsets of</u> E' .

<u>Proof</u>. Parts (a) and (b) follow from (i) of (1.2.4)(h), while part (d) is an immediate consequence of (1.2.4)(b) by making use of the following equality

$$K_{V,A} = Q_V \, j_A \, .$$

(c) It is known from (1.2.4)(b) that $Q_V \in L^p(E, E_V)$, hence there is an $W \in \mathcal{U}_E$ such that $Q_V(W)$ is a totally bounded subset of the normed space E_V , thus $Q_V(W)$ is separable in E_V . As $E_V = \bigcup_n n\, Q_V(W)$, we conclude that E_V is separable.

Suppose further that $(E, \mathcal{P})$ is metrizable. Then $\mathcal{P}$ has a countable neighbourhood basis $\{V_n : n \geq 1\}$ at 0. As each E_{V_n} is separable, there exists a sequence $\{x_k^{(n)}, k \geq 1\}$ such that $\{Q_{V_n}(x_k^{(n)}) : k \geq 1\}$ is dense in E_{V_n}. We then conclude from

$$p_{V_n}(x) = \hat{p}_{V_n}(Q_{V_n}(x)) \quad \text{for all} \quad x \in E \quad \text{and} \quad n \geq 1$$

that the double sequence $\{x_k^{(n)}, n, k \geq 1\}$ is dense in $(E, \mathcal{P})$.

(e) Let W, in $\mathcal{U}_E$, be such that $M \subset W^o$. In view of (1.2.4)(d), there exists an $V \in \mathcal{U}_E$ with $V \subset W$ such that the embedding map $E'(W^o) \to E'(V^o)$ is compact, hence W^o is compact in the Banach space $E'(V^o)$, thus M is relatively compact and separable in $E'(V^o)$. Finally, as $\beta(E', E)$ is coarser than the norm topology on $E'(V^o)$, it follows that M is separable for $\beta(E', E)$.

(f) Clearly $(E', \beta(E', E))$ is a Fréchet space. To prove that every $\beta(E', E)$-bounded subset of E' is $\beta(E', E)$-relatively compact, it suffices to verify that every $\beta(E', E)$-bounded sequence in E' has a convergent subsequence; but this is obvious since such a sequence is $\mathcal{P}$-equicontinuous by the σ-infrabarrelledness of E; and hence $\beta(E', E)$-relatively compact by part (a). Therefore $(E', \beta(E', E))$ is a Fréchet-Montel space.

Each $\beta(E', E)$-relatively compact subset B of E' is contained in the closed absolutely convex hull of some null sequence $\{f_n\}$ in $(E', \beta(E', E))$, hence there exists an $V \in \mathcal{U}_E$ such that $B \subset \overline{\Gamma}(\{f_n\}) \subset V^o$

by the σ-infrabarrelledness of E . Therefore $\mathcal{P}$ is the topology of uniform convergence on $\beta(E', E)$-relatively compact subsets of E' by part (a).

(1.2.7) Corollary. <u>If</u> $(E, \mathcal{P})$ <u>is a quasi-complete Schwartz space, then every</u> $\sigma(E, E')$-<u>convergent sequence in</u> E <u>is also convergent to the same limit</u>.

<u>Proof</u>. Let $\{x_n\}$ be $\sigma(E, E')$-convergent to x . Then the set $\{x, x_1, x_2, \ldots\}$ is bounded, hence there exists a closed bounded subset A of E such that $\{x, x_1, \ldots\} \subset B$. In view of (1.2.6)(b), B is compact, hence on B the topology $\sigma(E, E')$ coincides with $\mathcal{P}$, thus x_n is $\mathcal{P}$-convergent to x .

Recall that a locally convex space E is said to be σ-<u>barrelled</u> if every countable $\sigma(E', E)$-bounded subset of E' is equicontinuous.

(1.2.8) Corollary. <u>Let</u> E <u>be a Schwartz</u> σ-<u>barrelled space, and let</u> $\{f_n\}$ <u>be a sequence in</u> E' <u>which is</u> $\sigma(E', E)$-<u>convergent to</u> $f \in E'$. <u>Then there exists an</u> $V \in \mathcal{U}_E$ <u>such that</u> f_n <u>converges to</u> f <u>in the Banach space</u> $E'(V^o)$; <u>consequently</u> f_n <u>is</u> $\beta(E', E)$-<u>convergent to</u> f .

<u>Proof</u>. The set $\{f, f_1, f_2, \ldots\}$ is $\sigma(E', E)$-bounded, hence equicontinuous. In view of (1.2.6)(e), there exists an $V \in \mathcal{U}_E$ such that $\{f, f_1, f_2, \ldots\}$ is a relatively compact subset of the Banach space $E'(V^o)$, thus f_n converges to f in $E'(V^o)$; consequently f_n is $\beta(E', E)$-convergent to f .

1.3 Precompact-bounded linear mappings

An $T \in L^*(E, F)$ is said to be precompact-bounded (quassi-Schwartz in the terminology of Randtke [1]) if there is a precompact seminorm p on E such that $\{Tx \in F : p(x) \leq 1\}$ is bounded in F.

Clearly $T \in L^*(E, F)$ is precompact-bounded if and only if there is a precompact seminorm p on E such that for any continuous seminorm q on F it is possible to find an $\mu_q \geq 0$ such that

$$q(Tx) \leq \mu_q p(x) \quad \text{for all} \quad x \in E .$$

In view of (1.2.3), it is easily seen that the set consisting of all precompact-bounded linear maps from E into F, denoted by $L^{pb}(E, F)$, is a vector subspace of $L^{\ell b}(E, F)$. Furthermore, the composition of two continuous linear maps, in which one of them is precompact-bounded, is precompact-bounded.

The following result gives the relationship between precompact-bounded linear maps and precompact linear maps.

(1.3.1) Lemma. For any locally convex spaces E and F, we have

$$L^{pb}(E, F) \subseteq L^{p}(E, F) .$$

Moreover, if in addition, F is metrizable, then $L^{pb}(E, F) = L^{p}(E, F)$.

Proof. Let $T \in L^{pb}(E, F)$ and let p be a precompact seminorm on

E such that $B = \{Tx \in F : p(x) \leq 1\}$ is a bounded subset of F. The boundedness of B ensures that $p^{-1}(0) \subseteq \operatorname{Ker} T$, hence there exists an $S \in L(E_p, F)$ such that $T = S\,Q_p$. In view of (1.2.1), the precompactness of p implies that $Q_p \in L^p(E, E_p)$, so that $T \in L^p(E, F)$.

Suppose further that F is metrizable and that $T \in L^p(E, F)$. Then (1.1.6) shows that there exist an $(\zeta_n) \in c_0$, an equicontinuous sequence $\{x'_n\}$ in E' and an absolutely convex bounded subset B of F such that

$$r_B(Tu) \leq \sup\{|\zeta_n \langle u, x'_n\rangle| : n \geq 1\} \quad \text{for all} \quad u \in E,$$

where r_B is the gauge of B defined on $F(B)$. The seminorm p, defined by

$$p(x) = r_B(Tx) \quad \text{for all} \quad x \in E,$$

is a precompact seminorm on E such that $\{Tx \in F : p(x) \leq 1\} \subseteq B$, thus $T \in L^{pb}(E, F)$.

We shall show in the next section that if the strong dual of F is a Schwartz space, then $L^{pb}(E, F) = L^p(E, F)$. From this result, it is easy to give an example of precompact linear maps which are not precompact-bounded.

In terms of the notion of precompact-bounded linear maps, we present the following characterization of Schwartz spaces which is due to Randtke [1].

(1.3.2) Theorem. <u>The following statements are equivalent</u>.

(a) E <u>is a Schwartz space</u>.

(b) $L^{pb}(E, F) = L^{\ell b}(E, F)$ <u>for any locally convex space</u> F.

(c) $L^{pb}(E, F) = L^{p}(E, F)$ <u>for any locally convex space</u> F.

(d) $L(E, Y) = L^{pb}(E, Y)$ <u>for any normed (or Banach) space</u> Y.

(e) <u>For any locally convex (or normed) space</u> F <u>and any continuous bilinear form</u> ψ <u>on</u> $E \times F$, <u>there exist a precompact seminorm</u> p <u>on</u> E <u>and a continuous seminorm</u> q <u>on</u> F <u>such that</u>

$$|\psi(x, y)| \leq p(x)q(y) \quad \text{for all} \quad (x, y) \in E \times F .$$

(f) $L^{eq}(E, F(\mathcal{J})') \subset L^{pb}(E, F'_{\beta})$ <u>for any locally convex (or normed) space</u> $(F, \mathcal{J})$.

<u>Proof</u>. The implications (a) $\Rightarrow$ (b) and (a) $\Rightarrow$ (e) are obvious. As precompact maps are bounded, it follows that (b) implies (c). The implication (d) $\Rightarrow$ (a) follows from (1.2.4)(e). Therefore we complete the proof by showing the implications (c) $\Rightarrow$ (d) and (e) $\Rightarrow$ (f) $\Rightarrow$ (d).

(c) $\Rightarrow$ (d) : It suffices to show that $L(E, Y) \subseteq L^{pb}(E, Y)$. To do this, let $T \in L(E, Y)$ and $F = (Y, \sigma(Y, Y'))$. Then $T \in L^{p}(E, F)$ and thus $T \in L^{pb}(E, F)$. Therefore there is a precompact seminorm p on E such that $\{Tx \in Y : p(x) \leq 1\}$ is $\sigma(Y, Y')$-bounded and surely norm bounded.

(e) $\Rightarrow$ (f) : Let $T \in L^{eq}(E, F(\mathcal{J})')$, and let b_T be the associated bilinear form of T; that is,

$$b_T(x, y) = \langle y, Tx \rangle \quad \text{for all} \quad (x, y) \in E \times F .$$

As the map $T \mapsto b_T$ being an algebraic isomorphism from $L^{eq}(E, F(\mathfrak{J})')$ onto $B(E, F)$, it follows that there exist a precompact seminorm p on E and a continuous seminorm q on F such that

$$|\langle y, Tx\rangle| = |b_T(x, y)| \leq p(x)q(y) \quad \text{for all} \quad (x, y) \in E \times F . \qquad (3.1)$$

Let $U = \{y \in F : q(y) \leq 1\}$ and $N = \{Tx \in F : p(x) \leq 1\}$. For any absolutely convex bounded subset B of F there is an $\mu > 0$ such that $B \subset \mu U$. It then follows from (3.1) that $N \subset \mu B^{o}$, and hence that $T \in L^{pb}(E, F'_{\beta})$.

(f) ⇒ (d) : Let $T \in L(E, Y)$ and let $e_Y : Y \to Y''$ be the evaluation map. Then $S = e_Y \circ T \in L(E, Y'')$, hence $S \in L^{pb}(E, Y'')$ because of $L^{eq}(E, (Y'_{\beta})')) = L(E, Y'')$. Therefore there exist a precompact seminorm p on E and $\mu > 0$ such that

$$\{Sx \in Y'' : p(x) \leq 1\} \subset \mu\Sigma^{oo} , \qquad (3.2)$$

where Σ is the closed unit ball in Y and Σ^{oo} is the bipolar of Σ taken in Y''. Since $T(E) \subset Y$ and e_Y is an isometry from Y into Y'', it follows from (3.2) that $\{Tx \in Y : p(x) \leq 1\} \subset \mu\Sigma$, and hence that $T \in L^{pb}(E, Y)$.

From the equivalence between (a), (e) and (f) of (1.3.2), we obtain immediately the following result which can be viewed, in some sense, as the abstract kernel theorem on Schwartz spaces.

(1.3.3) Corollary. Let E be a Schwartz space and F a locally convex space. For any continuous bilinear form ψ on $E \times F$ there exists

<u>a unique</u> $T \in L^{pb}(E, F'_\beta)$ <u>such that</u>

$$\psi(x, y) = \langle y, Tx\rangle \quad \text{for all} \quad (x, y) \in E \times F .$$

We are now going to verify the factorization of precompact-bounded maps which should be compared with (1.1.2).

(1.3.4) Proposition. <u>For any</u> $T \in L(E, F)$, <u>the following statements are equivalent</u>.

(a) $T \in L^{pb}(E, F)$.

(b) <u>There exist an</u> $(\zeta_n) \in c_o$ <u>and an equicontinuous sequence</u> $\{x'_n\}$ <u>in</u> E' <u>such that for any continuous seminorm</u> q <u>on</u> F <u>it is possible to find an</u> $\beta_q \geq 0$ <u>for which</u>

$$q(Tx) \leq \beta_q \sup\{|\zeta_n \langle x, x'_n\rangle| : n \geq 1\} \quad \text{for all} \quad x \in E .$$

(c) T <u>is the composition of the following continuous linear maps</u>

$$E \xrightarrow{\ Q\ } X \xrightarrow{\ \widetilde{T}\ } Y \xrightarrow{\ J\ } F ,$$

<u>where</u> X <u>and</u> Y <u>are normed spaces and</u> $\widetilde{T} \in L^p(X, Y)$.

<u>Proof</u>. The equivalence of (a) and (b) is obvious. If one of $T \in L(E, F)$ and $S \in L(F, G)$ is precompact-bounded, then $ST \in L^{pb}(E, G)$, hence the implication (c) $\Rightarrow$ (a) follows. To prove the implication (a) $\Rightarrow$ (c), let $T \in L^{pb}(E, F)$ and let p be a precompact seminorm on E such that $B = \{Tx \in F : p(x) \leq 1\}$ is bounded in F . Then $p^{-1}(0) \subset \operatorname{Ker} T$, hence

there is an $S \in L(E_p, F(B))$ such that $T = j_B S Q_p$. In view of (1.2.1), the precompactness of p ensures that there is a continuous seminorm r on E with $p \leq r$ such that $Q_{p,r} \in L^p(E_r, E_p)$. Therefore we have

$$T = j_B \, S \, Q_{p,r} Q_r \quad . \tag{3.3}$$

As $SQ_{p,r} \in L^p(E_r, F(B))$, the implication (a) $\Rightarrow$ (c) follows from (3.3).

1.4 Properties of Schwartz spaces

We begin the investigations of this section with the following result concerning the stability properties of Schwartz spaces.

(1.4.1) Proposition. (a) Subspaces of a Schwartz space are Schwartz spaces; the product of an arbitrary family of Schwartz spaces is a Schwartz space. Consequently, the topological projective limit of an arbitrary family of Schwartz spaces is a Schwartz space.

(b) Every separated quotient space of a Schwartz space by a closed vector subspace is a Schwartz space; the locally convex direct sum of a countable family of Schwartz spaces is a Schwartz space. Consequently, the topological inductive limit of a countable family of Schwartz spaces is a Schwartz space.

Proof. Part (a) follows from (1.2.2) and the definition of Schwartz spaces. Let $(E, \mathcal{P})$ be a Schwartz space, let N be a closed vector subspace of E and $\hat{\mathcal{P}}$ the quotient topology of $\mathcal{P}$ by N. For any absolutely convex

$\hat{\mathcal{P}}$-neighbourhood $\hat{U}$ of 0 in $E/_N$, there exists, by (1.2.4)(f), an $V \in \mathcal{U}_E$ such that for any $\mu > 0$ it is possible to find a finite subset $\{x_1, \ldots, x_m\}$ of E such that

$$V \subset \bigcup_{j=1}^{m} (x_j + \mu (Q_N)^{-1} (\hat{U})) .$$

Clearly, $\hat{V} = Q_N(V)$ is a o-neighbourhood in $(E/_N, \hat{\mathcal{P}})$ such that

$$\hat{V} \subset \bigcup_{j=1}^{m} (Q_N(x_j) + \mu \hat{U}) ,$$

hence $(E/_N, \hat{\mathcal{P}})$ is a Schwartz space by (1.2.4)(f).

The proof for countably direct sum will be based on two conditions of (1.2.4)(h). Let $E = \bigoplus_{i=1}^{\infty} E_i$, where each E_i is a Schwartz space. Since a subset B of E is bounded (resp. precompact) if and only if $\pi_i(B)$ is bounded (resp. precompact) in E_i and $\Delta = \{i \in \mathbb{N} : \pi_i(B) \neq \{0\}\}$ is finite, it follows from (1.2.4)(h)(i) that bounded subsets of E are precompact. In order to verify that E satisfies the condition (ii) of (1.2.4)(h), let W be an absolutely convex o-neighbourhood in E and $j_n : E_n \to E$ the canonical embedding map. Then we identify E_n with the closed subspace $j_n(E_n)$ of E, hence each $W_n = j_n^{-1}(W) = W \cap E_n$ is an absolutely convex o-neighbourhood in E_n, and thus there exists an absolutely convex o-neighbourhood V_n in E_n with $V_n \subset W_n$ such that V_n satisfies the condition (ii) of (1.2.4)(h) (with respect to W_n). The absolutely convex hull of $\bigcup_{n\geq 1} (n+1)^{-1} V_n$, denoted by V, is a o-neighbourhood in E. Now for any $\mu > 0$,

$$(n+1)^{-1} V_n \subset \mu W_n \subset \mu W \text{ for all } n > 0 \text{ with } n + 1 \geq \mu^{-1} .$$

For any n with $0 < n \leq \mu^{-1}$, there exists a bounded subset B_n of E_n such that

$$(n+1)^{-1}V_n \subset V_n \subset B_n + \mu^{-1}W_n .$$

The absolutely convex hull of $\cup\{B_n : 0 < n \leq \mu^{-1}\}$, denoted by B , is clearly a bounded subset of E , and satisfies

$$(n+1)^{-1}V_n \subset B + \mu W \text{ for all } n \geq 1 .$$

Since $B + \mu W$ is absolutely convex, it follows that $V \subset B + \mu W$, as asserted.

(1.4.2) Corollary. <u>The separated quotient space of a Fréchet-Schwartz space by a closed vector subspace is a Fréchet-Schwartz space and a fortiori a Montel space</u>.

<u>Proof</u>. Since the separated quotient space of a Fréchet space by a closed vector subspace is a Fréchet space, the result follows from (1.4.1)(b) and the condition (i) of (1.2.4)(h).

Denote by $\underleftarrow{\lim}\, T_{\alpha,\mu}(X_\mu)$ the topological projective limit of a family $\{X_\mu : \mu \in \Lambda\}$ of normed spaces X_μ with respective to a family of continuous linear maps $T_{\alpha,\mu} : X_\mu \to X_\alpha$ ($\mu \geq \alpha$ in Λ) . Randtke [5] calls $\underleftarrow{\lim}\, T_{\alpha,\mu}(X_\mu)$ a <u>precompact projective limit</u> if for any $\alpha \in \Lambda$ there is an $\mu \in \Lambda$ with $\mu \geq \alpha$ such that

$$T_{\alpha,\mu} : X_\mu \to X_\alpha \text{ is a precompact linear map.}$$

As the range of every precompact linear map is separable, it follows that if $\underleftarrow{\lim}\, T_{\alpha,\mu}(X_\alpha)$ is a precompact projective limit of $\{X_\mu : \mu \in \Lambda\}$ with respective to a family of continuous linear maps $T_{\alpha,\mu} : X_\mu \to X_\alpha$, then each X_μ must be separable.

Suppose now that $(E, \mathcal{P})$ is a Schwartz space and that $\{p_\mu : \mu \in \Lambda\}$ is the family of all precompact seminorms on E. Then each $(E_\mu, \hat{p}_\mu)$ is a separable normed space by (1.2.5), hence (1.2.1) and (1.2.4)(d), together with a well-known result show that $(E, \mathcal{P})$ is topologically isomorphic to the precompact projective limit $\varprojlim Q_{\alpha,\mu}(E_\mu)$ of a family $\{(E_\mu, \hat{p}_\mu) : \mu \in \Lambda\}$ of separable normed spaces $(E_\mu, \hat{p}_\mu)$ with respective to a family of canonical maps $Q_{\alpha,\mu} : E_\mu \to E_\alpha$ ($\mu \geq \alpha$ in Λ), where $E_\mu = E/_\mu^{-1}(0)$. The converse still holds in view of (1.2.2) and the following well-known fact

$$\varprojlim T_{\alpha,\mu}(X_\mu) = \bigcap_{\alpha \leq \mu} \mathrm{Ker}(\pi_\alpha - T_{\alpha,\mu} \circ \pi_\mu),$$

where $\pi_\alpha : \prod X_\mu \to X_\alpha$ is the α-th projection. Therefore we have proved the following result, due to Randtke [5].

(1.4.3) Proposition. <u>A locally convex space</u> $(E, \mathcal{P})$ <u>is a Schwartz space if and only if it is topologically isomorphic to a precompact projective limit</u> $\varprojlim T_{\alpha,\mu}(X_\mu)$ <u>of a family</u> $\{X_\mu : \mu \in \Lambda\}$ <u>of separable normed spaces</u> X_μ <u>with respect to a family of continuous linear maps</u> $T_{\alpha,\mu} : X_\mu \to X_\alpha$ ($\mu \geq \alpha$ <u>in</u> Λ).

Let $\mathfrak{S}$ be a topologizing family for E' consisting of closed, absolutely convex bounded subsets of E such that $\cup \mathfrak{S} = E$. Then the family $\{P(A, U) : A \in \mathfrak{S}$ and $U \in \mathcal{U}_F\}$ is a neighbourhood basis at 0 for the $\mathfrak{S}$-topology on $L(E, F)$, where

$$P(A, U) = \{T \in L(E, F) : T(A) \subset U\}.$$

For any non-zero $y \in F$, the map $x' \longmapsto T_{x'}$, defined by

$$T_{x'}(x) = \langle x, x'\rangle y \quad \text{for all} \quad x \in E ,$$

is a topological isomorphism from E' onto a subspace of $L_{\mathfrak{S}}(E, F)$. Also for any non-zero $x' \in E'$, the map $y \longmapsto S_y$, defined by

$$S_y(x) = \langle x, x'\rangle y \quad \text{for all} \quad x \in E ,$$

is a topological isomorphism from F onto a subspace of $L_{\mathfrak{S}}(E, F)$. If $L_{\mathfrak{S}}(E, F)$ is a Schwartz space, then so do $E'_{\mathfrak{S}}$ and F by (1.4.1). The converse also holds as shown by the following result, due to Randtke [1].

(1.4.4) Theorem. $L_{\mathfrak{S}}(E, F)$ <u>is a Schwartz space if and only if both</u> $E'_{\mathfrak{S}}$ <u>and</u> F <u>are Schwartz spaces</u>. <u>In particular</u>, $L_\beta(E, F)$ <u>is a Schwartz space if and only if</u> $(E', \beta(E', E))$ <u>and</u> F <u>are Schwartz spaces</u>.

<u>Proof</u>. The necessity has been shown before this theorem. To prove the sufficiency, it is sufficient to show that each gauge $p_{(A,U)}$ of $P(A, U)$ is a precompact seminorm on $L(E, F)$.

In fact, let p_{A^0} be the gauge of A^0 defined on E' and q_U the gauge of U defined on F. Then the assumptions imply that there are (ζ_n), (μ_n) in c_0 and two equicontinuous sequences $\{x''_n\}$ and $\{y'_n\}$ in $(E'_{\mathfrak{S}})'$ and F' respectively such that

$$p_{A^0}(u') \leq \sup\{|\zeta_n\langle u', x''_n\rangle| : n \geq 1\} \quad \text{for all} \quad u' \in E' \tag{4.1}$$

and

$$q_U(v) \leq \sup\{|\mu_k\langle v, y'_k\rangle| : k \geq 1\} \quad \text{for all} \quad v \in F \tag{4.2}$$

As $T'(y'_k) \in E'$ for all $T \in L(E, F)$ and k, it follows that the maps $f_{k,n}$, defined by

$$f_{k,n}(T) = \langle T'(y'_k), x''_n \rangle \quad \text{for all} \quad T \in L(E, F),$$

are linear functionals on $L(E, F)$. We claim that $\{f_{k,n} : k, n \geq 1\}$ is an equicontinuous sequence in $(L_{\mathcal{B}}(E, F))'$. To this end, let $W \in \mathcal{U}_F$ and let $B \in \mathcal{B}$ be such that

$$y'_k \in W^o \quad \text{and} \quad x''_n \in B^{oo} \quad \text{for all} \quad k \quad \text{and} \quad n,$$

where B^{oo} is the polar of B^o taken in $(E'_{\mathcal{B}})'$. For any $T \in P(B, W)$ we have $T'(W^o) \subset B^o$, thus

$$|f_{k,n}(T)| = |\langle T'(y'_k), x''_n \rangle| \leq 1 \quad \text{for all} \quad T \in P(B, W)$$

which obtains our assertion.

If $1 \leq j \leq k$, let us put

$$g_{(k-1)(k-1)+j} = f_{j,k} \quad \text{and} \quad \beta_{(k-1)(k-1)+j} = \zeta_j \mu_k ;$$

if $1 \leq j < k$, let us put

$$g_{(k-1)(k-1)+k+j} = f_{k,k-j} \quad \text{and} \quad \beta_{(k-1)(k-1)+k+j} = \zeta_k \mu_{k-j} .$$

Then $\{g_j\}$ is an equicontinuous sequence in $(L_{\mathcal{B}}(F, F))'$, and $(\beta_j) \in c_o$ since $(\zeta_n), (\mu_n) \in c_o$. Furthermore, for any $T \in L(E, F)$ we obtain from (4.1) and (4.2) that

$$\begin{aligned} p_{(A,U)}(T) &= \sup\{|\langle Tx, y' \rangle| : x \in A \text{ and } y' \in U^o\} \\ &= \sup\{q_U(Tx) : x \in A\} \end{aligned}$$

$$\begin{aligned}
&\leq \sup_k \sup\{|\mu_k \langle Tx, y'_k\rangle| \ : x \in A\} \\
&= \sup_k \{|\mu_k| \, p_{A^o}(T'y'_k)\} \\
&\leq \sup\{|\mu_k \zeta_n \langle T'y'_k, x'_n\rangle| \ : k, n \geq 1\} \\
&= \sup\{|\mu_k \zeta_n \, f_{k,n}(T)| \ : k, n \geq 1\} \\
&= \sup\{|\beta_j \langle T, g_j\rangle| \ : j \geq 1\} ,
\end{aligned}$$

thus $p_{(A,U)}$ is a precompact seminorm.

Let $\mathfrak{S}$ be a topologizing family for E' consisting of closed, absolutely convex bounded subsets of E such that $\cup\mathfrak{S} = E$. Then $E'_{\mathfrak{S}}$ is a Schwartz space if and only if for any $B \in \mathfrak{S}$ there is an $A \in \mathfrak{S}$ with $A^o \subset B^o$ such that $Q_{B^o,A^o} \in L^p(E'_{A^o}, E'_{B^o})$. We are going to verify that $E'_{\mathfrak{S}}$ is a Schwartz space if and only if for any $B \in \mathfrak{S}$ there exists an $A \in \mathfrak{S}$ with $B \subset A$ such that the canonical embedding $j_{A,B} : E(B) \to E(A)$ is precompact. To this end, we first notice that $\widehat{j'_A} : E'_{A^o} \to E(A)'$ is an into isometry, where $\widehat{j'_A}$ is the injection associated with the adjoint map j'_A of the embedding map $j_A : E(A) \to E$. It is trivial that the evaluation map $e_A : E(A) \to E(A)''$ is an into isometry. It is also easy to show that the following diagrams are commutative:

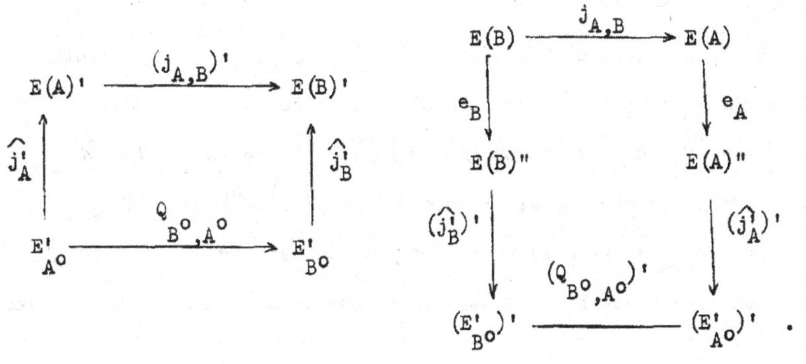

Furthermore, we claim that

$$(\hat{j}_B')' \circ e_B : E(B) \to (E'_{B^o})' \quad \text{and}$$

$$(\hat{j}_A')' \circ e_A : E(A) \to (E'_{A^o})'$$

are into isometries. Indeed, let r_B be the gauge of B defined on $E(B)$, let p_{B^o} be the gauge of B^o defined on E', and let $(\hat{p}_{B^o})^*$ be the dual norm of the quotient norm $\hat{p}_{B^o}$ of p_{B^o}. Then we have, for any $x \in E(B)$, that

$$\begin{aligned}(\hat{p}_{B^o})^*((\hat{j}_B')'(e_B x)) &= \sup\{|\langle Q_{B^o}(x'), (\hat{j}_B')'(e_B x)\rangle| : x' \in B^o\} \\ &= \sup\{|\langle x, \hat{j}_B'(Q_{B^o}(x'))\rangle| : x' \in B^o\} \\ &= \sup\{|\langle x, j_B'(x')\rangle| : x' \in B^o\} \\ &= \sup\{|\langle x, x'\rangle| : x' \in B^o\} = r_B(x) ,\end{aligned}$$

which obtains our assertion. Therefore we have verified the following result, due to Randtke [1].

(1.4.5) Proposition. $E'_{\mathcal{B}}$ __is a Schwartz space if and only if for any__ $B \in \mathcal{B}$ __there exists an__ $A \in \mathcal{B}$ __with__ $B \subset A$ __such that the canonical embedding map__ $j_{A,B} : E(B) \to E(A)$ __is precompact__.

If E is infrabarrelled, then $\{V^o : V \in \mathcal{U}_E\}$ is a fundamental family of $\beta(E', E)$-bounded subsets of E'. In view of (1.2.4)(d), E is a Schwartz space if and only if for any $W \in \mathcal{U}_E$ there is an $V \in \mathcal{U}_E$ with $V \subset W$ such that the canonical embedding map $E'(W^o) \to E'(V^o)$ is precompact, and this is the case if and only if the strong bidual $(E'', \beta(E'', E'))$ of E is a Schwartz space by (1.4.5). Therefore we have proved the following result.

(1.4.6) Corollary. <u>An infrabarrelled space</u> E <u>is a Schwartz space if and only if</u> $(E'', \beta(E'', E'))$ <u>is a Schwartz space</u>.

A locally convex space E is called a <u>co-Schwartz</u> (or <u>dual Schwartz</u>) <u>space</u> if its strong dual $(E', \beta(E', E))$ is a Schwartz space.

Schwartz spaces are, in general, not co-Schwartz spaces, and co-Schwartz spaces are, in general, not Schwartz spaces. In fact, we shall show below (see (1.4.13)) that Fréchet co-Schwartz spaces need not be Schwartz spaces, and that Schwartz (DF)-spaces are, in general, not co-Schwartz spaces.

(1.4.7) Proposition. <u>If</u> F <u>is a co-Schwartz space, then</u>

$$L^{pb}(E, F) = L^{p}(E, F) = L^{\ell b}(E, F)$$

<u>for any locally convex space</u> E .

<u>Proof</u>. It suffices to verify that $L^{\ell b}(E, F) \subseteq L^{pb}(E, F)$. To this end, let $T \in L^{\ell b}(E, F)$, let $V \in \mathcal{U}_E$ be such that $B = T(V)$ is a bounded subset of F , and $T_{(V,B)}$ be the induced map of T . Then $T_{(V,B)} \in L(E_V, F(B))$ and

$$T = j_B \, T_{(V,B)} \, Q_V \; . \tag{4.3}$$

Since F is a co-Schwartz space, there exists, by (1.4.5), a closed absolutely convex bounded subset A of F with $B \subset A$ such that the canonical embedding map $j_{A,B} : F(B) \to F(A)$ is precompact. As $j_B = j_A \, j_{A,B}$, it follows from (4.3) that

$$T = j_A(j_{A,B}\, T_{(V,B)})\, Q_V .$$

As $j_{A,B}\, T_{(V,B)} \in L^p(E_V, F(A))$, we conclude from (1.3.4)(c) that $T \in L^{pb}(E, F)$.

From the preceding result it is not hard to construct an example of precompact linear map which is not precompact-bounded since an uncountable product of $\mathbb{R}$ is not a co-Schwartz space.

Co-Schwartz spaces have the following important properties (compared with (1.2.6), (1.2.7) and (1.2.8)) which are similar to those of Schwartz spaces.

(1.4.8) Proposition. <u>For a co-Schwartz space</u> $(E, \mathcal{P})$, <u>the following assertions hold</u>:

(a) <u>Bounded subsets of</u> E <u>are precompact and separable</u>, <u>hence</u> $\beta(E', E)$ <u>coincides with the topology of precompact convergence</u>, <u>and equicontinuous subsets of</u> E' <u>are</u> $\beta(E', E)$-<u>relatively compact</u>.

(b) E <u>is quasi-complete if and only if it is a semi-Montel space</u>.

(c) <u>For any absolutely convex bounded subset</u> B <u>of</u> E <u>there exists an absolutely convex bounded subset</u> A <u>of</u> E <u>such that the normed space</u> E(A) <u>is separable and</u> B <u>is precompact and separable in</u> E(A) .

(d) <u>If</u>, <u>in addition</u>, E <u>is a dual metric space</u>, <u>then</u> $(E', \beta(E', E))$ <u>is a Fréchet-Montel space</u>, $\mathcal{P}$ <u>is the topology of uniform convergence on</u> $\beta(E', E)$-<u>relatively compact subsets of</u> E' , <u>and</u> E <u>is separable</u>.

(e) *If* E *is quasi-complete, then a sequence* $\{x_n\}$ *in* E *is* $\sigma(E, E')$*-convergent to* x *if and only if* x_n *is* $\mathcal{P}$*-convergent to* x .

(f) *If* E *is* σ*-barrelled, then every* $\sigma(E', E)$*-convergent sequence* $\{f_n\}$ *in* E' *is also* $\beta(E', E)$*-convergent to the same limit.*

Proof. Part (c) follows from (1.4.5) and the fact $E(A)$ is the image of the precompact linear map $j_{A,B}$, while part (b) is a consequence of (a) . On $E(A)$, $\mathcal{P}$ is coarser than the norm topology, part (c) implies (a) . Clearly (e) is a consequence of (b). It remains to verify parts (d) and (f).

(d) We first show that E is separable. Indeed, let $\{A_n : n \geq 1\}$ be a fundamental sequence of bounded sets in E . Then each A_n is separable by (a), hence there exists a sequence $\{x_k^{(n)}, k \geq 1\}$ in A_n which is dense in A_n . Clearly the double sequence $\{x_k^{(n)}, n, k \geq 1\}$ is dense in E .

An almost-verbation transcription of the proof of (f) of (1.2.6) shows that the other parts are valid.

(f) Let f_n be $\sigma(E', E)$-convergent to f . Then the σ-barrelledness ensures that there exists an $V \in \mathcal{U}_E$ such that $f_n, f \in V^o$ for all $n \geq 1$. In view of (a), $\beta(E', E)$ and $\sigma(E', E)$ coincide on V^o , we conclude that f_n is $\beta(E', E)$-convergent to f .

By virtue of (a) of (1.4.8), every co-Schwartz normed space is finite-dimensional. We shall see in Chapter 3 that the converse is also true.

The following result should be compared with (1.2.6).

(1.4.9) Proposition. E is a co-Schwartz space if and only if it satisfies the following two conditions:

(i) for any $V \in \mathcal{U}_E$ and any absolutely convex bounded subset A of E, the canonical map $K_{V,A} : E(A) \to E_V$ is precompact, and

(ii) for any precompact subset C of E there is an absolutely convex bounded subset A of E with $C \subset A$ such that C is a precompact subset of the normed space $E(A)$.

Proof. Necessity. Since precompact subsets of E are bounded, (ii) is a consequence of (c) of (1.4.8). To prove (i), let D be an absolutely convex bounded subset of E such that $A \subset D$ and the canonical embedding map $j_{D,A} : E(A) \to E(D)$ is precompact. Clearly

$$K_{V,A} = K_{V,D}\, j_{D,A} ,$$

we conclude that $K_{V,A}$ is precompact.

Sufficiency. We first claim that every bounded subset A of E is precompact. Indeed, we may assume that A is absolutely convex. Then for any $V \in \mathcal{U}_E$, the condition (i) implies that $K_{V,A}(A) = Q_V(A)$ is precompact in the normed space E_V, hence there exists a finite subset $\{a_1, \ldots, a_n\}$ of A such that

$$Q_V(A) \subset \bigcup_{i=1}^{n} (2^{-1} Q_V(V) + Q_V(a_i)) ,$$

thus

$$A \subset \bigcup_{i=1}^{n} (2^{-1}V + a_i) + p_V^{-1}(0) \subset \bigcup_{i=1}^{n} (V + a_i) \quad ,$$

which obtains our assertion.

For any absolutely convex bounded subset C of E, C is precompact, hence the condition (ii) ensures that there exists an absolutely convex bounded subset A of E with $C \subset A$ such that C is precompact in the normed space $E(A)$, thus the canonical embedding $j_{A,C} : E(C) \to E(A)$ is precompact since C is a o-neighbourhood in the normed space $E(C)$. Therefore E is a co-Schwartz space by (1.4.5).

(1.4.10) Proposition. <u>Let</u> E <u>be a</u> (DF)-<u>space which is either a Schwartz space or a co-Schwartz space</u>. <u>Then</u> E <u>is infrabarrelled</u>; <u>consequently quasi-complete Schwartz</u> (<u>or co-Schwartz</u>) (DF)-<u>spaces are Montel spaces</u>.

<u>Proof</u>. We first notice that (DF)-spaces are dual metric spaces. If E is a co-Schwartz space, then E is separable by (d) of (1.4.8), thus E is infrabarrelled by a result of Grothendieck (see Köthe [1, p.399]). If E is a Schwartz space, then E is infrabarrelled by (1.2.6)(f).

Following Köthe [1], the <u>polar topology</u> $\mathscr{P}^{o}$ of $\mathscr{P}$ is defined to be the topology on E' of uniform convergence on $\mathscr{P}$-precompact subsets of E.

(1.4.11) Proposition. <u>For a</u> (DF)-<u>space</u> $(E, \mathscr{P})$, <u>the following statements are equivalent</u>.

(a) E <u>is a Schwartz space</u>.

(b) E *is infrabarrelled and bounded subsets of* E *are precompact.*

(c) *Bounded subsets of* E *are precompact, and for any* $\beta(E', E)$-*convergent sequence* $\{x'_n\}$ *in* E' *there is an* $V \in \mathcal{U}_E$ *such that* x'_n *converges to the same limit in the Banach space* $E'(V^o)$.

In particular, a (DF)-*space which is either a Montel space or a co-Schwartz space is a Schwartz space.*

Proof. (a) $\Rightarrow$ (b) : Follows from (1.2.6)(a) and (1.4.10).

(b) $\Rightarrow$ (c) : Let x'_n be $\beta(E', E)$-convergent to x' . Then the set $\{x', x'_n\}$ is $\beta(E', E)$-compact because of $\beta(E', E) = \mathcal{P}^o$ and the infrabarrelledness. As $(E', \beta(E', E))$ is a Fréchet space, there is an $V \in \mathcal{U}_E$ such that $\{x', x'_n\}$ is a relatively compact subset of the Banach space $E'(V^o)$, thus x'_n converges to x' in $E'(V^o)$.

(c) $\Rightarrow$ (a) : We first notice that $(E', \beta(E', E))$ is a Fréchet space and $\beta(E', E) = \mathcal{P}^o$, W^o is $\beta(E', E)$-compact for any $W \in \mathcal{U}_E$. Now, for any $W \in \mathcal{U}_E$ there is a null sequence $\{x'_n\}$ in $(E', \beta(E', E))$ such that each $x' \in W^o$ is of the form

$$\sum_{n=1}^{\infty} \lambda_n x'_n \quad \text{with} \quad \sum_{n=1}^{\infty} |\lambda_n| \leq 1 .$$

By the hypothese, there is an $V \in \mathcal{U}_E$ such that x'_n converges to 0 in $E'(V^o)$. We claim that $\sum_{n=1}^{\infty} \lambda_n x'_n$ converges to x' in $E'(V^o)$. Indeed, let p_{V^o} be the gauge of V^o defined on $E'(V^o)$. There is an $k > 0$ such that $p_{V^o}(x'_n) \leq 1$ for all $n \geq k$, hence

$$p_{V^o}(\lambda_n x'_n + \dots + \lambda_{n+m} x'_{n+m}) \leq \Sigma_{j=1}^{m} |\lambda_{n+j}| \quad \text{for all} \quad n \geq k \quad \text{and} \quad m \geq 0 ,$$

and thus $\Sigma_{n=1}^{\infty} \lambda_n x'_n$ converges to x' in $E'(V^o)$ since $\Sigma_{n=1}^{\infty} |\lambda_n| \leq 1$, $E'(V^o)$ is complete and the relative topology on $E'(V^o)$ induced by $\beta(E', E)$ is coarser than the p_{V^o}-topology. Therefore W^o is a compact subset of the Banach space $E'(V^o)$, and thus E is a Schwartz space by (1.2.4)(d).

According to (1.4.8) and (1.2.6), Fréchet-Schwartz spaces and Fréchet-co-Schwartz spaces are Montel spaces and surely separable because Fréchet-Montel spaces are separable. Furthermore, we have the following result, due to Terzioğlu [3].

(1.4.12) Proposition. *For a Fréchet space* E, *the following statements are equivalent.*

(a) E *is a co-Schwartz space.*

(b) E *is a Montel space.*

(c) *Bounded subsets of* E *are precompact.*

In particular, Fréchet-Schwartz spaces are co-Schwartz spaces.

Proof. The implication (a) ⇒ (b) follows from (1.4.8)(b), and the implication (b) ⇒ (c) is obvious. To prove the implication (c) ⇒ (a), let B be an absolutely convex bounded subset of E. Then B is precompact, hence there is an absolutely convex compact subset A of E with $B \subset A$ such that B is a relatively compact subset of the Banach space $E(A)$ by (1.4.9), thus $j_{A,B} \in L^p(E(B), E(A))$. Therefore E is a co-Schwartz space by (1.4.5).

(1.4.13) Examples. (a) <u>Fréchet co-Schwartz spaces are, in general, not Schwartz spaces</u>.

(b) <u>Schwartz</u> (DF)-<u>spaces are, in general, not co-Schwartz spaces</u>.

<u>Proof</u>. It is known from Köthe [1, §31.5] that there exists a Fréchet-Montel space G in which there is a closed subspace N such that $G/_N$ is <u>not</u> a Montel space. According to (1.4.12) and (1.4.1)(b), G is a Fréchet co-Schwartz space which is <u>not</u> a Schwartz space.

To prove (b), we suppose, on the contrary, that each Schwartz (DF)-space is a co-Schwartz space. Then G'_β is a Schwartz (DF)-space, where G is the space mentioned in the above. Therefore the strong bidual G''_β of G is a Fréchet-Schwartz by the assumption, and thus G must be a Schwartz space by (1.4.1)(a) since G is a subspace of G''_β, which gives a contradiction.

Let N be a closed vector subspace of $(E, \mathscr{P})$. Then we identify $(E/_N)'$ with $N^\perp$ and $E'/_{N^\perp}$ with N'. It is well-known that

$$\beta(E', E)|_{N^\perp} \leq \beta(N^\perp, E/_N) \quad \text{and} \quad \beta(E'/_{N^\perp}, N) \leq \hat{\beta}(E', E) ,$$

where $\beta(E', E)|_{N^\perp}$ is the relative topology on $N^\perp$ induced by $\beta(E', E)$, and $\hat{\beta}(E', E)$ is the quotient topology of $\beta(E', E)$ by $N^\perp$. If either E is semi-reflexive or the subspace N is an σ-infrabarrelled space for which its strong dual is bornological, then $\beta(E'/_{N^\perp}, N) = \hat{\beta}(E', E)$. On the other hand, it is easily seen that the equality

$$\beta(E', E)|_{N^\perp} = \beta(N^\perp, E/_N) \tag{4.4}$$

is equivalent to the fact that every bounded subset of the quotient space $(E/_N, \hat{\mathcal{P}})$ is contained in the image, under the quotient map Q_N, of some bounded subset of E. The last statement holds for E being a (DF)-space as shown by Grothendieck. We shall show that the formula (4.4) holds whenever E is a Fréchet-Schwartz space. To do this we need the following

(1.4.14) Lemma. Let E and F be complete metrizable topological vector spaces and let $T \in L(E, F)$ be surjective. Then every compact subset K of F is the image, under T, of some compact subset of E.

Proof. Let $\{V_n\}$ be a countable basis of neighbourhoods of 0 in E such that $V_{n+1} + V_{n+1} \subset V_n$. Then $\{T(V_n)\}$ is a countable basis of neighbourhoods of 0 in F by the Banach open mapping theorem. For each n, the compactness of K ensures that there is a finite subset K_n of K such that $K \subset K_n + T(V_n)$, so that

$$K \subset T(C_n + V_n) ,$$

where C_n is a finite subset of E such that $K_n = T(C_n)$ because T is surjective. From this and $K_{n+1} \subset K$, we can choose $C_{n+1} \subset C_n + V_n$. Since $V_{n+1} + \dots + V_{n+j} \subset V_n$ for all $j \geq 1$, it follows that

$$C_{n+1+j} \subset C_{n+j} + V_{n+j} \subset C_{n+1} + V_n \quad \text{for all } j \geq 1 ,$$

and hence that $C = \bigcup_n C_n$ is a precompact subset of E because of $C \subset (\bigcup_{j=1}^{n+1} C_j) + V_n$. The completeness of E implies that $\overline{C}$ is compact and

$$K \subset T(C_n) + T(V_n) \subset T(\overline{C}) + T(V_n) \quad \text{for all } n .$$

From this it follows that $K \subset \bigcap_n (T(\overline{C}) + T(V_n)) = T(\overline{C})$ because $T(\overline{C})$ is compact. Therefore the set B, defined by

$$B = \overline{C} \cap (T^{-1}(K)) ,$$

has the required properties.

(1.4.15) Theorem. Let $(E, \mathcal{P})$ be a Fréchet-Schwartz space and N a closed vector subspace of E. Then

$$\beta(E', E)|_{N^\perp} = \beta(N^\perp, E/_N) \text{ and } \beta(E'/_{N^\perp}, N) = \hat{\beta}(E', E) .$$

Proof. The equality $\beta(E'/_{N^\perp}, N) = \hat{\beta}(E', E)$ is obvious because E is a Fréchet-Montel space by (1.2.6). To prove the equality $\beta(E', E)|_{N^\perp} = \beta(N^\perp, E/_N)$, we notice that the quotient space $(E/_N, \hat{\mathcal{P}})$ is a Fréchet-Schwartz space, hence bounded subsets of $E/_N$ are $\hat{\mathcal{P}}$-relatively compact by (1.2.6). Therefore every bounded subset of $E/_N$ is contained in the image, under Q_N, of some compact subset of E by (1.4.14), thus every bounded subset of $E/_N$ is contained in the image, under Q_N, of some bounded subset of E since bounded subsets of E are relatively compact.

(1.4.16) Theorem. Let $(E, \mathcal{P})$ and $(F, \mathcal{J})$ be Fréchet-Schwartz spaces and $T \in L(E, F)$. Then the following statements are equivalent.

(a) T is a topological homomorphism for $\mathcal{P}$ and $\mathcal{J}$.

(b) T is a weak homomorphism.

(c) $T'(F')$ is $\sigma(E', E)$-closed in F'.

(d) $T(E)$ <u>is</u> $\mathcal{J}$ -<u>closed in</u> F .

(e) $T(E)$ <u>is</u> $\sigma(E, E')$-<u>closed in</u> F .

(f) T' <u>is a topological homomorphism for</u> $\sigma(F', F)$ <u>and</u> $\sigma(E', E)$.

(g) T' <u>is a topological homomorphism for</u> $\beta(F', F)$ <u>and</u> $\beta(E', E)$.

(h) $T'(F')$ <u>is</u> $\beta(E', E)$-<u>closed in</u> E' .

<u>Proof</u>. In view of Dieudonné-Schwartz's homomorphism theorem (see Horváth [1, p.308]), the statements (a) through (f) are mutually equivalent. Since Fréchet-Schwartz spaces are Montel spaces, it follows that $\beta(E', E)$ is consistent with the dual pair $\langle E, E'\rangle$, and hence that (h) and (c) are equivalent. On the other hand, Köthe's homomorphism theorem (see Horváth [2, 5.7]) shows that (g) implies (b) because E and F are reflexive. Therefore we complete the proof by showing the implication (a) $\Rightarrow$ (g).

To do this we first notice that the bijection $\check{T}$ associated with T is a topological isomorphism from the quotient space $E/_{\text{Ker } T}$ onto the subspace $T(E)$ of F . Hence $\check{T}$ gives a one-to-one correspondence between bounded subsets of $E/_{\text{Ker } T}$ and of $T(E)$, thus $(\check{T})'$ is a topological isomorphism from $((T(E))', \beta((T(E))', T(E)))$ onto $(E/_{\text{Ker } T})'_\beta$. As $(T(E))'$ being algebraically isomorphic to $F'/_{\text{Ker } T'}$, and $(E/_{\text{Ker } T})'$ being algebraically isomorphic to $T'(F') = (\text{Ker } T)^{\perp}$, it follows that the bijection $\check{T}'$ associated with T' can be identified with $(\check{T})'$. Since $T(E)$ is closed in F , we conclude from (1.4.15) that

$$\beta((T(E))', T(E)) = \hat{\beta}(F', F) \quad \text{and} \quad \beta(T'(F'), E/_{\text{Ker } T}) = \beta(E', E)\big|_{T'(E')} ,$$

and hence that T' is a topological homomorphism from $(F', \beta(F', F))$ onto $(E', \beta(E', E))$.

1.5 Universal Schwartz spaces

Let us call a locally convex topology $\mathcal{T}$ on G a *Schwartz topology* if $(G, \mathcal{T})$ is a Schwartz space. We first show that the existence of the finest Schwartz topology on G which is consistent with the dual pair $\langle G, G'\rangle$ as follows.

(1.5.1) Lemma. *For a locally convex space* G, *the family of all* $\tau(G, G')$-*precompact seminorms on* G *determines the finest Schwartz topology* $\mathcal{T}$ *on* G *which is consistent with the dual pair* $\langle G, G'\rangle$.

Proof. Clearly $\mathcal{T}$ is consistent with $\langle G, G'\rangle$. To verify that $(G, \mathcal{T})$ is a Schwartz space, let q be a $\mathcal{T}$-continuous seminorm on G. Then there exist an $(\zeta_n) \in c_o$ and an $\tau(G, G')$-equicontinuous sequence $\{u_n'\}$ in G' such that

$$q(x) \leq \sup\{|\zeta_n \langle x, u_n'\rangle| : n \geq 1\} \quad \text{for all} \quad x \in G . \tag{5.1}$$

We now choose (μ_n) and (τ_n) in c_o such that $\zeta_m = \mu_m \tau_m$ for all $m \geq 1$, and define

$$f_m = \mu_m u_m' \quad \text{for all} \quad m \geq 1 . \tag{5.2}$$

Then $\sup\{|\mu_n \langle x, u_n'\rangle| : n \geq 1\}$ defines a $\mathcal{T}$-continuous seminorm on G, and

$$|\langle x, f_m\rangle| \leq \sup\{|\mu_n \langle x, u_n'\rangle| : n \geq 1\} \quad \text{for all} \quad x \in G \quad \text{and} \quad m \geq 1 .$$

Therefore $\{f_m\}$ is an $\mathcal{T}$-equicontinuous sequence in G' such that

$$q(x) \leq \sup\{|\tau_n \langle x, f_n\rangle| : n \geq 1\} \quad \text{for all} \quad x \in G$$

by virtue of (5.1) and (5.2), and thus q is a precompact seminorm on $(G, \mathfrak{L})$; consequently $(G, \mathfrak{L})$ is a Schwartz space.

Finally, if $\mathscr{P}$ is a Schwartz topology on G consistent with $\langle G, G'\rangle$, then every $\mathscr{P}$-equicontinuous sequence in G' must be $\tau(G, G')$-equicontinuous, hence $\mathscr{P}$ is coarser than $\mathfrak{L}$ by the definition of precompact seminorms.

Observe that if X is a separable normed space, then the closed unit ball Σ' in X' is metrizable and compact for $\sigma(X', X)$, and surely separable for $\sigma(X', X)$, hence there is a $\sigma(X', X)$-dense sequence $\{g_n\}$ in Σ' ; consequently X is isometrically isomorphic to a subspace of the Banach space $(\ell^\infty, \|\cdot\|_\infty)$ under the map $x \mapsto (\langle x, g_n\rangle)$.

Suppose now that E is a Schwartz space and that $\{p_\alpha : \alpha \in \Lambda\}$ is the family of all precompact seminorms on E. Then for each $\alpha \in \Lambda$ there exists an isometric isomorphism S_α from E_α into $(\ell^\infty, \|\cdot\|_\infty)$, therefore we identify E_α with a subspace of $(\ell^\infty, \|\cdot\|_\infty)$. For clearity of the notation, let us put

$$X_\alpha = \ell^\infty \quad \text{and} \quad \|\cdot\|_\alpha = \|\cdot\|_\infty \quad \text{for all } \alpha \in \Lambda .$$

Suppose now that μ, in Λ, is such that $\mu \geq \alpha$. Then $S_\alpha Q_{\alpha,\mu} : E_\mu \to (X_\alpha, \|\cdot\|_\alpha)$ is compact, and hence by a result of Lindenstrauss and Rosenthal [1, (4.1)] there exists a compact linear map $T_{\alpha,\mu} : (X_\mu, \|\cdot\|_\mu) \to (X_\alpha, \|\cdot\|_\alpha)$ such that

$$T_{\alpha,\mu} S_\mu = S_\alpha Q_{\alpha,\mu} .$$

As S_α are isometric isomorphisms, it follows from (1.4.3) that E is topologically isomorphic to the subspace

$$F = \{x = (x_\alpha) \in \prod X_\alpha : \pi_\alpha(x) = T_{\alpha,\mu}(\pi_\mu(x)) \text{ for all } \mu \geq \alpha \text{ in } \Lambda\} \qquad (5.3)$$

of the product space $\prod(X_\alpha, \|\cdot\|_\alpha)$, equipped with the relative topology $\mathcal{P}$ induced by the product topology of the norm topologies, where $\pi_\alpha : \prod X_\mu \to X_\alpha$ is the α-th projection. Therefore we identify E with $(F, \mathcal{P})$. On the other hand, as the norm topology on ℓ^∞ coincides with $\tau(\ell^\infty, (\ell^\infty)')$, it follows from (1.5.1) that the topology $\mathcal{L}$ on ℓ^∞ determined by all $\|\cdot\|_\infty$-precompact seminorms on ℓ^∞ is the finest Schwartz topology consistent with $\langle \ell^\infty, (\ell^\infty)' \rangle$, and hence from (1.4.1) that the relative topology on F induced by $\mathcal{L}^\wedge$, denoted by $\mathcal{T}$, is a Schwartz topology. As $\mathcal{L}$ is coarser than the $\|\cdot\|_\infty$-topology, it follows that $\mathcal{T}$ is coarser than $\mathcal{P}$. Moreover, we show that

$$\mathcal{P} \leq \mathcal{T} .$$

In fact, let $\alpha \in \Lambda$ and let $W = \Sigma_\alpha \times \prod_{\mu \neq \alpha} X_\mu$, where $\Sigma_\alpha = \{x_\alpha \in X_\alpha : \|x_\alpha\|_\alpha \leq 1\}$. Then $W \cap F$ is a $\mathcal{P}$-neighbourhood of 0 in F. Let μ, in Λ, be such that $\mu \geq \alpha$, and let $T_{\alpha,\mu} : (X_\mu, \|\cdot\|_\mu) \to (X_\alpha, \|\cdot\|_\alpha)$ be a compact linear map. Then there exists a $\|\cdot\|_\mu$-precompact seminorm q_μ on X_μ such that

$$\|T_{\alpha,\mu}(x_\mu)\|_\alpha \leq q_\mu(x_\mu) \text{ for all } x_\mu \in X_\mu , \qquad (5.4)$$

hence $V_\mu = \{x_\mu \in X_\mu : q_\mu(x_\mu) \leq 1\}$ is a $\mathcal{L}$-neighbourhood of 0 in X_μ. Denote by $V = V_\mu \times \prod_{\tau \neq \mu} X_\tau$. Then $V \cap F$ is a $\mathcal{T}$-neighbourhood of 0 in

F . Furthermore,

$$V \cap F \subset W \cap F$$

Indeed, if $x = (x_\alpha) \in V \cap F$, then $q_\mu(x_\mu) \leq 1$ and

$$x_\alpha = \pi_\alpha(x) = T_{\alpha,\mu}(\pi_\mu(x)) = T_{\alpha,\mu}(x_\mu)$$

by (5.3), hence (5.4) shows that $\|x_\alpha\|_\alpha \leq q_\mu(x_\mu) \leq 1$; consequently $x \in W \cap F$.

Therefore $\mathcal{P}$ and $\mathcal{J}$ coincide, and thus we obtain the following

(1.5.2) Theorem (Randtke). Let $\mathcal{L}$ be the locally convex topology on ℓ^∞ determined by the family of all $\|\cdot\|_\infty$-precompact seminorms on ℓ^∞ . Then a locally convex space E is a Schwartz space if and only if it is topologically isomorphic to a subspace of the product space $((\ell^\infty)^\Lambda, \mathcal{L}^\Lambda)$ for some index set Λ . Consequently, every metrizable Schwartz space is topologically isomorphic to a subspace of $((\ell^\infty)^{\mathbb{N}}, \mathcal{L}^{\mathbb{N}})$.

If G denotes either the Banach space c_0 or the Banach space C[0, 1] , then the topology $\mathcal{L}$ on G determined by all precompact seminorms is the finest Schwartz topology on G which is consistent with $\langle G, G'\rangle$ by (1.5.1), and $(G, \mathcal{L})$ is also a universal Schwartz space (see Randtke [4]).

CHAPTER 2. VECTOR SEQUENCE SPACES AND ABSOLUTELY SUMMING MAPPINGS

2.1 Locally convex topologies on vector sequence spaces

Throughout these notes $(E, \mathscr{P})$ will denote a locally convex space whose Hausdorff locally convex topology $\mathscr{P}$ is determined by a family $\mathbb{P}$ of seminorms, Λ will denote a non-empty index set and $\mathscr{F}(\Lambda)$ will denote the direct set consisting of all non-empty finite subsets of Λ ordered by the set inclusion. Elements in $\mathscr{F}(\Lambda)$ will be denoted by α, β, γ etc..

Denote by E^{Λ} (resp. $E^{(\Lambda)}$) the algebraic product (resp. algebraic direct sum) of E with Λ times, and by $\mathscr{P}^{\Lambda}$ (resp. $\mathscr{P}^{(\Lambda)}$) the product (resp. locally convex direct sum) topology. Elements in E^{Λ} will be denoted by $[x_i, \Lambda]$ which are called <u>families</u> (with index set Λ) in E, and elements in $E^{(\Lambda)}$ will be sometimes denoted by $[x_i, (\Lambda)]$.

For each $\alpha \in \mathscr{F}(\Lambda)$, we define the map $P_\alpha : [x_i, \Lambda] \mapsto [x_i^{(\alpha)}, (\Lambda)]$ by setting

$$x_i^{(\alpha)} = \begin{cases} x_i & \text{if } i \in \alpha \\ 0 & \text{if } i \notin \alpha \end{cases}.$$

Then P_α is clearly a linear map from E^{Λ} into $E^{(\Lambda)}$, and $\{[x_i^{(\alpha)}, (\Lambda)], \alpha \in \mathscr{F}(\Lambda)\}$ is a net in $E^{(\Lambda)}$. $[x_i^{(\alpha)}, (\Lambda)]$ is called the <u>α-family associated with</u> $[x_i, \Lambda]$ and $\{[x_i^{(\alpha)}, (\Lambda)], \alpha \in \mathscr{F}(\Lambda)\}$ is called <u>the net</u>

<u>in</u> $E^{(\Lambda)}$ <u>associated with</u> $[x_\iota, \Lambda]$. We also define the map $J_\alpha : E \to E^{(\Lambda)}$ by setting

$$\pi_\iota(J_\alpha(x)) = \begin{cases} x & \text{if } \iota \in \alpha \\ 0 & \text{if } \iota \notin \alpha \end{cases} ,$$

where π_ι is the ι-th projection; in particular, if $\iota \in \Lambda$ then we write J_ι for $J_{\{\iota\}}$. Clearly J_α is linear and injective. J_α is referred to as the α-<u>embedding map</u>, and J_ι is called simply the <u>embedding map</u>.

Recall that a family $[\zeta_\iota, \Lambda]$ in $\mathbb{C}$ is <u>summable</u> if the net $\{\Sigma_{\iota\in\alpha}\zeta_\iota, \alpha \in \mathfrak{F}(\Lambda)\}$ converges. The uniquely determined limit ζ is called the <u>sum</u> of $[\zeta_\iota, \Lambda]$ and we write

$$\zeta = \Sigma_\Lambda\zeta_\iota \quad \text{or} \quad \zeta = \Sigma_\iota\zeta_\iota \ .$$

If Λ is finite, then ζ obviously coincides with the ordinary sum. Each summable family of numbers contains at most countably many non-zero terms. On the other hand, if each $\zeta_\iota \geq 0$, then $[\zeta_\iota, \Lambda]$ is summable if and only if the net $\{\Sigma_{\iota\in\alpha}\zeta_\iota, \alpha \in \mathfrak{F}(\Lambda)\}$ is bounded; in this case,

$$\Sigma_\Lambda\zeta_\iota = \sup\{\Sigma_{\iota\in\alpha}\zeta_\iota : \alpha \in \mathfrak{F}(\Lambda)\} \ .$$

Moreover, it is not hard to show that a family $[\zeta_\iota, \Lambda]$ of numbers is summable if and only if $[|\zeta_\iota|, \Lambda]$ is summable; hence we can express the summability of $[\zeta_\iota, \Lambda]$ by the inequality

$$\Sigma_\Lambda|\zeta_\iota| < +\infty \ .$$

It is also easily seen that if there is a constant $C > 0$ such that

$$|\Sigma_{\iota\in\alpha}\zeta_\iota| \leq C \quad \text{for all} \quad \alpha \in \mathcal{F}(\Lambda) ,$$

then

$$\Sigma_{\iota\in\alpha}|\zeta_\iota| \leq 4C \quad \text{for all} \quad \alpha \in \mathcal{F}(\Lambda) .$$

(2.1.1) Lemma. Let ψ be a functional (not necessarily linear) defined on E such that

$$\psi(0) = 0 \quad \text{and} \quad \psi(x) \geq 0 \quad \text{for all} \quad x \in E ,$$

and let $[x_\iota, \Lambda]$ be a family in E. For each $\alpha \in \mathcal{F}(\Lambda)$, the family $[\psi(x_\iota), \Lambda\backslash\alpha]$ of numbers (with index set $\Lambda\backslash\alpha$) is summable if and only if the family $[\psi(x_\iota - x_\iota^{(\alpha)}), \Lambda]$ of numbers (with index set Λ) is summable; in this case,

$$\Sigma_\Lambda \psi(x_\iota - x_\iota^{(\alpha)}) = \Sigma_{\Lambda\backslash\alpha}\psi(x_\iota) . \tag{1.1}$$

Furthermore, the following inequality holds:

$$\Sigma_\Lambda\psi(x_\iota - x_\iota^{(\beta)}) \leq \Sigma_\Lambda\psi(x_\iota - x_\iota^{(\alpha)}) \quad \text{for all} \quad \beta \in \mathcal{F}(\Lambda) \quad \text{with} \quad \beta \geq \alpha .$$

Proof. Assume that $\Sigma_{\Lambda\backslash\alpha}\psi(x_\iota) < +\infty$. For any $\beta \in \mathcal{F}(\Lambda)$ we have either $\beta \cap \alpha = \phi$ or $\beta \cap \alpha \neq \phi$. If $\beta \cap \alpha = \phi$, then $\beta \in \mathcal{F}(\Lambda\backslash\alpha)$ and $x_\iota^{(\alpha)} = 0$ for all $\iota \in \beta$, hence

$$\Sigma_{\iota\in\beta}\psi(x_\iota - x_\iota^{(\alpha)}) = \Sigma_{\iota\in\beta}\psi(x_\iota) \leq \Sigma_{\Lambda\backslash\alpha}\psi(x_\iota) .$$

If $\beta \cap \alpha \neq \phi$, then

$$\Sigma_{\iota\in\beta}\psi(x_\iota - x_\iota^{(\alpha)}) \leq \Sigma_{\iota\in\beta\setminus\alpha}\psi(x_\iota - x_\iota^{(\alpha)}) + \Sigma_{\iota\in\alpha}\psi(x_\iota - x_\iota^{(\alpha)})$$
$$= \Sigma_{\iota\in\beta\setminus\alpha}\psi(x_\iota - x_\iota^{(\alpha)}) \leq \Sigma_{\Lambda\setminus\alpha}\psi(x_\iota) \ .$$

As β was arbitrary, we see that $[\psi(x_\iota - x_\iota^{(\alpha)}), \Lambda]$ is summable and

$$\Sigma_\Lambda\psi(x_\iota - x_\iota^{(\alpha)}) \leq \Sigma_{\Lambda\setminus\alpha}\psi(x_\iota) \tag{1.2}$$

Conversely, if $\Sigma_\Lambda\psi(x_\iota - x_\iota^{(\alpha)}) < +\infty$, then any $\gamma \in \mathcal{F}(\Lambda\setminus\alpha)$ is disjoint from α , hence

$$\Sigma_{\iota\in\gamma}\psi(x_\iota) = \Sigma_{\iota\in\gamma}\psi(x_\iota - x_\iota^{(\alpha)}) \leq \Sigma_\Lambda\psi(x_\iota - x_\iota^{(\alpha)})$$

which shows that $[\psi(x_\iota), \Lambda\setminus\alpha]$ is summable and

$$\Sigma_{\Lambda\setminus\alpha}\psi(x_\iota) \leq \Sigma_\Lambda\psi(x_\iota - x_\iota^{(\alpha)}) \ . \tag{1.3}$$

Combining (1.2) and (1.3) we obtain the required equality (1.1). The final conclusion follows from (1.1) and the fact that $\mathcal{F}(\Lambda\setminus\beta) \subset \mathcal{F}(\Lambda\setminus\alpha)$.

Following Pietsch [1], a family $[x_\iota, \Lambda]$ in $(E, \mathcal{P})$ is said to be

(i) <u>weakly summable</u> if for any $f \in E'$, the family $[\langle x_\iota, f\rangle, \Lambda]$ of numbers is summable;

(ii) <u>summable</u> (or <u>unconditionally Cauchy</u>) if the net $\{\Sigma_{\iota\in\alpha}x_\iota, \alpha \in \mathcal{F}(\Lambda)\}$ is a $\mathcal{P}$-Cauchy net in E ;

(iii) <u>absolutely summable</u> if for any $p \in \mathbb{P}$, the family $[p(x_\iota), \Lambda]$ of positive numbers is summable.

The sum of a summable family $[x_i, \Lambda]$ in E is defined to be the limit x of the $\mathcal{P}$-Cauchy net $\{\Sigma_{i\in\alpha} x_i, \alpha \in \mathcal{F}(\Lambda)\}$ which belongs to the completion of E, and we write

$$x = \Sigma_\Lambda x_i \quad \text{or} \quad x = \Sigma_i x_i .$$

Denote by $\ell^1_w(\Lambda, E)$ the set consisting of all weakly summable families in E with index set Λ, and by $\ell^1(\Lambda, E)$ (resp. $\ell^1[\Lambda, E]$) the set of all summable families (resp. absolutely summable families) in E with index set Λ. Then it is easily seen that they are all vector subspaces of E^Λ and

$$E^{(\Lambda)} \subset \ell^1[\Lambda, E] \subset \ell^1(\Lambda, E) \subset \ell^1_w(\Lambda, E) \subset E^\Lambda .$$

If $\Lambda = \mathbb{N}$, then we write $\ell^1_w(E)$ for $\ell^1_w(\mathbb{N}, E)$, $\ell^1(E)$ for $\ell^1(\mathbb{N}, E)$, $\ell^1[E]$ for $\ell^1[\mathbb{N}, E]$, and similarly $[x_i]$ for $[x_i, \mathbb{N}]$.

As the topology $\sigma(E, E')$ is consistent with the dual pair $\langle E, E'\rangle$, it follows that a family $[x_i, \Lambda]$ in E is weakly summable if and only if the net $\{\Sigma_{i\in\alpha} x_i, \alpha \in \mathcal{F}(\Lambda)\}$ is bounded in E. Therefore we define for any $p \in \mathbb{P}$

(a) $\quad p_w([x_i, \Lambda]) = \sup\{\Sigma_\Lambda |\langle x_i, f\rangle| : f \in V_p^o\} \qquad [x_i, \Lambda] \in \ell^1_w(\Lambda, E)$;

(a)* $\quad p_\varepsilon([x_i, \Lambda]) = p_w([x_i, \Lambda]) \qquad [x_i, \Lambda] \in \ell^1(\Lambda, E)$;

(b) $\quad p_\pi([x_i, \Lambda]) = \Sigma_\Lambda p(x_i) \qquad [x_i, \Lambda] \in \ell^1[\Lambda, E]$,

where $V_p = \{x \in E : p(x) \leq 1\}$ and V_p^o is the polar of V_p.

Clearly, p_w, p_ε and p_π are seminorms and

$$p_\varepsilon([x_i, \Lambda]) \leq p_\pi([x_i, \Lambda]) \quad \text{for all} \quad [x_i, \Lambda] \in \ell^1[\Lambda, E] .$$

Furthermore, for any $\alpha \in \mathfrak{F}(\Lambda)$, Lemma (2.1.1) shows that

$$p_\varepsilon([x_i, \Lambda] - [x_i^{(\alpha)}, (\Lambda)]) = p_\varepsilon([x_i, \Lambda\backslash\alpha]) \quad \text{for all} \quad [x_i, \Lambda] \in \ell^1(\Lambda, E) \qquad (1.4)$$

and

$$p_\pi([x_i, \Lambda] - [x_i^{(\alpha)}, (\Lambda)]) = p_\pi([x_i, \Lambda\backslash\alpha]) \quad \text{for all} \quad [x_i, \Lambda] \in \ell^1[\Lambda, E] ,$$

where $[x_i^{(\alpha)}, (\Lambda)]$ is the α-family associated with $[x_i, \Lambda]$.

Denote by $\mathcal{P}_W$ the locally convex topology on $\ell^1_W(\Lambda, E)$ determined by $\{p_W : p \in \mathbb{P}\}$, by $\mathcal{P}_\varepsilon$ the locally convex topology on $\ell^1(\Lambda, E)$ determined by $\{p_\varepsilon : p \in \mathbb{P}\}$, and by $\mathcal{P}_\pi$ the locally convex topology on $\ell^1[\Lambda, E]$ determined by $\{p_\pi : p \in \mathbb{P}\}$. $\mathcal{P}_\varepsilon$ is called the ε-topology and $\mathcal{P}_\pi$ is called the π-topology. We also write $\ell^1_\varepsilon(\Lambda, E)$ for the locally convex space $(\ell^1(\Lambda, E), \mathcal{P}_\varepsilon)$ and $\ell^1_\pi[\Lambda, E]$ for the locally convex space $(\ell^1[\Lambda, E], \mathcal{P}_\pi)$.

Clearly, if $\mathcal{P}$ is metrizable (resp. normable), then so do $\mathcal{P}_\pi$ and $\mathcal{P}_W$. On the other hand, it is easily seen that $\mathcal{P}_\varepsilon$ is the relative topology on $\ell^1(\Lambda, E)$ induced by $\mathcal{P}_W$, and that

$$\mathcal{P}_\varepsilon \leq \mathcal{P}_\pi \quad \text{on} \quad \ell^1[\Lambda, E] .$$

Therefore the canonical embedding map from $\ell^1[\Lambda, E]$ into $\ell^1(\Lambda, E)$ is a continuous linear map from $(\ell^1[\Lambda, E], \mathcal{P}_\pi)$ into $(\ell^1(\Lambda, E), \mathcal{P}_\varepsilon)$. Furthermore, the algebraic and topological relationship between $(\ell^1[\Lambda, E], \mathcal{P}_\pi)$ and $(\ell^1(\Lambda, E), \mathcal{P}_\varepsilon)$ are determined, for any infinite index set Λ , by the corresponding relationship between $\Lambda = \mathbb{N}$ (see Schaefer [1]) . On the other hand, each j-th projection $\pi_j : (\ell^1_w(\Lambda, E), \mathcal{P}_w) \to (E, \mathcal{P})$ is clearly continuous; and we also have for any $p \in \mathbb{P}$ that

$$\Gamma(\bigcup_{j\in\Lambda} J_j(V_p)) \subset \{[x_i, (\Lambda)] \in E^{(\Lambda)} : p_\pi([x_i, (\Lambda)]) \leq 1\} ,$$

hence the embedding map from $(E^{(\Lambda)}, \mathcal{P}^{(\Lambda)})$ into $(\ell^1[\Lambda, E], \mathcal{P}_\pi)$ is continuous. Therefore we conclude that

$$\mathcal{P}^\Lambda \leq \mathcal{P}_w \quad \text{on} \quad \ell^1_w(\Lambda, E) \quad \text{and} \quad \mathcal{P}_\pi \leq \mathcal{P}^{(\Lambda)} \quad \text{on} \quad E^{(\Lambda)} .$$

It is known that summable families are weakly summable, but the converse need not be true. However, we have the following

(2.1.2) Lemma. <u>An</u> $[x_i, \Lambda] \in \ell^1_w(\Lambda, E)$ <u>is summable if and only if its associated net</u> $\{[x_i^{(\alpha)}, (\Lambda)], \alpha \in \mathcal{F}(\Lambda)\}$ (in $E^{(\Lambda)}$) <u>is</u> $\mathcal{P}_w$<u>-convergent to</u> $[x_i, \Lambda]$.

<u>Proof</u>. <u>Necessity</u>. Let $p \in \mathbb{P}$ and $V_p = \{x \in E : p(x) \leq 1\}$. For any $\delta > 0$ there exists an $\alpha \in \mathcal{F}(\Lambda)$ such that

$$p(\Sigma_{i\in\beta} x_i - \Sigma_{i\in\gamma} x_i) \leq \frac{\delta}{4} \quad \text{for all } \beta, \gamma \in \mathcal{F}(\Lambda) \text{ with } \beta, \gamma \geq \alpha .$$

In particular, for any $f \in V_p^o$ and $\gamma \in \mathcal{F}(\Lambda\backslash\alpha)$, we have

$$|\Sigma_{i\in\gamma} \langle x_i, f\rangle| = |f(\Sigma_{i\in\gamma\cup\alpha} x_i) - f(\Sigma_{i\in\alpha} x_i)| \leq \frac{\delta}{4} ,$$

and hence

$$\Sigma_{i\in\gamma} |\langle x_i, f\rangle| \leq 4 \cdot \frac{\delta}{4} = \delta .$$

As $\gamma \in \mathfrak{F}(\Lambda\backslash\alpha)$ was arbitrary, it follows that

$$\Sigma_{\Lambda\backslash\alpha}|\langle x_i , f\rangle| \leq \delta \quad \text{for all} \quad f \in V_p^o$$

or, equivalently, $p_\varepsilon([x_i , \Lambda\backslash\alpha]) \leq \delta$. For any $\beta \geq \alpha$, Lemma (2.1.1) shows that

$$\Sigma_\Lambda|\langle x_i - x_i^{(\beta)}, f\rangle| \leq \Sigma_{\Lambda\backslash\alpha}|\langle x_i , f\rangle| \leq \delta \quad \text{for all} \quad f \in V_p^o .$$

In view of the definition of p_W , we obtain

$$p_W([x_i , \Lambda] - [x_i^{(\beta)}, (\Lambda)]) \leq \delta \quad \text{for all} \quad \beta \in \mathfrak{F}(\Lambda) \quad \text{with} \quad \beta \geq \alpha ,$$

which shows that $\{[x_i^{(\beta)}, (\Lambda)], \beta \in \mathfrak{F}(\Lambda)\}$ is $\mathcal{P}_W$-convergent to $[x_i , \Lambda]$.

<u>Sufficiency</u>. For any $p \in \mathbb{P}$ and $\delta > 0$, there exists an $\alpha_o \in \mathfrak{F}(\Lambda)$ such that

$$\Sigma_\Lambda|f(x_i^{(\beta)} - x_i^{(\gamma)})| \leq p_W([x_i^{(\beta)}, (\Lambda)] - [x_i^{(\gamma)}, (\Lambda)]) \leq \delta \quad \text{for all} \quad \beta, \gamma \in \mathfrak{F}(\Lambda) \quad \text{with} \quad \beta, \gamma \geq \alpha_o ,$$

whenever $f \in V_p^o$. As elements in $\mathfrak{F}(\Lambda)$ are finite, it follows that

$$|f(\Sigma_{i\in\beta}x_i - \Sigma_{i\in\gamma}x_i)| = |f(\Sigma_\Lambda x_i^{(\beta)} - \Sigma_\Lambda x_i^{(\gamma)})|$$
$$\leq \Sigma_\Lambda|f(x_i^{(\beta)} - x_i^{(\gamma)})| \leq \delta \quad \text{for all} \quad \beta, \gamma \geq \alpha_o \quad \text{and} \quad f \in V_p^o ,$$

and hence that

$$p(\Sigma_{i\in\beta}x_i - \Sigma_{i\in\gamma}x_i) \leq \delta \quad \text{for all} \quad \beta, \gamma \geq \alpha_o .$$

Therefore $\{\Sigma_{i\in\beta}x_i , \beta \in \mathfrak{F}(\Lambda)\}$ is a Cauchy net in E , and hence $[x_i , \Lambda]$ is summable.

(2.1.3) Theorem (Pietsch). $\ell^1(\Lambda, E)$ <u>is $\mathcal{P}_W$-closed in $\ell^1_W(\Lambda, E)$</u>. <u>Furthermore</u>, <u>if in addition</u>, $(E, \mathcal{P})$ <u>is complete</u>, <u>so do</u> $(\ell^1_W(\Lambda, E), \mathcal{P}_W)$, $(\ell^1(\Lambda, E), \mathcal{P}_\varepsilon)$ <u>and</u> $(\ell^1[\Lambda, E], \mathcal{P}_\pi)$.

<u>Proof</u>. Let $[x_i, \Lambda]$, in $\ell^1_W(\Lambda, E)$, belong to the $\mathcal{P}_W$-closure of $\ell^1(\Lambda, E)$ and $p \in \mathbb{P}$. For any $\delta > 0$ there exists an $[y_i, \Lambda]$ in $\ell^1(\Lambda, E)$ such that

$$p_W([x_i, \Lambda] - [y_i, \Lambda]) \leq \frac{\delta}{3} . \tag{1.5}$$

From this it then follows that

$$p_W([x_i^{(\beta)}, (\Lambda)] - [y_i^{(\beta)}, (\Lambda)]) \leq \frac{\delta}{3} \text{ for all } \beta \in \mathfrak{F}(\Lambda) . \tag{1.6}$$

According to (2.1.2), the net $\{[y_i^{(\alpha)}, \Lambda], \alpha \in \mathfrak{F}(\Lambda)\}$ in $E^{(\Lambda)}$ associated with $[y_i, \Lambda]$ is $\mathcal{P}_W$-convergent to $[y_i, \Lambda]$, hence there is an $\alpha_0 \in \mathfrak{F}(\Lambda)$ such that

$$p_W([y_i, \Lambda] - [y_i^{(\beta)}, (\Lambda)]) \leq \frac{\delta}{3} \text{ for all } \beta \in \mathfrak{F}(\Lambda) \text{ with } \beta \geq \alpha_0 . \tag{1.7}$$

By using the triangle inequality, formulae (1.5), (1.6) and (1.7) show that the net $\{[x_i^{(\beta)}, (\Lambda)], \beta \in \mathfrak{F}(\Lambda)\}$ associated with $[x_i, \Lambda]$ is $\mathcal{P}_W$-convergent to $[x_i, \Lambda]$, hence $[x_i, \Lambda] \in \ell^1(\Lambda, E)$ by (2.1.2), and thus $\ell^1(\Lambda, E)$ is $\mathcal{P}_W$-closed in $\ell^1_W(\Lambda, E)$.

Suppose further that $(E, \mathcal{P})$ is complete. Then, in view of the first assertion, it is sufficient to show that $(\ell^1_W(\Lambda, E), \mathcal{P}_W)$ and $(\ell^1[\Lambda, E], \mathcal{P}_\pi)$ are complete. But the completeness of $(\ell^1[\Lambda, E], \mathcal{P}_\pi)$ can be shown by the same manner of that of $(\ell^1_W(\Lambda, E), \mathcal{P}_W)$. Therefore we complete the proof by showing the completeness of $(\ell^1_W(\Lambda, E), \mathcal{P}_W)$.

Let $\{[x_{\iota,\mu}, \Lambda], \mu \in D\}$ be a $\mathcal{P}_w$-Cauchy net in $\ell^1_w(\Lambda, E)$. Then for any fixed $\iota \in \Lambda$, $\{x_{\iota,\mu}, \mu \in D\}$ is a Cauchy net in E since $\mathcal{P}^\Lambda \leq \mathcal{P}_w$ on $\ell^1_w(\Lambda, E)$, hence there exists $x_\iota \in E$ such that $\lim_\mu x_{\iota,\mu} = x_\iota$. We claim that $[x_\iota, \Lambda] \in \ell^1_w(\Lambda, E)$ is the $\mathcal{P}_w$-limit of $\{[x_{\iota,\mu}, \Lambda], \mu \in D\}$. Indeed, for any $p \in \mathbb{P}$ there is an $\mu_0 \in D$ such that

$$\Sigma_\Lambda|<x_{\iota,\mu_1}, f> - <x_{\iota,\mu_2}, f>| \leq 1 \quad \text{for all } \mu_1, \mu_2 \geq \mu_0 \text{ and } f \in V_p^o, \tag{1.8}$$

where $V_p = \{x \in E : p(x) \leq 1\}$, since $\{[x_{\iota,\mu}, \Lambda], \mu \in D\}$ is a $\mathcal{P}$-Cauchy net. In view of $\lim_\mu x_{\iota,\mu} = x_\iota$, we obtain by passing to the limit in (1.8) that

$$\Sigma_\Lambda|<x_\iota, f> - <x_{\iota,\mu_2}, f>| \leq 1 \quad \text{for all } \mu_2 \geq \mu_0 \text{ and } f \in V_p^o. \tag{1.9}$$

From this and $[x_{\iota,\mu_2}, \Lambda] \in \ell^1_w(\Lambda, E)$, we obtain

$$\Sigma_\Lambda|<x_\iota, f>| \leq \Sigma_\Lambda|<x_\iota, f> - <x_{\iota,\mu_2}, f>| + \Sigma_\Lambda|<x_{\iota,\mu_2}, f>| < +\infty \quad \text{for all } f \in V_p^o,$$

which shows that $[x_\iota, \Lambda] \in \ell^1_w(\Lambda, E)$. As $f \in V_p^o$ was arbitrary, it follows from (1.9) that the net $\{[x_{\iota,\mu}, \Lambda], \mu \in D\}$ is $\mathcal{P}_w$-convergent to $[x_\iota, \Lambda]$. Therefore $(\ell^1_w(\Lambda, E), \mathcal{P}_w)$ is complete.

In order to study the topological duals of $\ell^1_\varepsilon(\Lambda, E)$ and $\ell^1_\pi[\Lambda, E]$, we need the following result, which can be easily verified by using (2.1.2) and (2.1.1).

(2.1.4) Lemma. <u>The following statements hold</u>:

(a) $E^{(\Lambda)}$ <u>is dense in</u> $(\ell^1(\Lambda, E), \mathcal{P}_\varepsilon)$.

(b) <u>If</u> $[x_\iota, \Lambda] \in \ell^1[\Lambda, E]$, <u>then the net</u>

$\{[x_i^{(\alpha)}, (\wedge)], \alpha \in \mathcal{F}(\wedge)\}$ in $E^{(\wedge)}$ associated with $[x_i, \wedge]$ is $\mathcal{P}_\pi$-convergent to $[x_i, \wedge]$; consequently $E^{(\wedge)}$ is dense in $(\ell^1[\wedge, E], \mathcal{P}_\pi)$.

As $\mathcal{P}_\pi \leq \mathcal{P}^{(\wedge)}$ on $E^{(\wedge)}$, it follows from a well-known result and the preceding lemma that

$$(E_\pi^{(\wedge)})' = (\ell^1[\wedge, E], \mathcal{P}_\pi)' \subset (E^{(\wedge)}, \mathcal{P}^{(\wedge)})' = (E')^{\wedge} \ ,$$

where $E_\pi^{(\wedge)}$ is the space $E^{(\wedge)}$ equipped with the relative topology induced by $\mathcal{P}_\pi$. In order to give the topological dual of $E_\pi^{(\wedge)}$, we require the following terminology: A family $[x_i', \wedge]$ in E' is said to be equicontinuous if the set $\{x_i' : \iota \in \wedge\}$ is a $\mathcal{P}$-equicontinuous subset of E' . Clearly the set consisting of all equicontinuous families in E' (with index set $\wedge$), denoted by $e(\wedge, E')$, is a vector subspace of $(E')^{\wedge}$.

Let us define a map ϕ from $e(\wedge, E')$ onto $(E_\pi^{(\wedge)})'$ as follows: Assume that $[x_i', \wedge] \in e(\wedge, E')$. For each $[x_i, (\wedge)] \in E^{(\wedge)}$, the sum $\Sigma_\wedge <x_i, x_i'>$ has a meaning because $x_i = 0$ except for finitely many indices ι , therefore the map $\phi([x_i', \wedge])$, defined by

$$[x_i, (\wedge)] \longmapsto \Sigma_\wedge <x_i, x_i'> \quad \text{for all} \quad [x_i, (\wedge)] \in E^{(\wedge)} \ ,$$

is well-defined and linear on $E^{(\wedge)}$. Clearly $\phi([x_i', \wedge])$ is continuous on $E_\pi^{(\wedge)}$ and ϕ is injective. Furthermore, if $f \in (E_\pi^{(\wedge)})'$, then $x_i' = f \circ J_\iota \in E'$ ($\iota \in \wedge$) , where J_ι is the embedding map, hence $[x_i', \wedge]$ is such that

$$\phi([x_i', \wedge]) = f \ .$$

The equicontinuity of $[x'_i, \Lambda]$ can be seen from the following facts: Let $p \in \mathbb{P}$ be such that

$$|\langle[x_i, (\Lambda)], f\rangle| = |\Sigma_\Lambda \langle x_i, x'_i\rangle| \leq p_\pi([x_i, (\Lambda)]) \quad \text{for all} \quad [x_i, (\Lambda)] \in E^{(\Lambda)} ,$$

and let $W = \{x \in E : p(x) \leq 1\}$. For any $i \in \Lambda$ and $x \in W$, we have

$$|\langle x, x'_i\rangle| = |\langle J_i(x), f\rangle| \leq p_\pi(J_i(x)) = p(x) \leq 1 ,$$

hence $x'_i \in W^o$ and thus $[x'_i, \Lambda]$ is an equicontinuous family in E' .

In view of the Hahn-Banach extension theorem and (2.1.4) we obtain:

(2.1.5) Theorem. <u>The topological dual of</u> $(\ell^1[\Lambda, E], \mathcal{P}_\pi)$ <u>can be identified with the vector subspace of</u> $(E')^\Lambda$ <u>consisting of all equicontinuous families</u> $[x'_i, \Lambda]$ <u>in</u> E' , <u>the canonical bilinear form is given by</u>

$$\langle[x_i, \Lambda], [x'_i, \Lambda]\rangle = \Sigma_\Lambda \langle x_i, x'_i\rangle .$$

To study the topological dual of $(\ell^1(\Lambda, E), \mathcal{P}_\varepsilon)$, we first prove the following two interesting lemmas.

(2.1.6) Lemma. <u>Let</u> p <u>be a continuous seminorm on</u> $(E, \mathcal{P})$ <u>and</u> $V_p = \{x \in E : p(x) \leq 1\}$. <u>Denote by</u> Δ <u>the unit disk in the scalar field of</u> E . <u>For any</u> $[x_i, \Lambda] \in \ell^1(\Lambda, E)$, <u>the map</u> g , <u>defined by</u>

$$g([\zeta_i, \Lambda], x') = \Sigma_\Lambda \zeta_i \langle x_i, x'\rangle \quad \text{for all} \quad [\zeta_i, \Lambda] \in \Delta^\Lambda \quad \text{and} \quad x' \in V_p^o , \qquad (1.10)$$

<u>is continuous on the compact Hausdorff space</u> $\Delta^\Lambda \times V_p^o$ <u>and</u>

$$\|g\| = p_\varepsilon([x_i, \Lambda]) , \tag{1.11}$$

<u>where</u> $\|\cdot\|$ <u>is the sup-norm on</u> $C(\Delta^\wedge \times V_p^o)$. <u>Consequently, the map</u> $[x_i, \Lambda] \mapsto g$ <u>is a continuous linear map from</u> $(\ell^1(\Lambda, E), \mathcal{P}_\varepsilon)$ <u>into</u> $(C(\Delta^\wedge \times V_p^o), \|\cdot\|)$. <u>In particular, if</u> (E, p) <u>is a normed space, then</u> $(\ell^1(\Lambda, E), \mathcal{P}_\varepsilon)$ <u>is isometrically isomorphic to a subspace of</u> $(C(\Delta^\wedge \times V_p^o), \|\cdot\|)$ <u>under the map</u> $[x_i, \Lambda] \mapsto g$.

<u>Proof</u>. It is known that the net $\{[x_i^{(\alpha)}, \Lambda], \alpha \in \mathcal{F}(\Lambda)\}$ in $E^{(\Lambda)}$ associated with $[x_i, \Lambda]$ is $\mathcal{P}_\varepsilon$-convergent to $[x_i, \Lambda]$, hence for any $\delta > 0$ there is an $\alpha_o \in \mathcal{F}(\Lambda)$ such that

$$\Sigma_{\Lambda\setminus\alpha} |\langle x_i , x'\rangle| \leq \delta \quad \text{for all } x' \in V_p^o \text{ and } \alpha \geq \alpha_o \tag{1.12}$$

by (2.1.2). For any $\alpha \in \mathcal{F}(\Lambda)$ the map g_α , defined by

$$g_\alpha([\zeta_i, \Lambda], x') = \Sigma_{i\in\alpha} \zeta_i \langle x_i , x'\rangle \quad \text{for all } [\zeta_i, \Lambda] \in \Delta^\wedge \text{ and } x' \in V_p^o ,$$

is clearly continuous on $\Delta^\wedge \times V_p^o$ and $\|g_\alpha\| \leq p_\varepsilon([x_i^{(\alpha)}, (\Lambda)])$ because of

$$g_\alpha([\zeta_i, \Lambda], x') = \Sigma_\Lambda \zeta_i \langle x_i^{(\alpha)}, x'\rangle \quad \text{for all } [\zeta_i, \Lambda] \in \Delta^\wedge \text{ and } x' \in V_p^o .$$

Moreover, we claim that

$$\|g_\alpha\| = p_\varepsilon([x_i^{(\alpha)}, (\Lambda)]) .$$

In fact, for any $x' \in V_p^o$ and x_i , there is a scalar ζ_i such that

$$|\langle x_i , x'\rangle| = \zeta_i \langle x_i , x'\rangle \quad \text{and} \quad |\zeta_i| = 1 ,$$

hence

$$\Sigma_\Lambda |\langle x_i^{(\alpha)}, x'\rangle| = \Sigma_\Lambda \zeta_i \langle x_i^{(\alpha)}, x'\rangle = \Sigma_{i\in\alpha} \zeta_i \langle x_i , x'\rangle$$
$$= g_\alpha([\lambda_i, \Lambda], x') \leq \|g_\alpha\| ;$$

consequently $p_\varepsilon([x_\iota^{(\alpha)}, (\Lambda)]) \leq \|g_\alpha\|$. Therefore we obtain the required equality.

Now formula (1.12) and the definition of g_α ensure that

$$|g([\zeta_\iota, \Lambda], x') - g_\alpha([\zeta_\iota, \Lambda], x')| = |\Sigma_\Lambda \zeta_\iota \langle x_\iota - x_\iota^{(\alpha)}, x'\rangle|$$

$$\leq \Sigma_\Lambda |\langle x_\iota - x_\iota^{(\alpha)}, x'\rangle| = \Sigma_{\Lambda\setminus\alpha} |\langle x_\iota, x'\rangle| \leq \delta \quad \text{for all} \quad \alpha \geq \alpha_0 ,$$

hence the net $\{g_\alpha, \alpha \in \mathcal{F}(\Lambda)\}$ converges to g uniformly on $\Delta^\Lambda \times V_p^o$. Therefore g is continuous on $\Delta^\Lambda \times V_p^o$ and

$$\|g\| = \lim_\alpha \|g_\alpha\| = \lim_\alpha p_\varepsilon([x_\iota^{(\alpha)}, (\Lambda)]) = p_\varepsilon([x_\iota, \Lambda])$$

because the net $\{[x_\iota^{(\alpha)}, (\Lambda)], \alpha \in \mathcal{F}(\Lambda)\}$ is $\mathcal{P}_\varepsilon$-convergent to $[x_\iota, \Lambda]$.

Finally, the map $[x_\iota, \Lambda] \mapsto g$ is clearly linear. Furthermore, if (E, p) is a normed space, then p_ε is a norm on $\ell^1(\Lambda, E)$. It then follows from (1.11) that the map $[x_\iota, \Lambda] \mapsto g$ is injective. Therefore $(\ell^1(\Lambda, E), \mathcal{P}_\varepsilon)$ is isometrically isomorphic to a subspace of $(C(\Delta^\Lambda \times V_p^o), \|\cdot\|)$.

(2.1.7) Lemma. <u>For any</u> $f \in (\ell^1(\Lambda, E), \mathcal{P}_\varepsilon)'$, <u>there exist a</u> $\sigma(E', E)$-<u>closed equicontinuous subset</u> B <u>of</u> E' <u>and a positive Radon measure</u> μ <u>on</u> B <u>such that</u>

$$|\langle J_\iota(x), f\rangle| \leq \int_B |\langle x, x'\rangle| \, d\mu(x') \quad \underline{\text{for all}} \quad x \in E \quad \underline{\text{and}} \quad \iota \in \Lambda , \qquad (1.13)$$

<u>where</u> $J_\iota : E \to E^{(\Lambda)}$ <u>is the embedding map</u>.

<u>Proof</u>. There is a continuous seminorm p on E such that

$$|\langle[x_i, \Lambda], f\rangle| \leq p_\varepsilon([x_i, \Lambda]) \text{ for all } [x_i, \Lambda] \in \ell^1(\Lambda, E) , \tag{1.14}$$

hence (2.1.6) shows that the map $\psi : [x_i, \Lambda] \mapsto g$, defined by

$$g([\zeta_i, \Lambda], x') = \Sigma_\Lambda \zeta_i \langle x_i, x'\rangle \text{ for all } [\zeta_i, \Lambda] \in \Delta^\Lambda \text{ and } x' \in V_p^o ,$$

is a continuous linear map from $(\ell^1(\Lambda, E), \mathcal{P}_\varepsilon)$ into $C(\Delta^\Lambda \times V_p^o)$ such that

$$p_\varepsilon([x_i, \Lambda]) = \|g\| = \|\psi([x_i, \Lambda])\| \text{ for all } [x_i, \Lambda] \in \ell^1(\Lambda, E) . \tag{1.15}$$

It follows from (1.14) and (1.15) that $\operatorname{Ker}\psi \subset \operatorname{Ker} f$, and hence from the Hahn-Banach extension theorem that there is a continuous linear functional $\tilde{f}$ on $C(\Delta^\Lambda \times V_p^o)$ with norm ≤ 1 (by (1.15)) such that

$$\tilde{f} \circ \psi([x_i, \Lambda]) = \langle[x_i, \Lambda], f\rangle \text{ for all } [x_i, \Lambda] \in \ell^1(\Lambda, E) .$$

The Riesz representation theorem ensures that there exists a Radon measure ν on $\Delta^\Lambda \times V_p^o$ such that

$$\begin{aligned}\langle[x_i, \Lambda], f\rangle = \nu(\psi([x_i, \Lambda])) &= \int_{\Delta^\Lambda \times V_p^o} \psi([x_i, \Lambda])([\zeta_i, \Lambda], x')d\nu \\ &= \int_{\Delta^\Lambda \times V_p^o} \Sigma_\Lambda \zeta_i \langle x_i, x'\rangle d\nu \text{ for all } [x_i, \Lambda] \in \ell^1(\Lambda, E).\end{aligned} \tag{1.16}$$

On the other hand, $|\nu|$ is clearly a positive Radon measure on $\Delta^\Lambda \times V_p^o$, and the map $\phi : C(V_p^o) \to C(\Delta^\Lambda \times V_p^o)$, defined by

$$\phi(h)([\zeta_i, \Lambda], x') = h(x') \text{ for all } [\zeta_i, \Lambda] \in \Delta^\Lambda \text{ and } x' \in V_p^o ,$$

is an isometry from $C(V_p^o)$ into $C(\Delta^\Lambda \times V_p^o)$. By the Riesz representation theorem there is a positive Radon measure μ on V_p^o such that

$$\int_{V_p^o} h(x')d\mu(x') = \mu(h) = |\nu| \circ \phi(h) = \int_{\Delta^\Lambda \times V_p^o} \phi(h)([\zeta_i, \Lambda], x')d|\nu|$$

$$= \int_{\Delta^\Lambda \times V_p^o} h(x')d|\nu| \quad \text{for all} \quad h \in C(V_p^o) \, . \tag{1.17}$$

For any $x \in E$, the map $x' \mapsto |\langle x, x'\rangle|$ is continuous on V_p^o, hence (1.16) and (1.17) show that

$$|\langle J_\iota(x), f\rangle| \leq \int_{\Delta^\Lambda \times V_p^o} |\langle x, x'\rangle| \, d|\nu| = \int_{V_p^o} |\langle x, x'\rangle| \, d\mu(x') \, ,$$

which obtains the required inequality (1.13).

Since J_ι is a continuous linear map from $(E, \mathcal{P})$ into $(\ell^1(\Lambda, E), \mathcal{P}_\varepsilon)$, it follows that

$$x'_\iota = f \circ J_\iota \in E' \quad \text{whenever} \quad f \in (\ell^1(\Lambda, E), \mathcal{P}_\varepsilon)' \, ,$$

and hence from (1.13) that $[x'_\iota, \Lambda]$ is a family in E' such that

$$|\langle x, x'_\iota\rangle| \leq \int_B |\langle x, x'\rangle| \, d\mu(x') \quad \text{for all} \quad x \in E \quad \text{and} \quad \iota \in \Lambda \, .$$

As suggested by the above observation, a subset M of E' is said to be <u>prenuclear</u> if there exist a $\sigma(E', E)$-closed equicontinuous subset B of E' and a positive Radon measure μ on B such that

$$|\langle x, m'\rangle| \leq \int_B |\langle x, x'\rangle| \, d\mu(x') \quad \text{for all} \quad x \in E \quad \text{and} \quad m' \in M \, .$$

A family $[x'_\iota, \Lambda]$ in E' is said to be <u>prenuclear</u> if the set $\{x'_\iota : \iota \in \Lambda\}$ is prenuclear.

Clearly prenuclear subsets of E' are equicontinuous. It is not hard to show that the set consisting of all prenuclear families in E' (with index set Λ), denoted by $p(\Lambda, E')$, is a vector subspace of $(E')^\Lambda$.

In terms of the prenuclearity, the preceding result shows that every $f \in (\ell^1(\Lambda, E), \mathscr{P}_\varepsilon)'$ defines a prenuclear family $[x'_i, \Lambda]$ in E'. The converse is true as shown by the following result.

(2.1.8) Theorem. <u>The topological dual of</u> $(\ell^1(\Lambda, E), \mathscr{P}_\varepsilon)$ <u>can be identified with the vector subspace of</u> $(E')^\Lambda$ <u>consisting of all prenuclear families</u> $[x'_i, \Lambda]$ <u>(with index set</u> Λ<u>) in</u> E', <u>the canonical bilinear form is given by</u>

$$<[x_i, \Lambda], [x'_i, \Lambda]> = \Sigma_\Lambda <x_i, x'_i> .$$

<u>Proof</u>. As $E^{(\Lambda)}$ is dense in $(\ell^1(\Lambda, E), \mathscr{P}_\varepsilon)$, it follows that $(E_\varepsilon^{(\Lambda)})' = (\ell^1(\Lambda, E), \mathscr{P}_\varepsilon)'$, where $E_\varepsilon^{(\Lambda)}$ is the space $E^{(\Lambda)}$ equipped with the relative topology induced by $\mathscr{P}_\varepsilon$. Let us define a map ψ from $p(\Lambda, E')$ onto $(E_\varepsilon^{(\Lambda)})'$ as follows: For a given prenuclear family $[x'_i, \Lambda]$ in E', the map $\psi([x'_i, \Lambda])$, defined by

$$[x_i, (\Lambda)] \mapsto \Sigma_\Lambda <x_i, x'_i> \quad \text{for all} \quad [x_i, (\Lambda)] \in E^{(\Lambda)},$$

is well-defined and linear because of $x_i = 0$ except for finitely many indices i. To see the $\mathscr{P}_\varepsilon$-continuity of $\psi([x'_i, \Lambda])$, let V be an absolutely convex 0-neighbourhood in E and let μ be a positive Radon measure μ on V^o such that

$$|<x, x'_i>| \leq \int_{V^o} |<x, x'>| \, d\mu(x') \quad \text{for all} \quad x \in E \text{ and } i \in \Lambda .$$

If p is the gauge of V and $[x_i, (\Lambda)] \in E^{(\Lambda)}$, then

$$\Sigma_\Lambda |<x_i, x'_i>| \leq \int_{V^o} \Sigma_\Lambda |<x_i, x'>| \, d\mu(x') \leq \|\mu\| p_\varepsilon([x_i, (\Lambda)]) ,$$

thus $\psi([x_i', \Lambda])$ is $\mathcal{P}_\varepsilon$-continuous. Therefore the theorem holds by making use of (2.1.7).

2.2. Absolutely summing mappings

Clearly every $T \in L^*(E, F)$ associates a linear map $T^\wedge$ from $E^\wedge$ into $F^\wedge$ obtained by the equation

$$T^\wedge([x_i, \Lambda]) = [Tx_i, \Lambda] \quad \text{for all} \quad [x_i, \Lambda] \in E^\wedge .$$

The restriction of $T^\wedge$ to $E^{(\Lambda)}$, denoted by $T^{(\Lambda)}$, is clearly a linear map from $E^{(\Lambda)}$ into $F^{(\Lambda)}$. For each $i \in \Lambda$ we have

$$\pi_i \, T^{(\Lambda)} J_i = T .$$

On the other hand, if $T \in L(E, F)$ then $T^\wedge$ is a continuous linear map from $(E^\wedge, \mathcal{P}^\wedge)$ into $(F^\wedge, \mathcal{J}^\wedge)$.

A linear map $T : E \to F$ is said to be <u>absolutely summing</u> if

$$T^{(\mathbb{N})} \in L(E_\varepsilon^{(\mathbb{N})}, F_\pi^{(\mathbb{N})}) .$$

Clearly, absolutely summing maps are continuous, and the set consisting of all absolutely summing maps, denoted by $\pi_{\ell^1}(E, F)$, is a vector subspace of $L(E, F)$. Furthermore, an $T \in L(E, F)$ is absolutely summing if and only if for any continuous seminorm q on F there is a continuous seminorm p on E such that the inequality

$$\sum_{k=1}^{n} q(Tx_k) \leq \sup\{\sum_{k=1}^{n} |\langle x_k, x'\rangle| : x' \in V_p^o\} = \sup\{p(\sum_{k=1}^{n} c_k x_k) : |c_k| = 1\} \quad (2.1)$$

holds for any finite subset $\{x_1, \ldots, x_n\}$ of E, where $V_p = \{x \in E : p(x) \leq 1\}$. From this we conclude that $T \in \pi_{\ell^1}(E, F)$ if and only if $T^{(\Lambda)} \in L(E_\varepsilon^{(\Lambda)}, F_\pi^{(\Lambda)})$.

In view of (2.1), it is easily verified the following

(2.2.1) Lemma. Let E, F and G be locally convex spaces and suppose that $T \in L(E, F)$ and $S \in L(E, G)$. If one of T and S is absolutely summing, then $S \circ T \in \pi_{\ell^1}(E, G)$.

We now present some alternative characterizations of absolutely summing maps which we shall need in what follows.

(2.2.2) Theorem. For an $T \in L(E, F)$, the following statements are equivalent.

(a) $T \in \pi_{\ell^1}(E, F)$.

(b) $T^\Lambda \in L(\ell_\varepsilon^1(\Lambda, E), \ell_\pi^1[\Lambda, F])$ for any non-empty index set Λ.

(c) The adjoint map T' of T sends equicontinuous subsets of F' into prenuclear subsets of E'.

(d) For any continuous seminorm q on F there are a $\sigma(E', E)$-closed equicontinuous subset B of E' and a positive Radon measure μ on B such that

$$q(Tx) \leq \int_B |\langle x, x'\rangle| \, d\mu(x') \quad \text{for all } x \in E. \quad (2.2)$$

(e) <u>For any</u> $U \in \mathcal{U}_F$ <u>there exists an</u> $V \in \mathcal{U}_E$ <u>such that the following inequality holds</u>

$$|\Sigma_{i=1}^{n} \langle Tx_i , y_i' \rangle| \leq \sup\{\Sigma_{i=1}^{n} |\langle x_i , x' \rangle q_{U^o}(y_i')| \ : x' \in V^o\}$$

<u>for any finite subsets</u> $\{x_1, \dots, x_n\}$ <u>and</u> $\{y_1', \dots, y_n'\}$ <u>of</u> E <u>and</u> $F'(U^o)$ <u>respectively, where</u> q_{U^o} <u>is the gauge of</u> U^o <u>defined on the vector subspace</u> $F'(U^o)$ <u>of</u> F' <u>generated by</u> U^o .

<u>Proof</u>. The implication (b) ⇒ (c) is a consequence of (2.1.5) and (2.1.8), and the implication (c) ⇒ (d) follows from the definition of prenuclearity. Therefore we complete the proof by showing the implications (a) ⇒ (b) and (d) ⇒ (e) ⇒ (a).

(a) ⇒ (b) : For any continuous seminorm q on F there is a continuous seminorm p on E such that the inequality

$$\Sigma_{k=1}^{n} q(Tz_k) \leq \sup\{\Sigma_{k=1}^{n} |\langle z_k , x' \rangle| \ : x' \in V_p^o\}$$

holds for any finite subset $\{z_1, \dots, z_n\}$ of E, where $V_p = \{x \in E : p(x) \leq 1\}$. For any $[x_i , \Lambda] \in \ell^1(\Lambda, E)$ and $\alpha \in \mathcal{F}(\Lambda)$, we have

$$\Sigma_{i\in\alpha} q(Tx_i) \leq \sup\{\Sigma_{i\in\alpha} |\langle x_i , x' \rangle| \ : x' \in V_p^o\} .$$

It then follows that

$$q_\pi([Tx_i , \Lambda]) = \sup_{\alpha\in\mathcal{F}(\Lambda)} \Sigma_{i\in\alpha} q(Tx_i) \leq \sup_{\alpha\in\mathcal{F}(\Lambda)} \sup\{\Sigma_{i\in\alpha} |\langle x_i , x' \rangle| \ : x' \in V_p^o\} .$$

Thus $T^\wedge([x_i , \Lambda]) = [Tx_i , \Lambda] \in \ell^1[\Lambda, F]$ and $T^\wedge \in L(\ell_\varepsilon^1(\Lambda, E), \ell_\pi^1[\Lambda, F])$.

(d) ⇒ (e) : Assume that the statement (d) holds. Let q be the gauge of U and $V = B^o$. Then V is a o-neighbourhood in E. Now for any finite subsets $\{x_1, \ldots, x_n\}$ and $\{y'_1, \ldots, y'_n\}$ of E and $F'(U^o)$ respectively, we may assume without loss of generality that $q_{U^o}(y'_i) \neq 0$ $(i = 1, \ldots, n)$, then we have from (2.2) that

$$|\Sigma_{i=1}^n \langle Tx_i, y'_i\rangle| \leq \Sigma_{i=1}^n q_{U^o}(y'_i) q(Tx_i) \leq \mu(B) \sup\{\Sigma_{i=1}^n |\langle x_i, x'\rangle q_{U^o}(y'_i)| : x' \in B\}$$
$$\leq \mu(B) \sup\{\Sigma_{i=1}^n |\langle x_i, x'\rangle q_{U^o}(y'_i)| : x' \in V^o\}.$$

(e) ⇒ (a) : Assume that the statement (e) holds. Then for any continuous seminorm q on F there is an $V \in \mathcal{U}_E$ such that the inequality

$$|\Sigma_{i=1}^n \langle Tx_i, y'_i\rangle| \leq \sup\{\Sigma_{i=1}^n |\langle x_i, x'\rangle q_{U^o}(y'_i)| : x' \in V^o\}$$

holds for any finite subsets $\{x_1, \ldots, x_n\}$ and $\{y'_1, \ldots, y'_n\}$ of E and $F'(U^o)$ respectively, where $U = \{y \in F : q(y) \leq 1\}$. Replace y'_i by (sign $\langle Tx_i, y'_i\rangle)y'_i$ in the above inequality, we conclude that the inequality

$$\Sigma_{i=1}^n |\langle Tx_i, y'_i\rangle| \leq \sup\{\Sigma_{i=1}^n |\langle x_i, x'\rangle q_{U^o}(y'_i)| : x' \in V^o\} \tag{2.3}$$

holds for any finite subsets $\{x_1, \ldots, x_n\}$ and $\{y'_1, \ldots, y'_n\}$ of E and $F'(U^o)$ respectively.

Now for any finite subset $\{u_1, \ldots, u_n\}$ of F and $\delta > 0$, there are $v'_i \in U^o$ $(i = 1, 2, \ldots, n)$ such that

$$q(Tu_i) < |\langle Tu_i, v'_i\rangle| + n^{-1}\delta \quad (i = 1, 2, \ldots, n).$$

It then follows from (2.3) that

$$\Sigma_{i=1}^{n} q(Tu_i) < \Sigma_{i=1}^{n} |\langle Tu_i , v_i' \rangle| + \delta$$
$$\leq \sup\{\Sigma_{i=1}^{n} |\langle u_i , x' \rangle q_{U^o}(v_i')| : x' \in V^o\} + \delta$$
$$\leq \sup\{\Sigma_{i=1}^{n} |\langle u_i , x' \rangle| : x' \in V^o\} + \delta . \qquad (2.4)$$

As δ was arbitrary, we conclude from (2.4) and (2.1) that $T \in \pi_{\ell^1}(E, F)$.

(2.2.3) Corollary. For the identity map I on E , the following statements are equivalent:

(a) $I \in \pi_{\ell^1}(E)$.

(b) Every equicontinuous subset of E' is prenuclear.

(c) For any continuous seminorm p on E there are a $\sigma(E', E)$-closed equicontinuous subset B of E' and a positive Radon measure μ on B such that

$$p(x) \leq \int_B |\langle x, x' \rangle| \, d\mu(x') \quad \text{for all} \quad x \in E .$$

(d) The canonical embedding map ϕ of $\ell_\pi^1[E]$ into $\ell_\varepsilon^1(E)$ is a topological isomorphism from the first space onto the second.

Proof. In view of (2.2.2), the statements (a), (b) and (c) are mutually equivalent. To prove the implication (a) $\Rightarrow$ (d) , we see that $I^{\mathbb{N}} \in L(\ell_\varepsilon^1(E), \ell_\pi^1[E])$ and $\ell^1[E] \subset \ell^1(E)$, hence $I^{\mathbb{N}}$ is surjective. Clearly ϕ is the inverse map $(I^{\mathbb{N}})^{-1}$ of $I^{\mathbb{N}}$, hence ϕ is bijective and ϕ^{-1} is continuous. Therefore ϕ is a topological isomorphism since ϕ is always continuous. Finally, the implication (d) $\Rightarrow$ (a) is obvious since $\phi^{-1} = I^{\mathbb{N}}$.

If $(E, \mathcal{P})$ is a Fréchet space then so do $(\ell^1(E), \mathcal{P}_\varepsilon)$ and $(\ell^1[E], \mathcal{P}_\pi)$, thus $\ell^1(E) = \ell^1[E]$ implies that $\mathcal{P}_\varepsilon = \mathcal{P}_\pi$ by the Banach open mapping theorem. The assumption of the completeness of E can be dropped as shown by the following result.

(2.2.4) Corollary. Let $(E, \mathcal{P})$ be metrizable and $T \in L(E, F)$. Then $T \in \pi_{\ell^1}(E, F)$ if and only if $T^{\mathbb{N}} \in L^*(\ell^1(E), \ell^1[F])$, i.e., T sends every summable sequence in E into an absolutely summable sequence in F. In particular, $I \in \pi_{\ell^1}(E)$ if and only if $\ell^1(E) = \ell^1[E]$.

Proof. The necessity is obvious. Conversely, if $T^{\mathbb{N}} \in L^*(\ell^1(E), \ell^1[F])$, then it is not hard to show (see Pietsch [1, p.34-35]) that $T^{\mathbb{N}}$ sends bounded subsets of $(\ell^1(E), \mathcal{P}_\varepsilon)$ into bounded subsets of $(\ell^1[F], \mathcal{T}_\pi)$, thus $T^{\mathbb{N}} \in L(\ell^1(E), \ell^1[F])$ because $(\ell^1(E), \mathcal{P}_\varepsilon)$ is metrizable and a fortiori bornological.

Let B be a compact Hausdorff space and let μ be a positive Radon measure on B. Then $L^2_\mu(B)$ is a Hilbert space under the usual inner product defined by

$$[h, g] = \int_B h(t)\overline{g}(t)d\mu(t) \quad \text{for all } h, g \in L^2_\mu(B).$$

Furthermore, $C(B)$ is isometrically isomorphic to a subspace of $L^2_\mu(B)$ under the canonical embedding map.

In terms of Theorem (2.2.2), we are able to give a decomposition of absolutely summing maps as follows.

(2.2.5) Corollary. Let (Y, q) be a Banach space. Then every $T \in \pi_{\ell^1}(E, Y)$ is the composition of the following three continuous linear maps

$$E \xrightarrow{T_1} C(B) \xrightarrow{j_T} L^2_\mu(B) \xrightarrow{T_2} Y ,$$

where μ is a positive Radon measure on a compact Hausdorff space B and j_T is the canonical embedding map from $C(B)$ into $L^2_\mu(B)$.

Proof. We can assume that $T \neq 0$ otherwise the result is trivial. By (2.2.2), there are a $\sigma(E', E)$-closed equicontinuous subset B of E' and a positive Radon measure μ on B such that

$$q(Tx) \leq \int_B |\langle x, x'\rangle| \, d\mu(x') \quad \text{for all} \quad x \in E . \tag{2.2)*$$

The map $T_1 : E \to C(B)$, defined by

$$(T_1x)(x') = \langle x, x'\rangle \quad \text{for all} \quad x' \in B ,$$

is clearly a continuous linear map, hence $H = \{j_T(T_1x) : x \in E\}$ is a vector subspace of $L^2_\mu(B)$. Since q is a norm and j_T is injective, it follows from (2.2)* that $\text{Ker}\,(j_TT_1) \subset \text{Ker}\,T$, and hence that there exists a linear map $S_2 : H \to Y$ such that

$$S_2j_TT_1 = T .$$

If $\|\cdot\|_2$ denotes the usual norm on $L^2_\mu(B)$, then Cauchy-Schwarz's inequality and (2.2) show that

$$q(S_2j_TT_1(x)) = q(Tx) \leq \int_B |\langle x, x'\rangle| \, d\mu(x') = \int_B |(T_1x)(x')| \, d\mu(x')$$

$$= \mu(B) \left(\int_B |(j_TT_1(x)(x')|^2 d\mu(x')\right)^{\frac{1}{2}} = \mu(B) \| j_T(T_1x) \|_2 ,$$

hence $S_2 \in L(H, Y)$. Furthermore, the completeness of Y ensures that S_2 has a unique continuous extension to the closure $\overline{H}$ of H in $L^2_\mu(B)$, which is denoted by $\tilde{S}_2$. On the other hand, since $L^2_\mu(B)$ is a Hilbert space, there is a projection P from $L^2_\mu(B)$ onto $\overline{H}$. Therefore the map T_2 , defined by

$$T_2 = \tilde{S}_2 P ,$$

is a continuous linear map from $L^2_\mu(B)$ into Y such that

$$\begin{aligned} Tx &= S_2(j_T(T_1x)) = S_2(P(j_T(T_1x))) \\ &= T_2 j_T T_1(x) \quad \text{for all} \quad x \in E . \end{aligned}$$

(2.2.6) Corollary. <u>Let</u> E <u>be a locally convex space and</u> (Y, q) <u>a normed space</u>. <u>Then every</u> $T \in \pi_{\ell^1}(E, Y)$ <u>permits a factoring</u>

$$E \xrightarrow{S} C(B) \xrightarrow{J} N \xrightarrow{Q} Y ,$$

<u>where</u> B <u>is a compact Hausdorff space</u>, μ <u>a positive Radon measure on</u> B , N <u>a subspace of</u> $L^1_\mu(B)$, $J : C(B) \to L^1_\mu(B)$ <u>is the canonical embedding</u>, <u>and</u> S <u>and</u> Q <u>are continuous linear maps</u>.

<u>Proof</u>. By (2.2.2), there exist a $\sigma(E', E)$-closed equicontinuous subset B of E' and a positive Radon measure μ on B such that

$$q(Tx) \leq \int_B |\langle x, x'\rangle| \, d\mu(x') \quad \text{for all} \quad x \in E .$$

Let us define $S : E \to C(B)$ by

$$Sx = x|_B \quad \text{if} \quad x|_B(x') = \langle x, x'\rangle \quad \text{for all} \quad x' \in B ,$$

and denote by $\|\cdot\|_1$ the norm on $L^1_\mu(B)$. Then we have

$$q(Tx) \leq \int_B |\langle x, x'\rangle| \, d\mu(x') = \|J(Sx)\|_1 ,$$

hence $\operatorname{Ker}(JS) \subset \operatorname{Ker} T$ because q is a norm. Therefore there exists an $Q \in L(JS(E), Y)$ such that $T = QJS$. Clearly, $N = JS(E)$ is a subspace of $L^1_\mu(B)$, the result then follows.

The following example shows that absolutely summing mappings need not be precompact.

(2.2.7) Example. Let B be a compact Hausdorff space and μ a positive Radon measure on B . Then the canonical embedding $J : C(B) \to L^1_\mu(B)$ is an absolutely summing mapping which is not precompact.

Proof. We identify B with a $\sigma(C(B)', C(B))$-closed equicontinuous subset of $C(B)'$ by identifying each $b \in B$ with $\hat{b} \in C(B)'$ defined by

$$\langle f, \hat{b}\rangle = f(b) \quad \text{for all} \quad f \in C(B) .$$

If $\|\cdot\|_1$ denotes the usual norm on $L^1_\mu(B)$, then

$$\|Jf\|_1 = \int_B |f(b)| \, d\mu(b) = \int_B |\langle f, \hat{b}\rangle| \, d\mu(b) \quad \text{for all} \quad f \in C(B) ,$$

hence J is absolutely summing by (2.2.2)(d).

As $J(C(B))$ is dense in $L^1_\mu(B)$, it follows that the image of the closed unit ball in $C(B)$, under J , need not be a relatively compact subset of $L^1_\mu(B)$, and hence that J is not precompact.

In view of (2.2.1), the preceding example shows that the converse of (2.2.6) is still valid.

We shall see in §4.1 that the canonical embedding $J : C(B) \to L^1_\mu(B)$ is an integral linear map.

Suppose now that (X, p) and (Y, q) are normed spaces. Then an $T \in L(X, Y)$ is absolutely summing if and only if there is an $\zeta \geq 0$ such that the inequality

$$\Sigma_{k=1}^{n} q(Tx_k) \leq \zeta \sup\{\Sigma_{k=1}^{n} |\langle x_k, x'\rangle| : x' \in V_p^o\} \tag{2.5}$$

holds for any finite subset $\{x_1, \ldots, x_n\}$ of X, and this is the case if and only if $T^{\mathbb{N}} \in L(\ell_\varepsilon^1(X), \ell_\pi^1[Y])$. Therefore we define

$$\|T\|_{(s)} = |||T^{\mathbb{N}}||| \quad \text{for any} \quad T \in \pi_{\ell^1}(X, Y),$$

where $|||\cdot|||$ is the operator norm. Consequently, $\|\cdot\|_{(s)}$ is a norm on $\pi_{\ell^1}(X, Y)$ which is called the <u>absolutely summing norm</u>.

It is clear that $\|T\|_{(s)}$ is the infimum of all $\zeta \geq 0$ for which (2.5) holds for all finite subsets $\{x_1, \ldots, x_n\}$ of X.

If we determine an index set Λ such that a map $\iota \mapsto y'_\iota$ exists from Λ onto the closed unit ball Σ_Y^o in Y' and if $T \in \pi_{\ell^1}(X, Y)$, then (2.1.5) shows that $g = [y'_\iota, \Lambda] \in (\ell^1[\Lambda, Y], q_\pi)'$ with norm ≤ 1, hence $gT^\wedge = (T^\wedge)'(g) = [T'y'_\iota, \Lambda] \in (\ell^1(\Lambda, X), p_\varepsilon)'$ with

$$\langle [x_\iota, \Lambda], gT^\wedge\rangle = \Sigma_\Lambda \langle Tx_\iota, y'_\iota\rangle \quad \text{and} \quad \|gT^\wedge\| \leq \|T\|_{(s)} .$$

Since $(\ell^1(\Lambda, X), p_\varepsilon)$ is isometrically isomorphic to a subspace of $C(\Delta^\wedge \times \Sigma^0)$, where Σ^0 is the closed unit ball in X', a similar argument given in the proof of (2.1.7) shows that there is a positive Radon measure μ_0 on Σ^0 with $\mu_0(1) = \|gT^\wedge\|$ such that

$$|\langle J_\iota(x), gT^\wedge\rangle| \leq \int_{\Sigma^0} |\langle x, x'\rangle| \, d\mu_0(x') \quad \text{for all} \quad x \in X \quad \text{and} \quad \iota \in \Lambda .$$

Therefore, we obtain a positive Radon measure μ_0 on Σ^0 with $\mu_0(1) \leq \|T\|_{(s)}$ such that

$$q(Tx) \leq \int_{\Sigma^0} |\langle x, x'\rangle| \, d\mu_0(x') \quad \text{for all} \quad x \in X .$$

If $r(T)$ denotes the set of all positive Radon measures μ on Σ^0 such that

$$q(Tx) \leq \int_{\Sigma^0} |\langle x, x'\rangle| \, d\mu(x') \quad \text{for all} \quad x \in X ,$$

then $\|T\|_{(s)} \leq \mu(1)$ for all $\mu \in r(T)$. Consequently,

$$\|T\|_{(s)} = \inf\{\mu(1) : \mu \in r(T)\} = \mu_0(1)$$

for some $\mu_0 \in r(T)$.

Clearly, $p_\varepsilon(J_\iota x) = p(x)$ and $q_\pi(J_\iota(Tx)) = q(Tx)$, we obtain $|||T||| \leq \sup\{q_\pi([Tx_n]) : p_\varepsilon([x_n]) \leq 1\} = |||T^{\mathbb{N}}||| = \|T\|_{(s)}$ for all $T \in \pi_{\ell^1}(X, Y)$. Furthermore, if (Y, q) is complete, then a standard argument shows that $(\pi_{\ell^1}(X, Y), \|\cdot\|_{(s)})$ is a Banach space.

Further information on absolutely summing mappings from one normed space into another can be found in Pietsch [1].

The adjoint map of an absolutely summing map need not be absolutely summing (see Example (3.4.8)). Therefore Schaefer [2, p.264] induces the notion of hypermajorizing maps and then shows that an $T \in L(X, Y)$ is absolutely summing (resp. hypermajorizing) if and only if T' is hypermajorizing (resp. $T' \in \pi_{\ell^1}(Y', X')$). In view of (2.2.7), absolutely summing mappings need not be precompact; but for Hilbert spaces, we have the following

(2.2.8) Theorem. <u>For two Hilbert spaces</u> H_i $(i = 1, 2)$ <u>and</u> $T \in L(H_1, H_2)$, <u>the following statements are equivalent</u>.

(a) $\Sigma_\iota \|Te_\iota\|^2 < +\infty$ <u>for any orthonormal basis</u> $\{e_\iota : \iota \in \Lambda\}$ <u>in</u> H_1.

(b) <u>There is an orthonormal basis</u> $\{u_\iota : \iota \in \Lambda\}$ <u>in</u> H_1 <u>such that</u>

$$\Sigma_\iota \|Tu_\iota\|^2 < +\infty .$$

(c) T <u>is compact and admits a representation of the form</u>

$$Tx = \Sigma_n \mu_n [x, e_n] d_n = \lim_{\alpha \in \mathfrak{F}(\mathbb{N})} \Sigma_{n \in \alpha} \mu_n [x, e_n] d_n \quad \text{with} \quad \Sigma_n \mu_n^2 < +\infty ,$$

<u>where</u> $\mu_n \geq 0$, $\{e_n\}$ <u>and</u> $\{d_n\}$ <u>are orthonormal sequences in</u> H_1 <u>and</u> H_2 <u>respectively</u>.

(d) T <u>is compact and</u> $\Sigma_n \eta_n^2 < +\infty$, <u>where</u> η_n <u>are eigenvalues of the absolute value</u> $[T]$ <u>of</u> T.

(e) $T \in \pi_{\ell^1}(H_1, H_2)$.

<u>Proof</u>. The implication (a) $\Rightarrow$ (b) is obvious.

(b) ⇒ (a) : Let $\{d_j : j \in A\}$ be an orthonormal basis in H_2 . Then Parseval's equality ensures that

$$\|Tu_\iota\|^2 = \Sigma_j |[Tu_\iota , d_j]|^2 = \Sigma_j |[u_\iota , T'd_j]|^2 \quad \text{for all } \iota \in \Lambda$$

and

$$\|T'd_j\|^2 = \Sigma_\iota |[T'd_j , u_\iota]|^2 = \Sigma_\iota |[u_\iota , T'd_j]|^2 \quad \text{for all } j \in A ,$$

hence

$$\Sigma_\iota \|Tu_\iota\|^2 = \Sigma_\iota \Sigma_j |[u_\iota , T'd_j]|^2 = \Sigma_j \|T'd_j\|^2 . \qquad (2.6)$$

Since the sum of $\Sigma_j \|T'd_j\|^2$ does not depend upon the choice of the basis $\{u_\iota : \iota \in \Lambda\}$ in H_1 , we conclude that

$$\Sigma_\iota \|Te_\iota\|^2 = \Sigma_j \|T'd_j\|^2 < +\infty$$

for any orthonormal basis $\{e_\iota : \iota \in \Lambda\}$ in H_1 .

(a) ⇒ (c) : To prove the compactness of T , it suffices to verify that if a sequence $\{x_n\}$ in H_1 is $\sigma(H_1, H_1')$-convergent to x and if $\delta > 0$, then there is a natural number n_o such that

$$\|Tx_n - Tx\| < \delta \quad \text{for all} \quad n \geq n_o .$$

In fact, the sequence $\{x_n\}$ is bounded, hence there is a constant $C > 0$ such that

$$\|x\| \leq C \quad \text{and} \quad \|x_n\| \leq C \quad \text{for all} \quad n .$$

On the other hand, for any orthonormal basis $\{e_\iota : \iota \in \Lambda\}$ in H_1 , the hypothesis (a) implies that there exists $\alpha \in \mathcal{F}(\Lambda)$ such that

$$\Sigma_{\Lambda\setminus\alpha}\|Te_\iota\|^2 < \delta^2/16C^2 .$$

Since $\{e_\iota : \iota \in \Lambda\}$ is an orthonormal basis in H_1, we have

$$x_n - x = \Sigma_\iota [x_n - x, e_\iota] e_\iota ;$$

consequently, the continuity of T implies that

$$Tx_n - Tx = \Sigma_\iota [x_n - x, e_\iota] Te_\iota .$$

For each n, we have

$$\begin{aligned}\|Tx_n - Tx\|^2 &= \|\Sigma_{\iota\in\alpha}[x_n - x, e_\iota]Te_\iota + \Sigma_{\Lambda\setminus\alpha}[x_n - x, e_\iota]Te_\iota\|^2 \\ &\leq 2\|\Sigma_{\iota\in\alpha}[x_n - x, e_\iota]Te_\iota\|^2 + 2\|\Sigma_{\Lambda\setminus\alpha}[x_n - x, e_\iota]Te_\iota\|^2 . \qquad (2.7)\end{aligned}$$

Now we apply Cauchy-Schwarz's inequality and Bessel's inequality to get

$$\begin{aligned}2\|\Sigma_{\Lambda\setminus\alpha}[x_n - x, e_\iota]Te_\iota\|^2 &\leq 2(\Sigma_{\Lambda\setminus\alpha}|[x_n - x, e_\iota]|^2)(\Sigma_{\Lambda\setminus\alpha}\|Te_\iota\|^2) \\ &\leq 2\|x_n - x\|^2(\Sigma_{\Lambda\setminus\alpha}\|Te_\iota\|^2) < \frac{\delta^2}{2} \qquad (2.8)\end{aligned}$$

because of $\|x_n - x\|^2 \leq (\|x_n\| + \|x\|^2) \leq 4C^2$. On the other hand, since x_n is $\sigma(H_1, H_1')$-convergent to x and since α is a finite subset of Λ, we have $\lim_n \Sigma_{\iota\in\alpha}[x_n - x, e_\iota] = 0$, and thus $\lim_n \Sigma_{\iota\in\alpha}[x_n - x, e_\iota]Te_\iota = 0$ in H_2 by the continuity of the scalar multiplication. Therefore there is a natural number n_o such that

$$2\|\Sigma_{\iota\in\alpha}[x_n - x, e_\iota]Te_\iota\|^2 < \frac{\delta^2}{2} \quad \text{for all } n \geq n_o . \qquad (2.9)$$

Combining (2.8) and (2.9) we obtain from (2.7) that

$$\|Tx_n - Tx\|^2 < \delta^2 \quad \text{for all } n \geq n_o ,$$

which proves our assertion.

By the polar representation of a compact operator, we have

$$Tx = \Sigma_n \mu_n [x, e_n] d_n = W[T] ,$$

where $\{e_n\}$ and $\{d_n\}$ are orthonormal sequences in H_1 and H_2 respectively, $[T]e_n = \mu_n e_n$, $\mu_n \geq 0$, $\mu_n \downarrow 0$ and W is a partial isometry whose initial set is $\overline{\text{Im }[T]}$. Thus

$$||Te_n|| = ||[T]e_n|| = \mu_n \quad \text{for all} \quad n . \tag{2.10}$$

We extend $\{e_n\}$ to an orthonormal basis in H_1 by adding an orthonormal set $\{w_r\}$ for which $[T]w_r = 0$, then we have $Tw_r = 0$. According to (2.10), we obtain

$$\Sigma_n \mu_n^2 = \Sigma_n ||Te_n||^2 = \Sigma_n ||Te_n||^2 + \Sigma_r ||Tw_r||^2 < +\infty .$$

(c) $\Rightarrow$ (d) : By the hypothesis (c) and a well-known result, we have

$$[T]x = \Sigma_n \mu_n [x, e_n] e_n , \tag{2.11}$$

where $[T]$ is the absolute value of T . As $\{e_n\}$ is an orthonormal sequence in H_1 , it follows from (2.11) that

$$[T]e_n = \mu_n e_n \quad \text{for all} \quad n .$$

Therefore $\{\mu_n\} = \{\eta_n\}$ is the set of all eigenvalues of $[T]$; consequently,

$$\Sigma_n \eta_n^2 = \Sigma_n \mu_n^2 < +\infty .$$

(d) ⇒ (e) : In view of the polar representations of compact operators, we obtain

$$Tx = \Sigma_n \eta_n [x, e_n] d_n \quad \text{and} \quad \Sigma_n \eta_n^2 < +\infty ,$$

where $\{e_n\}$ and $\{d_n\}$ are orthonormal sequences in H_1 and H_2 respectively, $[T]e_n = \eta_n e_n$ and $\eta_n \geq 0$. Let us define two maps $T_1 \in L^*(H_1, \ell^1)$ and $T_2 \in L^*(\ell^2, H_2)$ by the following two equations

$$T_1 x = (\eta_n [x, e_n])_{n\geq 1} \quad \text{for all} \quad x \in H_1 ,$$
$$T_2((\zeta_n)) = \Sigma_n \zeta_n d_n \quad \text{for all} \quad (\zeta_n) \in \ell^2 .$$

In view of Cauchy-Schwarz's inequality and Bessel's inequality, we have

$$\|T_1 x\| = \Sigma_n |\eta_n [x, e_n]| \leq (\Sigma_{n=1}^{\infty} \eta_n^2)^{\frac{1}{2}} (\Sigma_n |[x, e_n]|^2)^{\frac{1}{2}} \leq (\Sigma_{n=1}^{\infty} \eta_n^2)^{\frac{1}{2}} \|x\| ,$$

and

$$\|T_2((\zeta_n))\| \leq (\Sigma_{n=1}^{\infty} |\zeta_n|^2)^{\frac{1}{2}} .$$

Therefore T_1 and T_2 are continuous and

$$|||T_1||| \leq (\Sigma_{n=1}^{\infty} \eta_n^2)^{\frac{1}{2}} \quad \text{and} \quad |||T_2||| \leq 1 .$$

Let J be the canonical embedding map from ℓ^1 into ℓ^2. Then $J \in \pi_{\ell^1}(\ell^1, \ell^2)$ by Pietsch [1, p.43]. As $T = T_2 J T_1$, it follows from (2.2.1) that $T \in \pi_{\ell^1}(H_1, H_2)$.

(d) ⇒ (a) : Let $\{e_\iota : \iota \in \Lambda\}$ be any orthonormal basis in H_1 and Σ the closed unit ball in H_1. For any $[\eta_\iota, \Lambda] \in \ell^2_\Lambda$ we have by the Cauchy-Schwarz inequality and Bessel's inequality that

$$\Sigma_i |\langle \eta_i e_i , x' \rangle| = (\Sigma_i |\eta_i|^2)^{\frac{1}{2}} (\Sigma_i |\langle e_i , x' \rangle|^2)^{\frac{1}{2}} \leq (\Sigma_i |\eta_i|^2)^{\frac{1}{2}} \|x\| \text{ for all } x \in H_1 .$$

It then follows from $T \in \pi_{\ell^1}(H_1, H_2)$ that

$$\|[T(\eta_i e_i), \Lambda]\|_\pi \leq \|T\|_{(s)} \|[\eta_i e_i , \Lambda]\|_\varepsilon = \|T\|_{(s)} \sup\{\Sigma_i |\langle \eta_i e_i , x' \rangle| : x' \in \Sigma^o\}$$
$$\leq \|T\|_{(s)} (\Sigma_i |\eta_i|^2)^{\frac{1}{2}} \text{ for all } [\eta_i , \Lambda] \in \ell^2_\Lambda ,$$

and hence from the definition of $\|\cdot\|_\pi$ that

$$\Sigma_i |\eta_i| \|Te_i\| = \|[T(\eta_i e_i), \Lambda]\|_\pi \leq \|T\|_{(s)} (\Sigma_i |\eta_i|^2)^{\frac{1}{2}} \text{ for all } [\eta_i , \Lambda] \in \ell^2_\Lambda .$$

Therefore $\Sigma_i \|Te_i\|^2 < +\infty$ since the above inequality holds for all $[\eta_i , \Lambda] \in \ell^2_\Lambda$.

An $T \in L(H_1, H_2)$ satisfying one of the preceding equivalent properties is called a Hilbert-Schmidt operator.

In view of (2.2.8) and Formula (2.6), an $T \in L(H_1, H_2)$ is a Hilbert-Schmidt operator if and only if its adjoint is a Hilbert-Schmidt operator.

(2.2.9) Example. Let K be a compact Hausdorff space, let μ be a positive Radon measure on K and let $j_K : C(K) \to L^2_\mu(K)$ be the canonical embedding. Suppose that H is a Hilbert space. Then

$$j_K \circ T \in \pi_{\ell^1}(H, L^2_\mu(K)) \text{ for all } T \in L(H, C(K)) .$$

Proof. Denote by p the norm on H defined by the inner product $[\cdot, \cdot]$, and by $\|\cdot\|_2$ the norm on $L^2_\mu(K)$, that is

$$\|h\|_2 = (\int_K |h(t)|^2 d\mu(t))^{\frac{1}{2}} \text{ for all } h \in L^2_\mu(K) .$$

For each $t \in K$, let δ_t be the Dirac measure on K defined by

$$\langle g, \delta_t \rangle = g(t) \quad \text{for all} \quad g \in C(K) .$$

Let $\{e_i, \Lambda\}$ be any orthonormal set in H. Then Bessel's inequality shows that

$$\Sigma_i |(Te_i)t|^2 = \Sigma_i |\langle Te_i, \delta_t \rangle|^2 \leq p^*(T'\delta_t)^2 \leq |||T|||^2 \quad \text{for all} \quad t \in K .$$

From this it follows that

$$\Sigma_i \|(j_K \circ T)e_i\|_2^2 = \int_K \Sigma_i |(Te_i)t|^2 d\mu(t) \leq |||T|||^2 \mu(K) ,$$

which obtains the desired result by (2.2.8).

2.3 Duality theorems in sequence spaces

The notions of summabilities and of absolutely summing mappings, studied in the previous two sections, depend upon on the Banach space ℓ^1 in one hand, and the locally convex topologies on the other hand. If ℓ^1 is replaced by a more general vector subspace λ of $\mathbb{K}^{\mathbb{N}}$ equipped with a suitable locally convex topology, then these two notions can be extended to those of λ-summabilities and of absolutely λ-summing mappings in a natural way. The purpose of the next section is devoted to such a study. To this aim, we first investigate some elementary properties of sequence spaces of numbers in this section.

We denote by ω the algebraic product $\mathbb{K}^{\mathbb{N}}$. Elements in ω will be denoted by $\zeta = [\zeta_i]$, $\eta = [\eta_i]$, $u = [u_i]$ etc., and e will denote the element in ω with all coordinates to be one. For any $n \geq 0$, we denote by e_n the element $[\delta_i^{(n)}]$ in ω, where $\delta_i^{(n)}$ is the Kronecker delta, and define the n-th section $\zeta^{(n)}$ of $\zeta = [\zeta_i]$ by

$$\zeta^{(n)} = \Sigma_{i=1}^{n} \zeta_i e_i \ .$$

The sequence $\{\zeta^{(n)}, n \geq 0\}$ in ω will be called the sectional sequence associated with ζ. Also we denote by $\zeta\eta$ the element $[\zeta_i \eta_i]$ in ω, by

$$AB = \{\zeta\eta : \zeta \in A \text{ and } \eta \in B\}$$

whenever A and B are subsets of ω, by $\frac{\zeta}{\eta}$ the element $[\frac{\zeta_i}{\eta_i}]$ in ω, here we use the convertion that $\frac{\zeta_i}{\eta_i} = 0$ if $\eta_i = 0$, and by $|\zeta|$ the element $[|\zeta_i|]$ in ω, where $|\zeta_i|$ is the modulus of ζ_i. For any two elements ζ and μ in ω, we define $|\zeta| \leq |\eta|$ by

$$|\zeta_i| \leq |\eta_i| \quad \text{for all} \quad i \in \mathbb{N} \ .$$

Vector subspaces of $\mathbb{K}^{\mathbb{N}}$ will be called sequence spaces.

Let λ be a sequence space and $A \subset \lambda$. We say that A is normal in λ (or a normal subset of λ) if

$$|u| \leq |\zeta| , \zeta \in A \text{ and } u \in \lambda \text{ imply that } u \in A \ .$$

In particular, a subset of ω, which is normal in ω, is said to be normal.

Clearly if λ is a normal sequence space in ω, then a subset A of λ is normal in λ if and only if it is normal, and

$$\zeta^{(n)} \in \lambda \quad \text{for all} \quad n \geq 0$$

whenever $\zeta \in \lambda$. If $\mathbb{K} = \mathbb{R}$, then ω is a Riesz space (i.e., vector lattice) under the usual ordering, hence every normal sequence space is an ℓ-ideal in ω ; if λ is a Riesz subspace of $\mathbb{R}^{\mathbb{N}}$, then a subset A of λ is normal in λ if and only if it is solid (see Wong and Ng [1, (10.4)]).

The intersection and the union of a family of normal subsets of λ are clearly normal in λ . Therefore, for any subset A of λ , the <u>normal hull of</u> A <u>in</u> λ , denoted by $h(A)$, is defined to be the intersection of all normal subsets of λ containing A , and the <u>normal kernel of</u> A <u>in</u> λ , denoted by $k(A)$, is defined to be the union of all normal subsets of λ contained in A .

The normal hull $h(A)$ of A in λ is the smallest normal subset of λ containing A and

$$h(A) = \{u \in \lambda : |u| \leq |\zeta| \ \text{ for some } \ \zeta \in A\} ; \tag{3.1}$$

the normal kernel $k(A)$ of A in λ is the largest normal subset of λ contained in A and

$$k(A) = \{\zeta \in A : |u| \leq |\zeta| \ \text{ and } \ u \in \lambda \ \text{ imply } \ u \in A\} . \tag{3.2}$$

Moreover, it is not hard to show that

(a) $k(A) \subset k(C)$ whenever $A \subset C \subset \lambda$,

(b) $k(\gamma A) = \gamma k(A)$ for any $\gamma \in \mathbb{K}$ with $\gamma \neq 0$,

(c) $k(A_1 \cap A_2) = k(A_1) \cap k(A_2)$.

We denote by $h_\omega(A)$ the normal hull of A in ω, and by $k_\omega(A)$ the normal kernel of A in ω. If λ is normal in ω, then

$$h_\omega(A) = h(A) \quad \text{and} \quad k_\omega(A) = k(A) .$$

If $\mathbb{K} = \mathbb{R}$ and λ is a Riesz subspace of ω, then $h(A)$ is the solid hull of A (see Peressini [1, p.102]), and $k(A)$ is the solid kernel of A (see Wong [2]).

For any sequence space λ, we define

$$\lambda^\times = \{\eta \in \omega : \langle|\zeta|, |\eta|\rangle = \Sigma_i |\zeta_i \eta_i| < +\infty \text{ for all } \zeta \in \lambda\} .$$

Then $\lambda^\times$ is a sequence space such that

$$\lambda \subset \lambda^{\times\times} \quad \text{and} \quad \lambda^\times = \lambda^{\times\times\times} . \tag{3.3}$$

$\lambda^\times$ is called the α-<u>dual</u> (or <u>Köthe dual</u>) of λ, and $\lambda^{\times\times}$ is called the α-<u>bidual</u> (or <u>Köthe bidual</u>) of λ.

Clearly if λ and μ are sequence spaces, then

$$\lambda \subset \mu \quad \text{implies} \quad \mu^\times \subset \lambda^\times .$$

Furthermore, we list several important sequence spaces together with their α-duals as follows:

(a) $(\mathbb{K}^{\mathbb{N}})^\times = \mathbb{K}^{(\mathbb{N})}$.

(b) $(\mathbb{K}^{(\mathbb{N})})^\times = \mathbb{K}^{\mathbb{N}}$.

(c) $(\ell^p)^\times = \ell^q$, where $p \geq 1$ and $\frac{1}{p} + \frac{1}{q} = 1$.

(d) $(\ell^\infty)^\times = \ell^1$.

(e) $(c_o)^\times = \ell^1$.

We see from (c) and (d) that $(\ell^p)^{\times\times} = \ell^p$ for all $1 \leq p \leq +\infty$, and from (d) and (e) that

$$c_o \neq (c_o)^{\times\times} = \ell^\infty .$$

A sequence space λ is said to be <u>perfect</u> if

$$\lambda = \lambda^{\times\times} .$$

The sequence spaces $\mathbb{K}^{\mathbb{N}}$, $\mathbb{K}^{(\mathbb{N})}$ and ℓ^p $(1 \leq p \leq +\infty)$ are all perfect, but the sequence space c_o is not perfect. In view of (3.3), the α-dual $\lambda^\times$ of λ is always perfect, hence $\lambda^{\times\times}$ is the smallest perfect sequence space containing λ . Furthermore, if the sequence space λ is perfect, then λ is normal and contains $\mathbb{K}^{(\mathbb{N})}$.

If λ is a sequence space containing $\mathbb{K}^{(\mathbb{N})}$ (in particular, λ is perfect), then λ and its α-dual $\lambda^\times$ form a dual pair $\langle\lambda, \lambda^\times\rangle$ under the natural bilinear form

$$\langle\zeta, \eta\rangle = \Sigma_i \zeta_i \eta_i \quad \text{for all} \quad \zeta \in \lambda \quad \text{and} \quad \eta \in \lambda^\times .$$

λ is perfect if and only if $(\lambda, \sigma(\lambda, \lambda^\times))$ is sequentially complete, as shown by Köthe [1, p.413]. Furthermore, we have the following important result whose proof can be found in Köthe [1, p.413-414].

(2.3.1) Theorem. Let λ be a normal sequence space containing $\mathbb{K}^{(\mathbb{N})}$. Then the following statements hold:

(a) The normal hull of every $\sigma(\lambda, \lambda^{\times})$-bounded subset of λ is $\sigma(\lambda, \lambda^{\times})$-bounded.

(b) Any $\zeta \in \lambda$ is the $\tau(\lambda, \lambda^{\times})$-limit of the sectional sequence $\{\zeta^{(n)}, n \geq 0\}$ associated with ζ.

(c) If, in addition, λ is perfect, then a subset of λ is $\sigma(\lambda, \lambda^{\times})$-bounded if and only if it is $\beta(\lambda, \lambda^{\times})$-bounded, and the normal hull of each $\sigma(\lambda, \lambda^{\times})$-compact subset is $\sigma(\lambda, \lambda^{\times})$-relatively compact.

Let λ be a normal sequence space containing $\mathbb{K}^{(\mathbb{N})}$. A locally convex topology $\mathfrak{L}$ on λ is said to be normal if it admits a neighbourhood basis at 0 consisting of normal sets in λ.

When $\mathbb{K} = \mathbb{R}$, a locally convex topology on λ is normal if and only if it is locally solid (see Wong and Ng [1]).

For any normal sequence space λ containing $\mathbb{K}^{(\mathbb{N})}$, there always exists a normal topology on λ as constructed by Köthe [1, p.407]. In fact, for any $\eta \in \lambda^{\times}$, the functional p_{η}, defined by

$$p_{\eta}(\zeta) = \Sigma_{i}|\zeta_{i}\eta_{i}| = \langle|\zeta|, |\eta|\rangle \quad \text{for all} \quad \zeta \in \lambda, \tag{3.4}$$

is a seminorm on λ such that

$$|\langle\zeta, \eta\rangle| \leq p_{\eta}(\zeta) \quad \text{for all} \quad \zeta \in \lambda, \tag{3.5}$$

and the unit ball

$$V_\eta = \{\zeta \in \lambda : p_\eta(\zeta) \leq 1\}$$

is a $\sigma(\lambda, \lambda^\times)$-closed, absolutely convex normal subset of λ. Hence, the family $\{p_\eta : \eta \in \lambda^\times\}$ of seminorms determines a unique normal topology on λ, which is called the Köthe topology and denoted by $\sigma_K(\lambda, \lambda^\times)$. In view of (3.5), $\sigma_K(\lambda, \lambda^\times)$ is finer than $\sigma(\lambda, \lambda^\times)$; actually $\sigma_K(\lambda, \lambda^\times)$ is consistent with the duality $\langle\lambda, \lambda^\times\rangle$ as shown by Köthe [1, p.409].

In view of (3.4), the norm topology on ℓ^1 coincides with $\sigma_K(\ell^1, \ell^\infty)$, hence the norm topology on ℓ^1 is the unique normal topology on ℓ^1 which is consistant with $\langle\ell^1, \ell^\infty\rangle$. On the other hand, if $\mathbb{K} = \mathbb{R}$, then the Köthe topology $\sigma_K(\lambda, \lambda^\times)$ coincides with the Dieudonné topology $\sigma_S(\lambda, \lambda^\times)$, that is, the topology of uniform convergence on order-bounded sets in $\lambda^\times$ or, equivalently, the locally solid topology associated with $\sigma(\lambda, \lambda^\times)$ (see Wong and Ng [1, p.160]); in this case, by means of the notions of cone-absolutely summing mappings and cone-prenuclear mappings, various characterizations for $\sigma_S(\lambda, \lambda^\times)$ are given by Wong [1].

The Köthe topology $\sigma_K(\lambda, \lambda^\times)$ has the following important properties as shown by Köthe [1, §30,5.(7) and §30,7.(6)].

(2.3.2) Theorem. Let λ be a normal sequence space containing $\mathbb{K}^{(\mathbb{N})}$ and $\sigma_K(\lambda, \lambda^\times)$ the Köthe topology.

(a) λ is $\sigma_K(\lambda, \lambda^\times)$-complete if and only if λ is perfect. Consequently $\lambda^{\times\times}$ is the $\sigma_K(\lambda, \lambda^\times)$-completion of λ.

(b) *If λ is perfect, then a subset A of λ is $\sigma(\lambda, \lambda^{\times})$-compact if and only if it is $\sigma_K(\lambda, \lambda^{\times})$-compact, and $\sigma(\lambda, \lambda^{\times})$-convergent and $\sigma_K(\lambda, \lambda^{\times})$-convergent sequences are always the same.*

It is clear that $\sigma(\ell^1, \ell^\infty)$ is not normal. But we shall show that any locally convex topology on λ, which is consistent with $\langle\lambda, \lambda^{\times}\rangle$, always associates a normal topology on λ which is also consistent with $\langle\lambda, \lambda^{\times}\rangle$. To this end, we need to establish duality relationship between normal hulls and normal kernels. The following technical lemma will be needed for the proof of our duality results.

(2.3.3) Lemma. *Let λ be a normal sequence space containing* $\mathbb{K}^{(\mathbb{N})}$. *For any fixed* $\zeta \in \lambda$, *we have*

$$\langle|\zeta|, |v|\rangle = \sup\{|\langle\zeta, \eta\rangle| : |\eta| \leq |v|\} \quad \text{for any } v \in \lambda^{\times}; \tag{3.6}$$

and, for any fixed $\eta \in \lambda^{\times}$, *we have*

$$\langle|u|, |\eta|\rangle = \sup\{|\langle\zeta, \eta\rangle| : \zeta \in \lambda \text{ and } |\zeta| \leq |u|\} \quad \text{for any } u \in \lambda. \tag{3.7}$$

Proof. The proof of (3.7) is similar to that of (3.6), and hence will be omitted. To prove (3.6), we first notice from the normality of $\lambda^{\times}$ that $\{\eta \in \omega : |\eta| \leq |v|\} \subset \lambda^{\times}$. For any $\eta \in \omega$ with $|\eta| \leq |v|$, we have

$$|\langle\zeta, \eta\rangle| \leq \Sigma_i |\zeta_i \eta_i| \leq \Sigma_i |\zeta_i v_i| = \langle|\zeta|, |v|\rangle ,$$

hence

$$\sup\{|\langle\zeta, \eta\rangle| : |\eta| \leq |v|\} \leq \langle|\zeta|, |v|\rangle .$$

In order to verify that they are equal, it suffices to show that if $M \geq 0$ is such that $|\langle\zeta, \eta\rangle| \leq M$ for all $\eta \in \omega$ with $|\eta| \leq |v|$, then

$$\langle|\zeta|, |v|\rangle \leq M .$$

In fact, for any $j \in \mathbb{N}$, let $\theta_j \in \mathbb{R}$ be such that $\zeta_j = |\zeta_j| e^{i\theta_j}$. Then we define

$$\eta_j = |v_j| e^{-i\theta_j} .$$

Clearly $\eta = [\eta_j] \in \omega$ is such that $|\eta| = |v|$, hence

$$M \geq |\langle\zeta, \eta\rangle| = |\Sigma_j \zeta_j \eta_j|$$

by the assumption. Therefore

$$M \geq |\langle\zeta, \eta\rangle| = |\Sigma_j \zeta_j \eta_j| = |\Sigma_j \zeta_j |v_j| e^{-i\theta_j}|$$
$$= \Sigma_j |\zeta_j v_j| = \langle|\zeta|, |v|\rangle$$

which obtains our assertion.

It should be noted that formula (3.7) holds for a sequence space λ with $\mathbb{K}^{(\mathbb{N})} \subset \lambda$ and satisfying the following conditions:

(i) $|\zeta| \in \lambda$ for all $\zeta \in \lambda$;

(ii) $\zeta a \in \lambda$ for all $\zeta \in \lambda$ and $a \in \omega$ with $|a| = e$.

Clearly the normality of λ implies that (i) and (ii) hold.

If $\mathbb{K} = \mathbb{R}$, then (2.3.3) is a well-known result in vector lattices (see Wong and Ng [1, (10.12) and (10.15)]).

(2.3.4) Corollary. Let λ be a normal sequence space containing $\mathbb{K}^{(\mathbb{N})}$.

(a) If $B \subset \lambda^{\times}$ is normal, then the absolute polar B° of B , taken in λ , is a normal subset of λ .

(b) If $A \subset \lambda$ is normal, then the absolute polar A° of A , taken in $\lambda^{\times}$, is a normal subset of $\lambda^{\times}$.

Proof. The proof of (b) is similar to that of (a), and hence will be omitted. To prove (a), let $u \in \lambda$ be such that $|u| \leq |\zeta|$, where $\zeta \in B^{\circ}$. For any $v \in B$, the normality of B ensures that $\{\eta \in \omega : |\eta| \leq |v|\} \subset B$. It then follows from $\zeta \in B^{\circ}$ and (3.6) that

$$|\langle u, v\rangle| \leq \langle |u|, |v|\rangle \leq \langle |\zeta|, |v|\rangle$$
$$= \sup\{|\langle\zeta, \eta\rangle| : |\eta| \leq |v|\} \leq 1 ,$$

and hence that $u \in B^{\circ}$.

A seminorm r , defined on a sequence space λ , is said to be regular if it follows from $|u| \leq |\zeta|$ with u, ζ in λ that $r(u) \leq r(\zeta)$.

If r is a regular seminorm on λ , then the unit ball

$$V = \{\zeta \in \lambda : r(\zeta) < 1\}$$

is clearly normal. The converse also holds as the following result shows.

(2.3.5) Corollary. Let λ be a normal sequence space containing $\mathbb{K}^{(\mathbb{N})}$, let r be a seminorm on λ and let

$$S = \{\zeta \in \lambda : r(\zeta) < 1\} \quad \underline{\text{and}} \quad S_1 = \{\zeta \in \lambda : r(\zeta) \leq 1\} .$$

Then the following statements are equivalent.

(a) r is regular.

(b) S is normal.

(b)' S_1 is normal.

(c) For any $\zeta \in \lambda$,

$$r(\zeta) = \sup\{<|\zeta|, |v|> : v \in S^o\} .$$

Proof. The implication (a) $\Rightarrow$ (b) has been noted before this result, the implications (a) $\Rightarrow$ (b)' and (c) $\Rightarrow$ (a) are obvious. If either S or S_1 is normal, then S^o is normal by (2.3.4)(b), thus

$$\sup\{<|\zeta|, |v|> : v \in S^o\} \leq \sup\{|<\zeta, \eta>| : \eta \in S^o\}$$

by (3.6). Consequently

$$\sup\{<|\zeta|, |v|> : v \in S^o\} = \sup\{|<\zeta, \eta>| : \eta \in S^o\}$$

since $|<\zeta, \eta>| \leq <|\zeta|, |\eta|>$. On the other hand, it is easily seen that

$$r(\zeta) = \sup\{|<\zeta, \eta>| : \eta \in S^o\} .$$

Therefore the implications (b) $\Rightarrow$ (c) and (b)' $\Rightarrow$ (c) hold.

Let $\mathfrak{S}$ be a saturated family (for definite, see, Schaefer [1, p.81]) consisting of $\sigma(\lambda^\times, \lambda)$-bounded subsets of $\lambda^\times$ which covers $\lambda^\times$. Rosier [1] calls $\mathfrak{S}$ a normal topologizing family for λ if it satisfies the following condition:

$$h(N) \in \mathcal{G} \qquad \text{for all} \quad N \in \mathcal{G} .$$

In view of (3.6), the normality of N ensures that

$$\sup\{|\langle\zeta, \eta\rangle| : \eta \in N\} = \sup\{\langle|\zeta|, |v|\rangle : v \in N\} \quad (\zeta \in \lambda) .$$

Therefore we obtain from (2.3.4)(a) the following result due to Rosier [1].

(2.3.6) Corollary. <u>Let λ be a normal sequence space. If $\mathcal{G}$ is a normal topologizing family for λ, then the $\mathcal{G}$-topology on λ, denoted by $\mathcal{I}_{\mathcal{G}}$, is a normal topology, and $\mathcal{I}_{\mathcal{G}}$ is determined by a family $\{r_{N^o} : N \in \mathcal{G}, N = h(N)\}$ of regular seminorms, where each r_{N^o} is the gauge of N^o, and is given by</u>

$$r_{N^o}(\zeta) = \sup\{|\langle\zeta, \eta\rangle| : \eta \in N\}$$
$$= \sup\{\langle|\zeta|, |v|\rangle : v \in N\} \quad (\zeta \in \lambda) .$$

We shall now establish duality relationship between normal hulls and normal kernels.

(2.3.7) Theorem. <u>Let λ be a normal sequence space containing</u> $\mathbb{K}^{(\mathbb{N})}$.

(a) $(h(B))^o = k(B^o)$ <u>for any</u> $B \subset \lambda^\times$.

(b) $(h(A))^o = k(A^o)$ <u>for any</u> $A \subset \lambda$.

<u>In other words, the absolute polar of the normal hull is the normal kernel of the absolute polar.</u>

Proof. The proof of (b) is similar to that of (a), and hence will be omitted. To prove (a), we first notice that $B \subset h(B)$, and from (2.3.4)(a) that $(h(B))^{\circ}$ is a normal subset of λ ; hence $(h(B))^{\circ} \subseteq k(B^{\circ})$. To prove that $k(B^{\circ}) \subseteq (h(B))^{\circ}$, let $u \in k(B^{\circ})$. Then formula (3.2) shows that

$$\{\zeta \in \lambda : |\zeta| \leq |u|\} \subset B^{\circ} . \tag{3.8}$$

For any $v \in h(B)$, there exists an $\eta \in B$ such that $|v| \leq |\eta|$ by (3.1). As $\eta \in B$, we conclude from (3.7) and (3.8) that

$$\begin{aligned} |\langle u, v\rangle| &\leq \langle |u|, |v|\rangle \leq \langle |u|, |\eta|\rangle \\ &= \sup\{|\langle\zeta, \eta\rangle| : \zeta \in \lambda \text{ and } |\zeta| \leq |u|\} \leq 1 . \end{aligned}$$

Therefore $u \in (h(B))^{\circ}$.

If $\mathbb{K} = \mathbb{R}$, (2.3.6) and (2.3.4) are well-known results in the theory of vector lattices (see Peressini [1, p.102]).

The preceding duality result has many interesting applications; we mention a few below.

(2.3.8) Corollary. Let λ be a normal sequence space containing $\mathbb{K}^{(\mathbb{N})}$.

(a) For any $\sigma(\lambda^{\times}, \lambda)$-closed absolutely convex subset N of $\lambda^{\times}$,

$$(k(N))^{\circ} = \overline{\Gamma}(h(N^{\circ})) .$$

(b) For any $\sigma(\lambda, \lambda^{\times})$-closed absolutely convex subset V of λ,

$$(k(V))^{\circ} = \overline{\Gamma}(h(V^{\circ})) .$$

In other words, the absolute polar of the normal kernel of a weakly closed absolutely convex set is the weakly closed absolutely convex hull of the normal hull of the absolute polar.

Proof. In view of (2.3.7)(a) and the bipolar theorem,

$$\bar{\Gamma}(h(N^{\circ})) = (h(N^{\circ}))^{\circ\circ} = (k(N^{\circ\circ}))^{\circ} = (k(N))^{\circ}$$

which obtains (a). In view of (2.3.7)(b) and the bipolar theorem, a similar argument shows that (b) holds.

(2.3.9) Corollary. *Let* λ *be a normal sequence space containing* $\mathbb{K}^{(\mathbb{N})}$.

(a) *If* N *is a* $\sigma(\lambda^{\times}, \lambda)$*-closed absolutely convex subset of* $\lambda^{\times}$, *then so does* $k(N)$, *and*

$$\bar{\Gamma}(h(B)) = \bar{\Gamma}(h(B^{\circ\circ})) \quad \text{for any } B \subset \lambda^{\times} . \tag{3.9}$$

(b) *If* V *is a* $\sigma(\lambda, \lambda^{\times})$*-closed absolutely convex subset of* λ, *then so does* $k(V)$, *and*

$$\bar{\Gamma}(h(A)) = \bar{\Gamma}(h(A^{\circ\circ})) \quad \text{for any } A \subset \lambda . \tag{3.10}$$

Proof. In view of (2.3.8)(a) and (2.3.7)(a), the bipolar theorem shows that

$$\bar{\Gamma}(k(N)) = (k(N))^{\circ\circ} = (\bar{\Gamma}(h(N^{\circ})))^{\circ} = (h(N^{\circ}))^{\circ}$$
$$= k(N^{\circ\circ}) = k(N)$$

which obtains the first assertion of (a).

As B^o is $\sigma(\lambda, \lambda^\times)$-closed absolutely convex, we conclude from (2.3.7)(b) and the bipolar theorem that

$$\bar{\Gamma}(h(B)) = (h(B))^{oo} = (k(B^o))^o = \bar{\Gamma}(h(B^{oo})) .$$

(2.3.10) Corollary. <u>Let</u> λ <u>be a perfect sequence space and</u> V <u>a</u> $\sigma(\lambda, \lambda^\times)$-<u>closed absolutely convex subset of</u> λ . <u>Then</u> V <u>is absorbing if and only if</u> $k(V)$ <u>is absorbing</u>.

<u>Proof</u>. The sufficiency is obvious. To prove the necessity, we first notice that V^o is $\sigma(\lambda^\times, \lambda)$-bounded, and hence $\sigma(\lambda^\times, \lambda^{\times\times})$-bounded by the perfectness of λ . In view of (2.3.1)(a), $h(V^o)$ is $\sigma(\lambda^\times, \lambda)$-bounded. It then follows from (2.3.7)(a) that

$$(h(V^o))^o = k(V^{oo}) = k(V) ,$$

and hence that $k(V)$ is absorbing.

The duality results for normal sets enable us to give a construction of a normal topology from a given locally convex topology.

(2.3.11) Theorem. <u>Let</u> λ <u>be a perfect sequence space, let</u> $\mathcal{T}$ <u>be a locally convex topology on</u> λ <u>which is consistent with</u> $\langle\lambda, \lambda^\times\rangle$, <u>and let</u> $\mathcal{W}$ <u>be a neighbourhood basis at</u> 0 <u>for</u> $\mathcal{T}$ <u>consisting of</u> $\sigma(\lambda, \lambda^\times)$-<u>closed absolutely convex sets in</u> λ . <u>Suppose further that</u>

$$k(\mathcal{W}) = \{k(W) : W \in \mathcal{W}\} .$$

<u>Then there exists a unique normal topology on</u> λ , <u>denoted by</u> $\mathcal{T}_K$, <u>such that</u>

$k(\mathcal{W})$ _is a neighbourhood basis at_ 0 _for this topology_ $\mathcal{L}_K$. _Furthermore, this topology has the following properties:_

(a) $\mathcal{L}_K$ _is the smallest normal topology on_ λ _which is finer than_ $\mathcal{L}$;

(b) $\mathcal{L}_K$ _is consistent with_ $\langle\lambda, \lambda^\times\rangle$;

(c) _if_ $r_{k(W)}$ _is the gauge of_ $k(W)$, _then_

$$\begin{aligned} r_{k(W)}(\zeta) &= \sup\{\langle|\zeta|, |v|\rangle : v \in W^o\} \\ &= \sup\{\Sigma_i |\zeta_i v_i| : v \in W^o\} \quad (\zeta \in \lambda) . \end{aligned} \tag{3.11}$$

Proof. We know from (2.3.9)(b) and (2.3.10) that each $k(W)$ is $\sigma(\lambda, \lambda^\times)$-closed, absolutely convex, normal and absorbing. Therefore, since $k(\gamma W) = \gamma k(W)$ for all $\gamma \in \mathbb{K}$ with $\gamma \neq 0$.

$$k(V_1 \cap V_2) = k(V_1) \cap k(V_2) ,$$

and since $k(W) \subseteq k(U)$ whenever $W \subseteq U$, there exists a unique normal topology $\mathcal{L}_K$ on λ such that $k(\mathcal{W})$ is a neighbourhood basis at 0 for $\mathcal{L}_K$. As $k(W) \subseteq W$, it follows that $\mathcal{L}_K$ is finer than $\mathcal{L}$; consequently $\mathcal{L}_K$ is Hausdorff.

Let $\mathcal{L}'$ be any normal topology on λ which is finer than $\mathcal{L}$, and let U be any $\mathcal{L}_K$-neighbourhood of 0. Then there exists an $W \in \mathcal{W}$ such that $k(W) \subseteq U$, hence there is a _normal_ $\mathcal{L}'$-neighbourhood V of 0 such that $V \subseteq W$. It then follows that $V \subseteq k(W)$, and hence that $V \subseteq U$. Therefore $\mathcal{L}'$ is finer than $\mathcal{L}_K$.

To prove (b), let $W \in \mathcal{U}$. Then W^o is $\sigma(\lambda^\times, \lambda)$-compact, and

$$(k(W))^o = \bar{\Gamma}(h(W^o)) \tag{3.12}$$

by (2.3.8)(b). In view of (2.3.2)(b) and (2.3.1)(c), the perfectness of λ ensures that the $\sigma(\lambda, \lambda^\times)$-closure $\overline{h(W^o)}$ of $h(W^o)$ is $\sigma_K(\lambda^\times, \lambda)$-compact, and hence $\bar{\Gamma}(\overline{h(W^o)})$ is $\sigma_K(\lambda^\times, \lambda)$-compact since $(\lambda^\times, \sigma_K(\lambda^\times, \lambda))$ is complete. Therefore $\bar{\Gamma}(\overline{h(W^o)})$ is $\sigma(\lambda^\times, \lambda)$-compact; consequently $\bar{\Gamma}(h(W^o))$ is $\sigma(\lambda^\times, \lambda)$-compact since $\bar{\Gamma}(h(W^o)) \subseteq \bar{\Gamma}(\overline{h(W^o)})$. As $k(\mathcal{U})$ is a neighbourhood basis at 0 for $\mathfrak{T}_K$, it follows from Mackey's theorem that $\mathfrak{T}_K$ is consistent with $<\lambda, \lambda^\times>$.

Finally, we know from (3.12) that

$$r_{k(W)}(\zeta) = \sup\{|<\zeta, \eta>| : \eta \in h(W^o)\} . \tag{3.13}$$

It then follows from (3.1) that

$$r_{k(W)}(\zeta) \leq \sup\{<|\zeta|, |v|> : v \in W^o\} . \tag{3.14}$$

Conversely for any $v \in W^o$, we have from (3.6) that

$$<|\zeta|, |v|> = \sup\{|<\zeta, \eta>| : |\eta| \leq |v|\} ,$$

and hence from (3.13) that

$$<|\zeta|, |v|> \leq r_{k(W)}(\zeta) . \tag{3.15}$$

As $v \in W^o$ was arbitrary, we obtain the formula (3.11) by making use of (3.14) and (3.15).

The foregoing argument leads to the following definition: Let λ be a perfect sequence space, and let $\mathcal{I}$ be a locally convex topology on λ which is consistent with $\langle\lambda, \lambda^{\times}\rangle$. Then the normal topology $\mathcal{I}_K$, constructed in the preceding result, is called the <u>normal topology on</u> λ <u>associated with</u> $\mathcal{I}$.

The construction of $\mathcal{I}_K$ is similar to that of the locally solid topology associated with a locally o-convex topology on a Riesz space (see Wong [2, Prop.3.2]).

As an immediate consequence of (2.3.9), we obtain some elementary properties about the associated normal topology $\mathcal{I}_K$ as follows.

(2.3.12) Corollary. <u>Let</u> λ <u>be a perfect sequence space, and let</u> $\mathcal{I}$ <u>and</u> $\mathcal{I}'$ <u>be locally convex topologies on</u> λ <u>which are consistent with</u> $\langle\lambda, \lambda^{\times}\rangle$. <u>Then the following assertions hold</u>:

(a) <u>If</u> $\mathcal{I}$ <u>is coarser than</u> $\mathcal{I}'$, <u>then</u> $\mathcal{I}_K$ <u>is coarser than</u> $\mathcal{I}'_K$.

(b) $\mathcal{I}$ <u>is normal if and only if</u> $\mathcal{I} = \mathcal{I}_K$; <u>consequently every normal topology on</u> λ <u>which is consistent with</u> $\langle\lambda, \lambda^{\times}\rangle$ <u>admits a neighbourhood basis at</u> 0 <u>consisting of</u> $\sigma(\lambda, \lambda^{\times})$-<u>closed, absolutely convex and normal sets in</u> λ <u>or, equivalently,</u> $\mathcal{I}$ <u>is determined by a family of regular seminorms</u>.

(c) <u>If</u> $\mathcal{I}$ <u>is metrizable (resp. normable), then</u> $\mathcal{I}_K$ <u>is metrizable (resp. normable)</u>.

(d) $(\lambda, \mathcal{I}_K)$ <u>is complete</u>.

(e) <u>Any</u> $\zeta \in \lambda$ <u>is the</u> $\mathcal{I}_K$<u>-limit of the sectional sequence</u> $\{\zeta^{(n)}, n \geq 0\}$ <u>associated with</u> ζ.

(f) <u>The Mackey topology</u> $\tau(\lambda, \lambda^\times)$ <u>is the finest normal topology on</u> λ <u>which is consistent with</u> $\langle\lambda, \lambda^\times\rangle$.

(g) <u>The Köthe topology</u> $\sigma_K(\lambda, \lambda^\times)$ <u>is the normal topology on</u> λ <u>associated with</u> $\sigma(\lambda, \lambda^\times)$, <u>and</u> $\sigma_K(\lambda, \lambda^\times)$ <u>is the coarsest normal topology on</u> λ <u>which is consistent with</u> $\langle\lambda, \lambda^\times\rangle$.

<u>Proof</u>. The statements (a), (c) and (f) are immediate consequences of (2.3.11), part (e) follows from (2.3.11) and (2.3.1)(b), (d) follows from (2.3.11) and (2.3.2)(a), while part (b) is a consequence of (2.3.11) and (2.3.5).

To prove part (g), we first notice from (3.11) and (3.4) that the Köthe topology $\sigma_K(\lambda, \lambda^\times)$ is the normal topology on λ associated with $\sigma(\lambda, \lambda^\times)$. Therefore (g) holds by making use of (2.3.11)(a).

Let λ be a perfect sequence space, and let $\mathcal{L}$ be any normal topology on λ (not necessarily consistent with $\langle\lambda, \lambda^\times\rangle$) . Denote by $[(\lambda, \mathcal{L})]$ the subspace of $(\lambda, \mathcal{L})$ consisting of all $\zeta \in \lambda$ which are the $\mathcal{L}$-limit of the sectional sequence $\{\zeta^{(n)}, n \geq 0\}$ associated with ζ . Then it is not hard to show that $[(\lambda, \mathcal{L})]$ is $\mathcal{L}$-closed. If, in addition, $\mathcal{L}$ is consistent with $\langle\lambda, \lambda^\times\rangle$, then (2.3.12)(e) shows that

$$[(\lambda, \mathcal{L})] = \lambda .$$

The $\beta(\lambda, \lambda^\times)$-closure of $\mathbb{K}^{(\mathbb{N})}$ in λ , denoted by λ_r , is called the <u>regular subspace</u> of λ ; and λ is said to be <u>regular</u> if $\lambda = \lambda_r$. It is observed by Crofts [1] that

$$\lambda_r = [(\lambda, \beta(\lambda, \lambda^{\times})] .$$

It then follows that if $(\lambda, \tau(\lambda, \lambda^{\times}))$ is infrabarrelled, then λ is regular.

We conclude this section with a precise formula for the diagonal mappings between two sequence spaces.

Let λ and μ be sequence spaces. An $u \in \omega$ is called a <u>diagonal map</u> from λ into μ if

$$u\lambda \subset \mu .$$

The set consisting of all diagonal maps from λ into μ, denoted by $D(\lambda, \mu)$, is obviously a sequence space. Members in $D(\lambda, \mu)$ can be represented as infinite diagonal matrices. Clearly

$$\ell^{\infty} \subset D(\lambda, \lambda) \text{ for any normal sequence space } \lambda \text{, and}$$
$$D(\lambda, c_0) \subset c_0 \text{ whenever } e \in \lambda . \tag{3.16}$$

It then follows from (3.16) that

$$D(\ell^{\infty}, c_0) = c_0 . \tag{3.17}$$

The following result, due to Crofts [1], gives a precise formula for $D(\lambda, \mu)$ plus showing the perfectness of $D(\lambda, \mu)$.

(2.3.13) Proposition. <u>Let</u> λ <u>and</u> μ <u>be two sequence spaces.</u> <u>Then</u>

$$D(\lambda, \mu) \subset (\lambda\mu^{\times})^{\times} .$$

<u>If, in addition,</u> μ <u>is perfect, then</u>

$$D(\lambda, \mu) = (\lambda \mu^{\times})^{\times} \; ; \tag{3.18}$$

<u>consequently</u> $D(\lambda, \mu)$ <u>is perfect</u>.

<u>Proof</u>. Let $u \in D(\lambda, \mu)$. For any $\zeta \in \lambda$ and $\eta \in \mu^{\times}$, we have that $u\zeta \in \mu$, and thus

$$|u||\zeta\eta| = |u\zeta||\eta| \in \ell^{1}$$

which implies that $u \in (\lambda \mu^{\times})^{\times}$.

If μ is perfect and $v \in (\lambda \mu^{\times})^{\times}$, then we have for any $\zeta \in \lambda$ and $\eta \in \mu^{\times}$ that

$$<|\eta|, |v\zeta|> = \Sigma_{i} |\eta_{i} v_{i} \zeta_{i}| = <|\zeta\eta|, |v|> < +\infty ,$$

so that $v\zeta \in \mu$ by the perfectness of μ . As ζ was arbitrary, we conclude that $v \in D(\lambda, \mu)$.

Notice that the α-dual of every sequence space is always perfect. The perfectness of $D(\lambda, \mu)$ then follows from $D(\lambda, \mu) = (\lambda \mu^{\times})^{\times}$.

We shall demonstrate that we cannot remove the perfectness of μ to obtain the equality (3.18) . However, the example requires the following result, which will be useful for the factoring result of nuclear maps.

(2.3.14) Lemma. <u>The following assertions hold</u>:

(a) $\ell^1 = \ell^1 c_o = \{u\eta : u \in \ell^1$ <u>and</u> $\eta \in c_o\}$.

(b) $\ell^1 = \ell^p \ell^q$, <u>where</u> $p \geq 1$ <u>and</u> $\frac{1}{p} + \frac{1}{q} = 1$.

(c) $\ell^r = \ell^t \ell^s$, <u>where</u> $1 \leq r \leq t$ <u>and</u> $\frac{1}{r} = \frac{1}{t} + \frac{1}{s}$.

<u>Proof</u>. (a) Clearly $\ell^1 c_o \subseteq \ell^1$ since every element in c_o is a bounded sequence of numbers. Conversely we shall show that for any $\zeta \in \ell^1$ there exist $u \in \ell^1$ and $\eta \in c_o$ such that

$$\zeta_j = u_j \eta_j \quad \text{and} \quad 0 < \eta_j \leq 1 \quad \text{for all} \quad j \geq 1 . \tag{3.19}$$

In fact, we can choose a sequence $\{n_k\}$ of positive integers with $n_1 < n_2 < \dots$ such that

$$\Sigma_{j=n_k}^{n_{k+1}} |\zeta_j| < 4^{-k} \quad \text{for all} \quad k \geq 1 .$$

Let us define

$$\eta_j = \begin{cases} 1 & \text{if} \quad 1 \leq j < n_2 \\ 2^{-k} & \text{if} \quad n_{k+1} \leq j < n_{k+2} . \end{cases}$$

Then $\eta = (\eta_j) \in c_o$ since $\lim_j \eta_j = 0$, and

$$\Sigma_{j=1}^{\infty} |\zeta_j| \eta_j^{-1} \leq \Sigma_{j=1}^{\infty} |\zeta_j| + \frac{1}{2} .$$

Thus the sequence $u = [u_j]$, defined by

$$u_j = \zeta_j \eta_j^{-1} \quad \text{for all} \quad j \geq 1 ,$$

has the required property.

(b) It is obvious that $\ell^{\infty}\ell^{1} \subseteq \ell^{1}$. As $c_o \subset \ell^{\infty}$, we conclude from part (a) that $\ell^{1} = \ell^{1}\ell^{\infty}$. Therefore the result holds for $p = 1$. If $p > 1$, then Holder's inequality shows that $\ell^{p}\ell^{q} \subseteq \ell^{1}$. Conversely, if $u = (u_n) \in \ell^{1}$, then the sequences $\zeta = [\zeta_n]$ and $\eta = [\eta_n]$, defined by

$$\zeta_n = u_n^{\frac{1}{p}} \quad \text{and} \quad \eta_n = u_n^{\frac{1}{q}} \quad \text{for all} \quad n \geqslant 1 ,$$

satisfy $\zeta \in \ell^{p}$, $\eta \in \ell^{q}$ and $u = \zeta\eta$, thus $u \in \ell^{p}\ell^{q}$. Therefore $\ell^{1} \subseteq \ell^{p}\ell^{q}$.

(c) For the proof of this part, we use the following notation: if $u = [u_n] \in \omega$ and q is any positive number, then we write u^q for the sequence $[u_n^q]$.

Since $1 = \frac{r}{t} + \frac{r}{s}$, it follows from part(b) that

$$\ell^{1} = \ell^{\frac{t}{r}}\ell^{\frac{s}{r}} . \tag{3.20}$$

If $u \in \ell^{r}$, then $u^{r} \in \ell^{1}$, hence there exist $\zeta \in \ell^{\frac{t}{r}}$ and $\eta \in \ell^{\frac{s}{r}}$ such that $u^{r} = \zeta\eta$; consequently

$$u = \zeta^{\frac{1}{r}}\eta^{\frac{1}{r}} . \tag{3.21}$$

As $\zeta \in \ell^{\frac{t}{r}}$ and $\eta \in \ell^{\frac{s}{r}}$, it follows that $\zeta^{\frac{t}{r}} \in \ell^{1}$ and $\eta^{\frac{s}{r}} \in \ell^{1}$, and hence that

$$\zeta^{\frac{1}{r}} \in \ell^{t} \quad \text{and} \quad \eta^{\frac{1}{r}} \in \ell^{s} .$$

Therefore $\eta \in \ell^{t}\ell^{s}$ by (3.2.1), thus $\ell^{r} \subseteq \ell^{t}\ell^{s}$.

Conversely, let $\zeta \in \ell^t$ and $\eta \in \ell^s$. Then

$$\zeta^{r \cdot \frac{t}{r}} = \zeta^t \in \ell^1 \quad \text{and} \quad \eta^{r \cdot \frac{s}{r}} = \eta^s \in \ell^1 ,$$

hence

$$\zeta^r \in \ell^{\frac{t}{r}} \quad \text{and} \quad \eta^r \in \ell^{\frac{s}{r}} .$$

It then follows from (3.20) that $\zeta^r \eta^r \in \ell^1$, and hence that $\zeta\eta \in \ell^r$. Therefore $\ell^t \ell^s \subseteq \ell^r$.

__Example.__ It is known that c_o is not perfect, and from (3.17) that $D(\ell^\infty, c_o) = c_o$. In view of (2.3.14)(b), $\ell^1 = \ell^\infty \ell^1$, hence

$$(\ell^\infty c_o^\times)^\times = (\ell^\infty \ell^1)^\times = \ell^\infty .$$

Therefore $D(\ell^\infty, c_o)$ is a proper vector subspace of $(\ell^\infty c_o^\times)^\times$.

(2.3.15) __Examples.__ (a) $D(\ell^1, \ell^\infty) = \ell^\infty$.

(b) $D(\ell^p, \ell^r) = \ell^1$ where $1 \leq r < p < \infty$ and $\frac{1}{s} = \frac{1}{r} - \frac{1}{p}$.

__Proof.__ (a) Clearly $\ell^\infty \subseteq D(\ell^1, \ell^\infty)$. To prove that $D(\ell^1, \ell^\infty) \subseteq \ell^\infty$, let $u = (u_n) \notin \ell^\infty$. Then there exists a subsequence (u_{n_k}) of (u_n) such that $|u_{n_k}| \geq k^3$ for all $k \geq 1$. For each $k \geq 1$, let

$$\eta_k = \begin{cases} k^{-2} & \text{if } k = n_k \\ 0 & \text{if } k \neq n_k . \end{cases}$$

Then $(\eta_k) \in \ell^1$ and $u\eta \notin \ell^1$ since

$$|u_{n_k} \eta_{n_k}| \geq k \quad \text{for all} \quad k \geq 1 .$$

Therefore $D(\ell^1, \ell^\infty) \subseteq \ell^\infty$, and thus $D(\ell^1, \ell^\infty) = \ell^\infty$.

(b) As $(\ell^r)^\times = \ell^{1/(1-r^{-1})} = \ell^{\frac{1}{r-1}}$, it follows from (2.3.12) that

$$D(\ell^p, \ell^r) = (\ell^p \ell^{\frac{r}{r-1}})^\times \tag{3.22}$$

and hence from (2.3.14)(c) that

$$\ell^p \ell^{\frac{r}{r-1}} = \ell^{\frac{pr}{r+p\ r-p}} . \tag{3.23}$$

Note that

$$1 - \frac{r+p\ r-p}{pr} = \frac{p-r}{pr} = \frac{1}{r} - \frac{1}{p} = \frac{1}{s} .$$

We conclude from (3.22) and (3.23) that

$$D(\ell^p, \ell^r) = \ell^s .$$

2.4 λ-summability of sequences

In the sequal, λ will be assumed to be a perfect sequence spaces, and $\mathfrak{T}$ will be assumed to be a normal topology on λ which is consistent with the dual pair $<\lambda, \lambda^\times>$. By virtue of (2.3.12)(b), $\mathfrak{T}$ admits a neighbourhood basis at 0 , denoted by $\mathcal{W}_\lambda$, consisting of $\sigma(\lambda, \lambda^\times)$-closed, absolutely convex and normal sets in λ or, equivalently, $\mathfrak{T}$ is determined by a family $\{r_S : S \in \mathcal{W}_\lambda\}$ of regular seminorms, where the gauge r_S of each $S \in \mathcal{W}_\lambda$ is given by

$$\begin{aligned} r_S(\zeta) &= \sup\{|<\zeta, \eta>| : \eta \in S^o\} \\ &= \sup\{<|\zeta|, |v|> : v \in S^o\} \quad (\zeta \in \lambda) \quad . \end{aligned} \tag{4.1}$$

For any element $x = [x_n]$ in $E^{\mathbb{N}}$, and any positive number k, the <u>k-th section of</u> x, denoted by $x^{(k)}$ or $[x_n]^{(k)}$, is defined by

$$x^{(k)} = [x_1, x_2, \ldots, x_k, 0, \ldots] .$$

The sequence $\{x^{(k)}, k \geq 1\}$ in $E^{\mathbb{N}}$ is called the <u>sectional sequence associated with</u> x. It is clear that $x^{(k)} \in E^{(\mathbb{N})}$ for all $k \geq 1$, hence the sectional sequence associated with x is a sequence in $E^{(\mathbb{N})}$.

An element $x = [x_n]$ in $E^{\mathbb{N}}$ is said to be

(i) <u>weakly</u> λ-<u>summable</u> if

$$[<x_n, f>] \in \lambda \quad \text{for all} \quad f \in E' ;$$

(ii) λ-<u>summable</u> if for any $\zeta \in \lambda^\times$, the sequence $\{\Sigma_{j=1}^n \zeta_j x_j, n \geq 1\}$ is a $\mathcal{P}$-Cauchy sequence in E ;

(iii) <u>absolutely</u> λ-<u>summable</u> if

$$[p_V(x_n)] \in \lambda \quad \text{for all} \quad V \in \mathcal{U}_E .$$

If $\lambda = \ell^1$, then $[x_n]$ is weakly ℓ^1-summable if and only if it is weakly summable, and $[x_n]$ is absolutely ℓ^1-summable if and only if it is absolutely summable. As $(\ell^1)^\times = \ell^\infty$, it follows that $[x_n]$ is ℓ^1-summable if and only if $\Sigma_{\mathbb{N}} x_n$ is bounded-multiplier $\mathscr{P}$-Cauchy (for definition, see Day [1, p.59]), and hence from Day [1, p.59] that $[x_n]$ is ℓ^1-summable if and only if it is summable in the sense of §2.1. Therefore the notions of weak λ-summability, of λ-summability, and of absolute λ-summability are respectively the generalizations of weak summability, of summability, and of absolute summability.

The concept of absolute λ-summability was introduced by Rosier [1] and De Grande-De Kimpe [1], and they gave some interesting properties of absolutely λ-summable sequences which will be studied in this section, while the concept of weak λ-summability was introduced by Ramanujan [2].

Denote by $\lambda_w(E)$ the set of all weakly λ-summable sequences in E, by $\lambda(E)$ the set of all λ-summable sequences in E, and by $\lambda[E]$ the set of all absolutely λ-summable sequences in E. Then it is easily seen that they are all vector subspaces of $E^{\mathbb{N}}$ and

$$E^{(\mathbb{N})} \subset \lambda[E] \subset \lambda(E) \subset \lambda_w(E) \subset E^{\mathbb{N}} .$$

If we write ϕ for the algebraic direct sum $\mathbb{K}^{(\mathbb{N})}$, then

$$E^{(\mathbb{N})} = \phi[E] = \phi(E) = \phi_w(E) .$$

For any $V \in \mathcal{U}_E$ and $S \in \mathcal{W}_\lambda$, let us define

$$\begin{aligned} \pi_{(S,V)}(x) &= r_S([p_V(x_n)]) \\ &= \sup\{\Sigma_n |\eta_n p_V(x_n)| : [\eta_n] \in S^o\} \end{aligned} \tag{4.2}$$

whenever $x = [x_n] \in \lambda[E]$. If $S = \{\eta\}$, then we write simply $\pi_{(\eta,V)}$ for

$\pi_{(\{n\},V)}$; in particular, we have for any $e_k = [\delta_n^{(k)}] \in \lambda^\times$ that

$$\pi_{(e_k,V)}(x) = r_{e_k}([p_V(x_n)]) = \Sigma_n |\delta_n^{(k)} p_V(x_n)|$$

$$= p_V(x_k) \quad \text{for all} \quad x = [x_n] \in \lambda[E] \ . \tag{4.3}$$

It is clear that each $\pi_{(S,V)}$ is a seminorm on $\lambda[E]$. The locally convex topology on $\lambda[E]$, determined by $\{\pi_{(S,V)} : S \in \mathcal{W}_\lambda, V \in \mathcal{U}_E\}$, is called the $(\pi, \mathcal{L})$-$\underline{\text{topology}}$, and denoted by $\pi_{(\wp,\mathcal{L})}$ (we shall see from (2.4.3) that this topology is always Hausdorff). If $\mathcal{L}$ is the Köthe topology $\sigma_K(\lambda, \lambda^\times)$, then we write simply $\pi_\wp$ for $\pi_{(\wp, \sigma_K(\lambda, \lambda^\times))}$; if $\mathcal{L}$ is the the Mackey topology $\tau(\lambda, \lambda^\times)$, then we write $\pi_{(\wp,\tau)}$ for $\pi_{(\wp, \tau(\lambda, \lambda^\times))}$. Clearly

$$\pi_\wp \leqslant \pi_{(\wp, \mathcal{L})} \leqslant \pi_{(\wp, \tau)} \ .$$

If $\wp$ and $\mathcal{L}$ are metrizable, then so does $\pi_{(\wp, \mathcal{L})}$, if $\wp$ and $\mathcal{L}$ are normable, then so does $\pi_{(\wp, \mathcal{L})}$.

On ℓ^1 , the normal topology which is consistent with $\langle \ell^1, \ell^\infty \rangle$ is the only normed topology on ℓ^1 , hence we obtain for $\lambda = \ell^1$ that

$$\wp_\pi = \pi_{(\wp, \mathcal{L})} = \pi_{(\wp, \tau)} = \pi_\wp \ .$$

The following result, due to De Grande-De Kimpe [1], is a generalization of (2.1.4)(b).

(2.4.1) Lemma. $\underline{\text{Let}}$ $\mathcal{L}$ $\underline{\text{be consistent with}}$ $\langle \lambda, \lambda^\times \rangle$. $\underline{\text{Then for}}$ $\underline{\text{any}}$ $x = [x_n] \in \lambda[E]$, x $\underline{\text{is the}}$ $\pi_{(\wp, \mathcal{L})}$-$\underline{\text{limit of the sectional sequence}}$

$\{x^{(k)}, k \geq 1\}$ <u>associated with</u> x . <u>Consequently</u>, $E^{(\mathbb{N})}$ <u>is</u> $\pi_{(\mathcal{P},\mathcal{L})}$-<u>dense in</u> $\lambda[E]$.

<u>Proof</u>. For any $V \in \mathcal{U}_E$ and $S \in \mathcal{U}_\lambda$, since $[p_V(x_n)] \in \lambda$, it follows from (2.3.12)(e) that $[p_V(x_n)]$ is the $\mathcal{L}$-limit of the sectional sequence $\{[p_V(x_n)]^{(k)}, k \geq 1\}$ associated with $[p_V(x_n)]$, and hence that

$$\lim_{k\to\infty} \pi_{(S,V)}(x - x^{(k)}) = \lim_{k\to\infty} r_S([p_V(x_n)] - [p_V(x_n)]^{(k)}) = 0 ,$$

which obtains our assertion.

In order to give a suitable topology on $\lambda_w(E)$, we first require the following result, due to Dubinsky and Ramanujan [1, p.15].

(2.4.2) Lemma. <u>For any</u> $x = [x_n] \in \lambda_w(E)$, <u>the map</u> L_x , <u>defined by</u>

$$L_x(f) = [\langle x_n, f\rangle] \quad \underline{\text{for all}} \quad f \in E' ,$$

<u>is a linear map from</u> E' <u>into</u> λ <u>such that</u> $L_x(V^o)$ <u>is</u> $\sigma_K(\lambda, \lambda^\times)$-<u>bounded for any</u> $V \in \mathcal{U}_E$.

<u>Proof</u>. Clearly L_x is linear. To prove the $\sigma_K(\lambda, \lambda^\times)$-boundedness of $L_x(V^o)$, it suffices to show that for any $\eta = [\eta_n] \in \lambda^\times$,

$$\sup\{\langle |L_x(f)|, |\eta|\rangle : f \in V^o\} < \infty . \tag{4.4}$$

To this end, let

$$M = \{f \in E' : \langle |L_x(f)|, |\eta|\rangle = \Sigma_n |\eta_n \langle x_n, f\rangle| \leq 1\} .$$

Then M is clearly absolutely convex, and absorbing since $x \in \lambda_w(E)$. We claim that M is $\sigma(E', E)$-closed. Indeed, let $f^{(\nu)}$ be a net in M which is $\sigma(E', E)$-convergent to f . For any $k \geq 1$,

$$\Sigma_{j=1}^{k} |\eta_j <x_j, f>| = \Sigma_{j=1}^{k} \lim_{\nu} |\eta_j <x_j, f^{(\nu)}>|$$

$$\leq \lim_{\nu} \Sigma_{j=1}^{\infty} |\eta_j <x_j, f^{(\nu)}>| \leq 1 ,$$

hence

$$\Sigma_{j=1}^{\infty} |\eta_j <x_j, f>| = \sup_{k} \Sigma_{j=1}^{k} |\eta_j <x_j, f>| \leq 1 ,$$

which shows that $f \in M$, thus M is $\sigma(E', E)$-closed.

As M being absorbing, it follows that M° is $\mathcal{P}$-bounded, and hence that there exists an $m > 0$ such that $M^{\circ} \subset m V$ since $V \in \mathcal{U}_E$. The bipolar theorem, together with $\sigma(E', E)$-closedness of M , shows that

$$V^{\circ} \subseteq m M$$

or, equivalently,

$$\Sigma_n |\eta_n <x_n, f>| \leq m \quad \text{for all} \quad f \in V^{\circ} ,$$

which obtains (4.4).

Since $\mathfrak{L}$ is consistent with $<\lambda, \lambda^{\times}>$, it follows that $L_x(V^{\circ})$ is $\mathfrak{L}$-bounded for any $V \in \mathcal{U}_E$, where $x = [x_n] \in \lambda_w(E)$. This leads to the following definitions.

For any $V \in \mathcal{U}_E$ and $S \in \mathcal{W}_{\lambda}$, let us define

$$w_{(S,V)}(x) = \sup\{r_S(L_x(f)) : f \in V^o\}$$
$$= \sup\{\Sigma_n |\eta_n \langle x_n, f\rangle| : [\eta_n] \in S^o, f \in V^o\} \qquad (4.5)$$

whenever $x = [x_n] \in \lambda_w(E)$, and

$$\varepsilon_{(S,V)}(x) = w_{(S,V)}(x) \quad \text{for all} \quad x = [x_n] \in \lambda(E) \qquad (4.6)$$

If $S = \{\eta\}$, then we write simply $\varepsilon_{(\eta,V)}$ and $w_{(\eta,V)}$ for $\varepsilon_{(\{\eta\},V)}$ and $w_{(\{\eta\},V)}$ respectively; in particular, for any $e_k = [\delta_n^{(k)}] \in \lambda^\times$ we have

$$w_{(e_k,V)}(x) = \sup\{\Sigma_n |\delta_n^{(k)} \langle x_n, f\rangle| : f \in V^o\}$$
$$= \sup\{|\langle x_k, f\rangle| : f \in V^o\}$$
$$= p_V(x_k) \quad \text{for all} \quad x = [x_n] \in \lambda_w(E) \qquad (4.7)$$

It is clear that $w_{(S,V)}$ and $\varepsilon_{(S,V)}$ are seminorms on $\lambda_w(E)$ and $\lambda(E)$ respectively. For any $f \in V^o$, since

$$|\langle x_n, f\rangle| \leq p_V(x_n) \quad \text{for all} \quad n \geq 1 ,$$

it follows from the regularity of the seminorm r_S that

$$\varepsilon_{(S,V)}(x) \leq \pi_{(S,V)}(x) \quad \text{for all} \quad x \in \lambda[E] \ . \qquad (4.8)$$

Denote by $w_{(\mathcal{P},\mathcal{L})}$ the locally convex topology on $\lambda_w(E)$ determined by $\{w_{(S,V)} : S \in \mathcal{W}_\lambda, V \in \mathcal{U}_E\}$, and by $\varepsilon_{(\mathcal{P},\mathcal{L})}$ the locally convex topology on $\lambda(E)$ determined by $\{\varepsilon_{(S,V)} : S \in \mathcal{W}_\lambda, V \in \mathcal{U}_E\}$. We call $\varepsilon_{(\mathcal{P},\mathcal{L})}$ the $(\varepsilon, \mathcal{L})$-<u>topology</u> on $\lambda(E)$. If $\mathcal{L}$ is the Köthe topology $\sigma_K(\lambda, \lambda^\times)$, we write simply $\varepsilon_{\mathcal{P}}$ for $\varepsilon_{(\mathcal{P}, \sigma_K(\lambda, \lambda^\times))}$, and $w_{\mathcal{P}}$ for $w_{(\mathcal{P}, \sigma_K(\lambda, \lambda^\times))}$; if $\mathcal{L}$ is the Mackey topology $\tau(\lambda, \lambda^\times)$, then we write

$\varepsilon_{(\wp, \tau)}$ and $w_{(\wp, \tau)}$ for $\varepsilon_{(\wp, \tau(\lambda, \lambda^\times))}$ and $w_{(\wp, \tau(\lambda, \lambda^\times))}$ respectively. Clearly

$$\varepsilon_\wp \leqslant \varepsilon_{(\wp, \mathcal{L})} \leqslant \varepsilon_{(\wp, \tau)} \quad \text{and} \quad w_\wp \leqslant w_{(\wp, \mathcal{L})} \leqslant w_{(\wp, \tau)} \ .$$

It is obvious that $\varepsilon_{(\wp, \mathcal{L})}$ is the relative topology on $\lambda(E)$ induced by $w_{(\wp, \mathcal{L})}$. On the other hand, (4.8) shows that

$$\varepsilon_{(\wp, \mathcal{L})} \leqslant \pi_{(\wp, \mathcal{L})} \quad \text{on} \ \lambda[E] \ .$$

We shall verify that on $\lambda(E)$ the product topology $\wp^{IN}$ is coarser than $\varepsilon_{(\wp, \mathcal{L})}$, hence $\varepsilon_{(\wp, \mathcal{L})}$ and $\pi_{(\wp, \mathcal{L})}$ are always Hausdorff. It is also trivial that if $\wp$ and $\mathcal{L}$ are metrizable, then so do $\varepsilon_{(\wp, \mathcal{L})}$ and $w_{(\wp, \mathcal{L})}$, and that if $\wp$ and $\mathcal{L}$ are normable, then so do $\varepsilon_{(\wp, \mathcal{L})}$ and $w_{(\wp, \mathcal{L})}$.

If $\lambda = \ell^1$, then it is clear that

$$\wp_\varepsilon = \varepsilon_{(\wp, \mathcal{L})} = \varepsilon_{(\wp, \tau)} = \varepsilon_\wp \ , \quad \text{and}$$
$$\wp_w = w_{(\wp, \mathcal{L})} = w_{(\wp, \tau)} = w_\wp \ .$$

(2.4.3) Lemma. For any $k \geqslant 1$, let $\pi_k : \lambda_w(E) \to E$ be the k-th projection, and let $J_k : E \to \lambda[E]$ be the k-th embedding map, that is

$$J_k(u) = [\delta_n^{(k)} u]_{n \geqslant 1} \quad \text{for all} \quad u \in E \ .$$

Then the following statements hold.

(a) For any $V \in \mathcal{U}_E$ and $e_k = [\delta_n^{(k)}] \in \lambda^\times$,

$$w_{(e_k, V)}(x) = \pi_{(e_k, V)}(x) = p_V(\pi_k(x)) = p_V(x_k) \quad \text{for all} \quad x = [x_n] \in \lambda_w(E), \quad (4.9)$$

<u>hence</u> $\mathcal{P}^{\mathbb{N}} \leq w_{\mathcal{P}}$ <u>on</u> $\lambda_w(E)$.

(b) <u>For any</u> $V \in \mathcal{U}_E$ <u>and</u> $S \in \mathcal{W}_\lambda$,

$$w_{(S,V)}(J_k(u)) = \pi_{(S,V)}(J_k(u)) = r_S(e_k)p_V(u) \quad \text{for all} \quad u \in E , \tag{4.10}$$

<u>hence</u> $\pi_{(\mathcal{P},\mathcal{L})} \leq \mathcal{P}^{(\mathbb{N})}$ <u>on</u> $E^{(\mathbb{N})}$.

<u>Proof</u>. As $x_k = \pi_k(x)$, formula (4.9) follows from (4.3) and (4.7), hence (a) holds by the continuity of each $\pi_k : (\lambda_w(E), w_{\mathcal{P}}) \to (E, \mathcal{P})$.

To prove part (b), we first notice that

$$\begin{aligned} r_S(e_k) &= \sup\{\Sigma_n |\delta_n^{(k)} \eta_n| : [\eta_n] \in S^o\} \\ &= \sup\{|\eta_k| : [\eta_n] \in S^o\} . \end{aligned}$$

Therefore we obtain

$$\begin{aligned} w_{(S,V)}(J_k(u)) &= \sup\{\Sigma_n |\eta_n \langle \delta_n^{(k)} u, f\rangle| : [\eta_n] \in S^o , f \in V^o\} \\ &= \sup\{|\eta_k \langle u, f\rangle| : [\eta_n] \in S^o, f \in V^o\} \\ &= r_S(e_k)p_V(u) , \end{aligned}$$

and

$$\begin{aligned} \pi_{(S,V)}(J_k(u)) &= \sup\{\Sigma_n |\eta_n p_V(\delta_n^{(k)} u)| : [\eta_n] \in S^o\} \\ &= \sup\{|\eta_k p_V(u)| : [\eta_n] \in S^o\} = r_S(e_k)p_V(x_k) , \end{aligned}$$

thus (b) holds by the continuity of each $J_k : (E, \mathcal{P}) \to (\lambda[E], \pi_{(\mathcal{P},\mathcal{L})})$.

The following result, which should be compared with (2.1.2), gives an alternative characterization of λ-summability for sequences.

(2.4.4) Lemma. <u>An</u> $x = [x_n] \in \lambda_w(E)$ <u>is</u> λ-<u>summable if and only if it is the</u> $w_\wp$-<u>limit of the sectional sequence</u> $\{x^{(k)}, k \geq 1\}$ <u>associated with</u> x .

<u>Proof</u>. <u>Necessity</u>. For any $\eta = [\eta_n] \in \lambda^\times$, the sequence $\{\Sigma_{j=1}^{n} \eta_j x_j , n \geq 1\}$ is a $\wp$-Cauchy sequence in E . For any $V \in \mathcal{U}_E$ and $\delta > 0$, there exists an $n_o \geq 1$ such that

$$p_V(\Sigma_{j=n+1}^{n+k} \eta_j x_j) \leq \frac{\delta}{4} \quad \text{for all} \quad n \geq n_o \quad \text{and} \quad k \geq 1, 2, \ldots .$$

In particular, for any $f \in V^o$ we have

$$|\Sigma_{j=k+1}^{k+q} \eta_j <x_j , f>| \leq \frac{\delta}{4} \quad \text{for all} \quad k \geq n_o \quad \text{and} \quad q = 1, 2, \ldots ,$$

thus

$$\Sigma_{j=k+1}^{k+q} |\eta_j <x_j , f>| \leq \delta \quad \text{for all} \quad k \geq n_o \quad \text{and} \quad q = 1, 2, \ldots .$$

Therefore

$$w_{(\eta,V)}(x - x^{(k)}) = \sup\{\Sigma_{j=k+1}^{\infty} |\eta_j <x_j , f>| \ : f \in V^o\} \leq \delta \quad \text{for all} \quad k \geq n_o ,$$

which shows that x is the $w_\wp$ -limit of $x^{(k)}$.

<u>Sufficiency</u>. Let x be the $w_\wp$-limit of $x^{(k)}$. Then $\{x^{(k)}, k \geq 1\}$ is a $w_\wp$-Cauchy sequence in $\lambda_w(E)$. For any $V \in \mathcal{U}_E$, $\eta = [\eta_n] \in \lambda^\times$ and $\delta > 0$ there exists an $n_o \geq 1$ such that

$$w_{(\eta,V)}(x^{(k+q)} - x^{(k)}) = \sup\{\Sigma_{j=k+1}^{k+q} |\eta_j <x_j , f>| \ : f \in V^o\}$$
$$\leq \delta \qquad \text{for all} \quad k \geq n_o \quad \text{and} \quad q = 1, 2, \ldots .$$

It then follows that

$$p_V\left(\sum_{j=1}^{k+q} \eta_j x_j - \sum_{j=1}^{k} \eta_j x_j\right) = \sup\{|\sum_{j=k+1}^{k+q} \langle \eta_j x_j , f\rangle| \ : f \in V^o\}$$

$$\leq \sup\{\sum_{j=k+1}^{k+q} |\eta_j \langle x_j , f\rangle| \ : f \in V^o\}$$

$$= w_{(\eta ,V)}(x^{(k+q)} - x^{(k)}) \leq \delta \quad \text{for all } k \geq n_o \text{ and } q = 1, 2, \ldots,$$

and hence that the sequence $\{\sum_{j=1}^{n} \eta_j x_j , n \geq 1\}$ is $\mathcal{P}$-Cauchy. Therefore $x \in \lambda(E)$.

Denote by $[\lambda_w(E), \mathcal{L}]$ the set of all $x \in \lambda_w(E)$ which are the $w_{(\mathcal{P}, \mathcal{L})}$-limits of sectional sequences associated with x , by $[\lambda_w(E)]$ the set $[\lambda_w(E), \sigma_k(\lambda, \lambda^x)]$, and by $[\lambda[E], \mathcal{L}]$ the set of all $x \in \lambda[E]$ which are the $\pi_{(\mathcal{P}, \mathcal{L})}$-limits of sectional sequences associated with x . Then (2.4.1) shows that

$$\lambda[E] = [\lambda[E], \mathcal{L}] = [\lambda[E], \sigma_K(\lambda, \lambda^x)] ,$$

while (2.4.4) ensures that

$$\lambda(E) = [\lambda_w(E)] .$$

As $w_{\mathcal{P}} \leq w_{(\mathcal{P}, \mathcal{L})}$, it follows that

$$[\lambda_w(E) , \mathcal{L}] \subset \lambda(E) .$$

Clearly $[\lambda_w(E), \mathcal{L}]$ is a vector subspace of $\lambda_w(E)$. To prove the $w_{(\mathcal{P}, \mathcal{L})}$-closedness of $[\lambda_w(E), \mathcal{L}]$ in $\lambda_w(E)$, we require the following notation: For any $f \in E'$, the map T_f , defined by

$$T_f x = [\langle x_n, f\rangle] \quad \text{for all} \quad x = [x_n] \in \lambda_w(E) \tag{4.11}$$

is a linear map from $\lambda_w(E)$ into λ such that

$$w_{(S,V)}(x) = \sup\{r_S(T_f x) : f \in V^o\} \quad \text{for all} \quad x \in \lambda_w(E) . \tag{4.12}$$

It then follows that for any $V \in \mathcal{U}_E$, the family $\{T_f : f \in V^o\}$ is equicontinuous from $(\lambda_w(E), w_{(\wp, \mathcal{L})})$ into $(\lambda, \mathcal{L})$. Clearly

$$T_f x^{(k)} = (T_f x)^{(k)} \quad \text{and} \quad |T_f x^{(k)}| \leq |T_f x| \quad \text{for all} \quad x \in \lambda_w(E) \quad \text{and} \quad k \geq 1 ,$$

it then follows from the regularily of r_S and (4.12) that

$$w_{(S,V)}(x^{(k)}) \leq w_{(S,V)}(x) \quad \text{for all} \quad x \in \lambda_w(E) \quad \text{and} \quad k \geq 1 . \tag{4.13}$$

On the other hand,

$$\begin{aligned} w_{(S,V)}(x - x^{(k)}) &= \sup\{r_S(T_f x - T_f x^{(k)}) : f \in V^o\} \\ &= \sup\{r_S(T_f x - (T_f x)^{(k)}) : f \in V^o\} \\ &= \sup\{\Sigma_{n=k+1}^{\infty} |\eta_n \langle x_n, f\rangle| : [\eta_n] \in S^o, f \in V^o\} ; \end{aligned}$$

we can give an equivalent definition for $[\lambda_w(E), \mathcal{L}]$ as follows: If $x \in \lambda_w(E)$, then $x \in [\lambda_w(E), \mathcal{L}]$ if and only if for any $V \in \mathcal{U}_E$ and $S \in \mathcal{U}_\lambda$,

$$\lim_{k \to \infty} r_S(T_f x - (T_f x)^{(k)}) = 0 \quad \text{uniformly with respect to} \quad f \in V^o .$$

(2.4.5) Lemma. $[\lambda_w(E), \mathcal{L}]$ <u>is always a</u> $w_{(\wp, \mathcal{L})}$<u>-closed vector subspace of</u> $\lambda_w(E)$.

<u>Proof.</u> Let $x \in \lambda_w(E)$ and let $x_{(\nu)}$ be a net in $[\lambda_w(E), \mathcal{L}]$

which is $w_{(\mathcal{P},\mathcal{L})}$-convergent to x . For any $V \in \mathcal{U}_E$ and $S \in \mathcal{W}_\lambda$, the linearity of T_f and (4.13) show that

$$\begin{aligned} w_{(S,V)}((x_{(\nu)})^{(k)} - x^{(k)}) &= \sup\{r_S((T_f(x_{(\nu)} - x))^{(k)}) : f \in V^o\} \\ &\leq \sup\{r_S(T_f(x_{(\nu)} - x)) : f \in V^o\} \\ &= w_{(S,V)}(x_{(\nu)} - x) \qquad \text{for all } k \geq 1 . \end{aligned}$$

For any $\delta > 0$, there exists an ν_o such that

$$w_{(S,V)}((x_{(\nu)})^{(k)} - x^{(k)}) \leq w_{(S,V)}(x_{(\nu)} - x) \leq \frac{\delta}{3} \text{ for all } \nu \geq \nu_o \text{ and } k \geq 1. \quad (4.14)$$

As $x_{(\nu_o)} \in [\lambda_w(E), \mathcal{L}]$, there is an $k_o \geq 1$ such that

$$w_{(S,V)}(x_{(\nu_o)} - (x_{(\nu_o)})^{(k)}) \leq \frac{\delta}{3} \text{ for all } k \geq k_o . \qquad (4.15)$$

For any $k \geq k_o$, we obtain from (4.14) and (4.15) that

$$\begin{aligned} w_{(S,V)}(x - x^{(k)}) &\leq w_{(S,V)}(x - x_{(\nu_o)}) + w_{(S,V)}(x_{(\nu_o)} - (x_{(\nu_o)})^{(k)}) \\ &\quad + w_{(S,V)}((x_{(\nu_o)})^{(k)} - x^{(k)}) \leq \delta , \end{aligned}$$

thus $x \in [\lambda_w(E), \mathcal{L}]$.

The following result is a generalizations of (2.1.3).

(2.4.6) Corollary. $[\lambda_w(E)]$ <u>is</u> $w_{(\mathcal{P},\mathcal{L})}$-<u>closed in</u> $\lambda_w(E)$.

<u>Proof</u>. In view of (2.4.5), $[\lambda_w(E)]$ is $w_{\mathcal{P}}$-closed in $\lambda_w(E)$. It then follows from $w_{\mathcal{P}} \leq w_{(\mathcal{P},\mathcal{L})}$ that $[\lambda_w(E)]$ is $w_{(\mathcal{P},\mathcal{L})}$-closed in $\lambda_w(E)$.

The remainder of this section is devoted to a study of the topological duals $\lambda[E]'$ and $\lambda(E)'$ of $(\lambda[E], \pi_{(\mathcal{P},\mathcal{L})})$ and $(\lambda(E), \varepsilon_{(\mathcal{P},\mathcal{L})})$ respectively.

We first note that if $\eta = [\eta_n] \in \lambda^\times$ and $u' = [u'_i]$ is an equicontinuous sequence in E', then there exist $S \in \mathcal{W}_\lambda$ and $V \in \mathcal{U}_E$ such that

$$\eta \in S^o, \text{ and } u'_i \in V^o \quad \text{for all } i \geq 1 ,$$

hence

$$\begin{aligned}\Sigma_{i=1}^{\infty} |\eta_i \langle x_i , u'_i\rangle| &\leq \sup\{\Sigma_{i=1}^{\infty} |v_i p_V(x_i)| : [v_i] \in S^o\} \\ &= \pi_{(S,V)}([x_i]) \quad \text{for all} \quad x = [x_i] \in \lambda[E] .\end{aligned}$$

Therefore the element $\eta u' = [\eta_i u'_i]$ in $(E')^{\mathbb{N}}$ defines a $\pi_{(\mathcal{P},\mathcal{L})}$-continuous linear functional on $\lambda[E]$ by the following equation

$$\langle x, \eta u'\rangle = \Sigma_{i=1}^{\infty} \eta_i \langle x_i , u'_i\rangle \quad \text{for all} \quad x = [x_i] \in \lambda[E] .$$

The converse is also true as shown by the following result, due to Rosier [1] and De Grande-De Kimpe [1], which is a generalization of (2.1.5) for $\Lambda = \mathbb{N}$.

(2.4.7) Theorem. <u>Let</u> $\mathcal{L}$ <u>be a normal topology on</u> λ <u>which is consistent with</u> $\langle\lambda, \lambda^\times\rangle$. <u>Then the topological dual</u> $\lambda[E]'$ <u>of</u> $(\lambda[E], \pi_{(\mathcal{P},\mathcal{L})})$ <u>can be identified with the vector subspace of</u> $(E')^{\mathbb{N}}$, <u>defined by</u>

$$\{\eta u' = [\eta_i u'_i] \in (E')^{\mathbb{N}} : \eta \in \lambda^\times, [u'_i] \text{ is equicontinuous in } E'\}, \qquad (4.16)$$

<u>the canonical bilinear form of the dual pair</u> $\langle\lambda[E], \lambda[E]'\rangle$ <u>is given by</u>

$$\langle x, \eta u' \rangle = \Sigma_{i=1}^{\infty} \eta_i \langle x_i , u'_i \rangle \quad ,$$

$\pi_{\wp}$ <u>and</u> $\pi_{(\wp,\tau)}$ <u>are consistent with</u> $\langle \lambda[E], \lambda[E]' \rangle$, <u>and</u>

$$\Sigma_{i=1}^{\infty} |\eta_i \langle x_i , u'_i \rangle| < \infty \tag{4.17}$$

<u>whenever</u> $[x_i] \in \lambda[E]$, $[\eta_i] \in \lambda^{\times}$ <u>and</u> $[u'_i]$ <u>is an equicontinuous sequence in</u> E' . <u>Furthermore, the</u> $\pi_{(\wp, \mathcal{I})}$<u>-equicontinuous subsets of</u> $\lambda[E]'$ <u>are the sets of the form</u>

$$\{[\eta_i u'_i] : [\eta_i] \in S^{o}, u'_i \in V^{o} \text{ for all } i \geq 1\} , \tag{4.18}$$

<u>where</u> $S \in \mathcal{W}_{\lambda}$ <u>and</u> $V \in \mathcal{U}_E$.

<u>Proof</u>. Let $f \in \lambda[E]'$, let $V \in \mathcal{U}_E$ and $S \in \mathcal{W}_{\lambda}$ be such that

$$|\langle [x_i], f \rangle| \leq \pi_{(S,V)}([x_i]) \text{ for all } x = [x_i] \in \lambda[E] .$$

For each $i \geq 1$, it is known from (2.4.3)(b) that the i-th embedding map $J_i : (E, \wp) \to (\lambda[E], \pi_{(\wp, \mathcal{I})})$ is continuous and

$$\pi_{(S,V)}(J_i u) = r_S(e_i) p_V(u) \text{ for all } u \in E ,$$

hence

$$x'_i = f \circ J_i \in E' \text{ and} \tag{4.19}$$
$$|\langle u, x'_i \rangle| = \langle J_i u, f \rangle| \leq r_S(e_i) p_V(u) \quad \text{for all } u \in E .$$

It then follows that

$$x'_i \in E'(V^{o}) \text{ and } q_{V^{o}}(x'_i) \leq r_S(e_i) \text{ for all } i \geq 1 , \tag{4.20}$$

where $q_{V^{o}}$ is the gauge of V^{o} defined on $E'(V^{o})$. On the other hand,

(2.4.1) shows that each $x = [x_i] \in \lambda[E]$ is the $\pi_{(\wp, \mathfrak{L})}$-limit of the sectional sequence $\{x^{(k)}, k \geq 1\}$ associated with x, hence

$$\langle x, f \rangle = \lim_k \langle x^{(k)}, f \rangle \tag{4.21}$$

by the $\pi_{(\wp, \mathfrak{L})}$-continuity of f. For each $k \geq 1$,

$$\langle x^{(k)}, f \rangle = \langle \Sigma_{i=1}^{k} J_i x_i , f \rangle = \Sigma_{i=1}^{k} \langle x_i , f \circ J_i \rangle = \Sigma_{i=1}^{k} \langle x_i , x'_i \rangle$$

we obtain from (4.21) that

$$\langle x, f \rangle = \Sigma_{i=1}^{\infty} \langle x_i , x'_i \rangle \ . \tag{4.22}$$

Let us now define

$$\eta_i = q_{V^o}(x'_i) \quad \text{and} \quad u'_i = \begin{cases} x'_i / q_{V^o}(x'_i) & \text{if } q_{V^o}(x'_i) \neq 0 \\ 0 & \text{if } q_{V^o}(x'_i) = 0 \ . \end{cases}$$

Then $u'_i \in V^o$ for all $i \geq 1$ and $[x'_i] = [\eta_i u'_i]$, hence (4.22) shows that

$$\langle x, f \rangle = \Sigma_{i=1}^{\infty} \eta_i \langle x_i , u'_i \rangle \qquad \text{for all} \quad x = [x_i] \in \lambda[E] \ .$$

Moreover, we claim that $[\eta_i] \in \lambda^{\times}$. Indeed, let $\zeta = [\zeta_i] \in \lambda$. Then

$$r_S(\zeta) = \sup\{\Sigma_{i=1}^{\infty} |v_i \zeta_i| \ : \ [v_i] \in S^o\} < +\infty \ . \tag{4.23}$$

For each $i \geq 1$, since

$$q_{V^o}(\zeta_i x'_i) = \sup\{|\langle u, \zeta_i x'_i \rangle| \ : \ u \in V\} ,$$

there exists an $u_i \in V$ such that

$$q_{V^o}(\zeta_i x'_i) < \langle u_i , \zeta_i x'_i \rangle + 2^{-i} = \langle \zeta_i u_i , x'_i \rangle + 2^{-i} \ . \tag{4.24}$$

For each $k \geq 1$, the element defined by

$$[\zeta_i u_i]^{(k)} = (\zeta_1 u_1, \ldots, \zeta_k u_k, 0, 0 \ldots .)$$

belongs to $\lambda[E]$, hence

$$\Sigma_{i=1}^{k} \langle u_i, \zeta_i x_i' \rangle \leq \pi_{(S,V)}([\zeta_i u_i]^{(k)}) = \sup\{\Sigma_{i=1}^{k} |v_i p_V(\zeta_i u_i)| : [v_i] \in S^o\}$$
$$\leq \sup\{\Sigma_{i=1}^{k} |v_i \zeta_i| : [v_i] \in S^o\} \leq r_S(\zeta) < \infty ,$$

thus $\Sigma_{i=1}^{\infty} \langle u_i, \zeta_i x_i' \rangle < \infty$ since k was arbitrary. Therefore (4.24) shows that

$$\Sigma_{i=1}^{\infty} |\zeta_i \eta_i| = \Sigma_{i=1}^{\infty} q_{V^o}(\zeta_i x_i') \leq \Sigma_{i=1}^{\infty} (\langle u_i, \zeta_i x_i' \rangle + 2^{-i}) \leq r_S(\zeta) , \qquad (4.25)$$

thus $\eta = [\eta_i] \in \lambda^{\times}$.

Therefore, the map $\eta u' \mapsto f$, defined by

$$\langle x, f \rangle = \Sigma_{i=1}^{\infty} \eta_i \langle x_i, u_i' \rangle \quad \text{for all} \quad x = [x_i] \in \lambda[E] ,$$

is a surjective map from the vector subspace $\{\eta u' \in (E')^{\mathbb{N}} : \eta \in \lambda^{\times}, [u_i'] \text{ is equicontinuous in } E'\}$ of $(E')^{\mathbb{N}}$ onto $\lambda[E]'$. Clearly this map is also injective and linear.

To prove (4.17), it suffices to show that

$$\Sigma_{i=1}^{\infty} |\gamma_i \eta_i \langle x_i, u_i' \rangle| < \infty \quad \text{for all} \quad [\gamma_i] \in c_o$$

since $\ell^1 = (c_o)^{\times}$. In fact, we first identify $[\eta_i u_i']$ with an $f \in \lambda[E]'$, and choose a sequence $[\alpha_i]$ such that

$$|\alpha_i| = 1 \quad \text{and} \quad |\gamma_i \eta_i \langle x_i, u_i' \rangle| = \alpha_i \gamma_i \eta_i \langle x_i, u_i' \rangle .$$

As $[\gamma_i] \in c_0$, it follows that $[\alpha_i \gamma_i x_i] \in \lambda[E]$, and hence that

$$\sum_{i=1}^{\infty} |\gamma_i \eta_i <x_i , u_i'>| = \sum_{i=1}^{\infty} \alpha_i \gamma_i \eta_i <x_i , u_i'> = <[\alpha_i \gamma_i x_i], f> < \infty$$

by the continuity of f .

Finally, the set $\{\eta u' : \eta \in S^o, u_i' \in V^o \text{ for all } i \geq 1\}$ is clearly an $\pi_{(\mathcal{P},\mathcal{L})}$-equicontinuous subset of $\lambda[E]'$, where $S \in \mathcal{U}_\lambda$ and $V \in \mathcal{U}_E$. Conversely, if M is an $\pi_{(\mathcal{P},\mathcal{L})}$-equicontinuous subset of $\lambda[E]'$, then there exist $S \in \mathcal{U}_\lambda$ and $V \in \mathcal{U}_E$ such that

$$|<x, f>| \leq \pi_{(S,V)}(x) \quad \text{for all } f \in M \text{ and } x \in \lambda[E] .$$

In view of the proof of the representation of f , we see from (4.25) and (4.23) that M is of the form described by (4.18).

Because of (4.17), Rosier [1] defines the α-dual of $\lambda[E]$, denoted by $\lambda[E]^\times$, as the vector space of all sequences $x' = [x_n']$ in $(E')^{\mathbb{N}}$ satisfying

$$\sum_{i=1}^{\infty} |<x_i , x_i'>| < \infty \quad \text{for all } [x_i] \in \lambda[E] .$$

In view of (4.17),

$$\lambda[E]' \subset \lambda[E]^\times .$$

Furthermore, it is easy to verify that

$$\lambda[E]^\times \subset \lambda^\times[E_\beta'] .$$

Therefore it is natural to ask under what condition on E or E_β' , either $\lambda[E]^\times = \lambda^\times[E_\beta']$ or $\lambda[E]' = \lambda^\times[E_\beta']$. In order to seeking some sufficient condition on E or E'_β, he introduces the following concept: A locally convex space E is said to be fundamentally λ-bounded if for any $\pi_{(\mathcal{P},\mathcal{L})}$-bounded subset $\mathbb{B}$ of $\lambda[E]$ there exist a $\sigma_K(\lambda, \lambda^\times)$-bounded subset R of λ and a closed absolutely convex bounded subset B of E such that

$$\mathbb{B} \subset [R, B] = \{x \in \lambda[E] : x_n \in E(B) \ (n \geq 1) \text{ and } [p_B(x_n)] \in R\} .$$

He also showed that if E is fundamentally λ-bounded, then $\lambda[E]^\times = \lambda^\times[E'_\beta]$, and that if E is σ-infrabarrelled and E'_β is fundamentally $\lambda^\times$-bounded, then $\lambda[E]' = \lambda^\times[E'_\beta]$. Every normed space is fundamentally λ-bounded; moreover, if $\lambda^\times$ has a fundamental sequence of bounded sets, then every metrizable locally convex space is fundamentally λ-bounded, and every (DF)-space is fundamentally $\lambda^\times$-bounded.

To study the topological dual $\lambda(E)'$ of $(\lambda(E), \varepsilon_{(\mathcal{P},\mathcal{L})})$, we first prove the following result which should be compared with (2.1.6).

(2.4.8) Lemma. Let $V \in \mathcal{U}_E$ and $\eta = [\eta_n] \in \lambda^\times$. Denote by Δ the unit disk in $\mathbb{C}$. For any $x = [x_n] \in \lambda(E)$, the map g, defined by

$$g([a_n], x') = \sum_{j=1}^{\infty} \eta_j a_j \langle x_j, x'\rangle \quad \text{for all } [a_n] \in \Delta^{\mathbb{N}} \text{ and } x' \in V^o ,$$

is continuous on the compact Hausdorff space $\Delta^{\mathbb{N}} \times V^o$ and

$$\|g\| = \varepsilon_{(\eta,V)}(x) ,$$

where $\|\cdot\|$ is the sup-norm on $C(\Delta^{\mathbb{N}} \times V^o)$. Consequently, the map $[x_n] \mapsto g$ is a continuous linear map from $(\lambda(E), \varepsilon_{\mathcal{P}})$ into $(C(\Delta^{\mathbb{N}} \times V^o), \|\cdot\|)$.

<u>Proof</u>. It is known from (2.4.4) that x is the $\varepsilon_{\wp}$-limit of the sectional sequence $\{x^{(k)}, k \geq 1\}$ associated with x. For any $\delta > 0$ there is an $n_o \geq 1$ such that

$$\Sigma_{k=1}^{\infty} |\eta_{k+n} \langle x_{k+n}, x' \rangle| \leq \delta \quad \text{for all} \quad n \geq n_o \quad \text{and} \quad x' \in V^o .$$

For each $k \geq 1$, the map g_k, defined by

$$g_k([a_n], x') = \Sigma_{j=1}^{k} \eta_j a_j \langle x_j, x' \rangle \quad \text{for all} \quad [a_n] \in \Delta^{\mathbb{N}} \quad \text{and} \quad x' \in V^o ,$$

is clearly continuous on $\Delta^{\mathbb{N}} \times V^o$, and

$$\|g_k\| \leq \varepsilon_{(\eta,V)}(x^{(k)}) .$$

Furthermore, we claim that they are equal. Indeed, for any $x' \in V^o$ and j there exists an α_j such that

$$|\alpha_j| = 1 \quad \text{and} \quad |\eta_j \langle x_j, x' \rangle| = \alpha_j \eta_j \langle x_j, x' \rangle .$$

Clearly $[\alpha_j] \in \Delta^{\mathbb{N}}$, hence we have

$$\Sigma_{j=1}^{k} |\eta_j \langle x_j, x' \rangle| = \Sigma_{j=1}^{k} \alpha_j \eta_j \langle x_j, x' \rangle = g_k([\alpha_j], x') \leq \|g_k\| ,$$

which implies that

$$\varepsilon_{(\eta,V)}(x^{(k)}) \leq \|g_k\| .$$

This proves our assertion.

As

$$|g([a_n], x') - g_k([a_n], x')| = |\Sigma_{j=k+1}^{\infty} \eta_j a_j \langle x_j, x' \rangle|$$

$$\leq \Sigma_{j=k+1}^{\infty} |\eta_j a_j \langle x_j', x' \rangle| \leq \delta \quad \text{for all} \quad k \geq n_o \quad \text{and} \quad x' \in V^o ,$$

it follows that g_k converges to g uniformly on $\Delta^{\mathbb{N}} \times V^o$, and hence that $g \in C(\Delta^{\mathbb{N}} \times V^o)$. Finally we have

$$\|g\| = \lim_k \|g_k\| = \lim_k \varepsilon_{(\eta,V)}(x^{(k)}) = \varepsilon_{(\eta,V)}(x) .$$

(2.4.9) Lemma. <u>For any</u> $f \in (\lambda(E), \varepsilon_{\wp})'$, <u>there exist an</u> $\eta = [\eta_i] \in \lambda^\times$, <u>a</u> $\sigma(E', E)$-<u>closed equicontinuous subset</u> B <u>of</u> E' <u>and a positive Radon measure</u> μ <u>on</u> B <u>such that</u>

$$|\langle J_i u, f\rangle| \leq \int_B |\eta_i \langle u, x'\rangle| d\mu(x') \quad \underline{\text{for all}} \quad u \in E \quad \underline{\text{and}} \quad i \in \mathbb{N} ,$$

<u>where</u> $J_i : E \to E^{(\mathbb{N})}$ <u>is the embedding map</u>.

<u>Proof</u>. There exist an $\eta = [\eta_i] \in \lambda^\times$ and $V \in \mathcal{U}_E$ such that

$$|\langle [x_i], f\rangle| \leq \varepsilon_{(\eta,V)}([x_i]) \quad \text{for all} \quad [x_i] \in \lambda(E) . \tag{4.26}$$

(2.4.8) shows that the map $\psi : [x_i] \mapsto g$, defined by

$$g([a_i], x') = \sum_{j=1}^{\infty} \eta_j a_j \langle x_j, x'\rangle \quad \text{for all} \quad [a_i] \in \Delta^{\mathbb{N}} \quad \text{and} \quad x' \in V^o ,$$

is a continuous linear map from $(\lambda(E), \varepsilon_{\wp})$ into $(C(\Delta^{\mathbb{N}} \times V^o), \|\cdot\|)$ such that

$$\|\psi([x_i])\| = \|g\| = \varepsilon_{(\eta,V)}([x_i]) \quad \text{for all} \quad [x_i] \in \lambda(E) .$$

It then follows from (4.26) that $\operatorname{Ker} \psi \subset \operatorname{Ker} f$, and hence from the Hahn-Banach extension theorem and Riesz's representation theorem that there exists a Radon measure ν on $\Delta^{\mathbb{N}} \times V^o$ such that

$$\langle [x_i], f\rangle = \nu(\psi([x_i])) = \int_{\Delta^{\mathbb{N}} \times V^o} \psi([x_i])([a_i], x')d\nu$$

$$= \int_{\Delta^{\mathbb{N}} \times V^o} \Sigma_{j=1}^{\infty} \eta_j a_j \langle x_j, x'\rangle d\nu \quad \text{for all} \quad [x_i] \in \lambda(E) .$$

In particular,

$$\langle J_i u, f\rangle = \int_{\Delta^{\mathbb{N}} \times V^o} \eta_i a_i \langle u, x'\rangle d\nu \quad \text{for all} \quad u \in E \quad \text{and} \quad i \in \mathbb{N} .$$

Thus the proof can be completed by using a similar argument given in the proof of (2.1.7).

(2.4.10) Theorem. The topological dual $\lambda(E)'$ of $(\lambda(E), \varepsilon_{\wp})$ can be identified with the vector subspace of $(E')^{\mathbb{N}}$ defined by

$$\{\eta u' = [\eta_i u_i'] \in (E')^{\mathbb{N}} : \eta \in \lambda^{\times} , [u_i'] \text{ is a prenuclear sequence in } E'\} ,$$

the canonical bilinear form of the dual pair $\langle\lambda(E), \lambda(E)'\rangle$ is given by

$$\langle x, \eta u'\rangle = \Sigma_{j=1}^{\infty} \eta_j \langle x_j , u_j'\rangle . \tag{4.27}$$

Proof. In view of (2.4.9), it suffices to show that if $\eta \in \lambda^{\times}$ and $[u_i']$ is a prenuclear sequence in E', then the equation (4.27) defines an $\varepsilon_{\wp}$-continuous linear functional on $\lambda(E)$. Indeed, let $V \in \mathcal{U}_E$ and let μ be a positive Radon measure on V^o such that

$$|\langle x, u_i'\rangle| \leq \int_{V^o} |\langle x, x'\rangle| d\mu(x') \quad \text{for all} \quad x \in E \quad \text{and} \quad i \geq 1 .$$

Then

$$\Sigma_{i=1}^{\infty} |\eta_i \langle x_i , u_i'\rangle| \leq \int_{V^o} \Sigma_{i=1}^{\infty} |\eta_i \langle x_i , x'\rangle| d\mu(x')$$

$$\leq \|\mu\| \sup\{\Sigma_{i=1}^{\infty} |\eta_i \langle x_i , x'\rangle| : x' \in V^o\}$$

.

$$= \|\mu\| \, \varepsilon_{(\eta,V)}([x_i]) \quad \text{for all} \quad [x_i] \in \lambda(E) ,$$

which obtains our assertion.

When $\lambda = \ell^1$, the preceding result is (2.1.8) for $\Lambda = \mathbb{N}$.

The concept of absolutely summing mappings depends on the sequence space ℓ^1 and its duality theory. When E and F are normed spaces, we see from (2.2.4) that $T \in L(E, F)$ is absolutely summing if and only if T sends every summable sequence in E into absolutely summable sequence in F . From this observation, Ramanujan [2] has generalized this concept to abstract sequence space λ , and called absolutely λ-summing mappings; namely an $T \in L(E, F)$ is said to be absolutely λ-summing if $[Tx_n] \in \lambda[F]$ whenever $[x_n] \in \lambda_w(E)$. Further information on absolutely λ-summing mappings can be found in Dubinsky and Ramanujan [1] and [2].

CHAPTER 3. NUCLEAR SPACES

3.1 Nuclear linear mappings

Let (X, p) and (Y, q) be normed spaces and p^* the dual norm of p. If $\{f_n\}$ and $\{y_n\}$ are two sequences in X' and Y respectively such that $\Sigma_{n=1}^{\infty} p^*(f_n)q(y_n) < +\infty$, then

$$q(\Sigma_{j=n}^{m} f_j(x)y_j) \leq (\Sigma_{j=n}^{m} p^*(f_j)q(y_j))\,p(x) \qquad \text{for all } x \in X \text{ and } m > n ,$$

hence the equation

$$Tx = \lim_n \Sigma_{j=1}^{n} f_j(x)y_j \qquad (x \in X)$$

defines a continuous linear map from X into the completion $\tilde{Y}$ of Y. Not every $T \in L(X, Y)$ has the above form. This leads Grothendieck to define the nuclear maps.

An $T \in L(X, Y)$ is said to be <u>nuclear</u> if there are two sequences $\{f_n\}$ and $\{y_n\}$ in X' and Y respectively such that

$$\Sigma_{n=1}^{\infty} p^*(f_n)q(y_n) < +\infty \quad \text{and} \quad Tx = \Sigma_{n=1}^{\infty} f_n(x)y_n \quad \text{for all } x \in X . \tag{3.1}$$

In this case we shall write symbolically

$$T = \Sigma_n f_n \otimes y_n . \tag{3.2}$$

Formula (3.1) (or (3.2)) is referred to as a <u>nuclear representation</u> of T. The set consisting of all nuclear maps from X into Y, denoted by

$N_{\ell^1}(X, Y)$, is a vector space. It is not hard to show that the functional $\|\cdot\|_{(n)}$ on $N_{\ell^1}(X, Y)$, defined by

$$\|T\|_{(n)} = \inf\{\Sigma_{n=1}^{\infty} p^*(f_n)q(y_n) : T = \Sigma f_n \otimes y_n , f_n \in X' \text{ and } y_n \in Y\} ,$$

is a norm on $N_{\ell^1}(X, Y)$ which is called the <u>nuclear norm</u>, and that

$L^f(X, Y) \subset N_{\ell^1}(X, Y) \subset \pi_{\ell^1}(X, Y)$ and $|||T||| \leq \|T\|_{(s)} \leq \|T\|_{(n)}$ for all $T \in N_{\ell^1}(X, Y)$ where $L^f(X, Y)$ is the vector subspace of $L(X, Y)$ of all elements of finite rank.

Nuclear maps are, in a sense, the only 'contructible' mappings, that is the mappings of the most elementary form that include mappings of finite rank. Many of difficulties in the theory of nuclear maps arise because the representation of a nuclear map $T : X \to Y$ depends on the range Y and not the image $T(X)$. To avoid this kind of difficulties, Stegall and Retherford [1] define fully nuclear operators.

(3.1.1) Proposition. $L^f(X, Y)$ <u>is dense in</u> $(N_{\ell^1}(X, Y), \|\cdot\|_{(n)})$. <u>Moreover</u>, <u>if in addition</u>, Y <u>is complete</u>, <u>then so does</u> $(N_{\ell^1}(X, Y), \|\cdot\|_{(n)})$.

<u>Proof</u>. Let $T \in N_{\ell^1}(X, Y)$, and let $f_n \in X'$ and $y_n \in Y$ be such that

$$\Sigma_{n=1}^{\infty} p^*(f_n)q(y_n) < \infty \quad \text{and} \quad T(x) = \Sigma_{n=1}^{\infty} f_n(x)y_n \quad \text{for all } x \in X .$$

For any $\delta > 0$, there exists n_o such that

$$\Sigma_{k=m}^{\infty} p^*(f_k)q(y_k) \leq \delta \quad \text{for all} \quad m \geq n_o . \tag{3.3}$$

For any positive integer m , we define

$$T_m(x) = \Sigma_{k=1}^{m} f_k(x)y_k \quad \text{for all} \quad x \in X .$$

Then $T_m \in L^f(X, Y)$ and

$$T(x) - T_m(x) = \Sigma_{k=m+1}^{\infty} f_k(x)y_k \quad \text{for all} \quad x \in X .$$

By the definition of the nuclear norm and (3.3), we obtain

$$\|T - T_m\|_{(n)} \leq \Sigma_{k=m+1}^{\infty} p^*(f_k)q(y_k) \leq \delta \quad \text{for all} \quad m \geq n_o ,$$

which implies the density of $L^f(X, Y)$ in $(L^n(X, Y), \|\cdot\|_{(n)})$.

Suppose now that Y is a Banach space. Then $(L(X, Y), |||\cdot|||)$ is complete. As

$$|||T||| \leq \|T\|_{(n)} \quad \text{for all} \quad T \in L^n(X, Y) ,$$

it follows that every Cauchy sequence in $(N_{\ell^1}(X, Y), \|\cdot\|_{(n)})$ is $|||\cdot|||$-Cauchy. Now the completeness of $(N_{\ell^1}(X, Y), \|\cdot\|_{(n)})$ can be easily verified by a standard argument.

From the preceding result, it is clear that every nuclear map is precompact. Therefore, Example (2.2.7) shows that there are absolutely summing mappings which need not be nuclear. But we shall show that the composition of three absolutely summing mappings is nuclear; in fact, this result is still valid when we replace three by two (one of the deepest results, due to Grothendieck (see Pietsch [1, p.66]), in the operator theory).

Clearly, the composition of two continuous linear maps, in which one of them is nuclear, must be nuclear and

$$\|T \circ S\|_{(n)} \leq \begin{cases} \|T\|_{(n)} \, |||S||| & \text{if } T \text{ is nuclear} \\ |||T||| \, \|S\|_{(n)} & \text{if } S \text{ is nuclear.} \end{cases}$$

On the other hand, if Y is a vector subspace of a normed space (Z, r) and if $T \in N_{\ell^1}(X, Y)$, then T, viewed as a continuous linear map from X into Z, is a nuclear map from X into Z because of $T = j_Y T$, where $j_Y : Y \to Z$ is the canonical embedding. A partial converse holds as shown by the following result.

(3.1.2) Proposition. _Let_ (X, p) _and_ (Z, r) _be normed spaces, let_ Y _be a dense vector subspace of_ Z _and let_ $T \in L(X, Y)$. _If_ $T \in N_{\ell^1}(X, Z)$, _then_ $T \in N_{\ell^1}(X, Y)$.

Proof. We first show that for any $z \in Z$ and $\delta > 0$, there exists a sequence $\{y_n\}$ in Y such that

$$z = \lim_n \Sigma_{k=1}^n y_k \quad \text{and} \quad \Sigma_{k=1}^\infty r(y_k) \leq (1 + \delta) r(z) .$$

In fact, for any $n \geq 1$ there exists $a_n \in Y$ such that

$$r(z - a_n) \leq (\delta/_{2^{n+1}}) r(z) .$$

Setting

$$y_1 = a_1 \quad \text{and} \quad y_n = a_n - a_{n-1} \quad \text{for } n > 1 .$$

Then we get $z = \lim_n a_n = \lim_n \Sigma_{k=1}^n y_k$ and

$$r(y_1) \leq (1 + \frac{\delta}{4})r(z) \quad \text{and} \quad r(y_n) \leq (\frac{\delta}{2^{n+1}} + \frac{\delta}{2^n})r(z) \quad \text{for} \quad n > 1 ,$$

thus we obtain

$$\Sigma_{n=1}^{\infty} r(y_n) \leq [1 + \frac{\delta}{4} + \delta \, \Sigma_{n=2}^{\infty} (\frac{1}{2^{n+1}} + \frac{1}{2^n})]r(z) \leq (1 + \delta)r(z) ,$$

as asserted.

Suppose now that $T \in N_{\ell^1}(X, Z)$. Then, for any $\delta > 0$, there exist sequences $\{f_n\}$ and $\{z_n\}$ in X' and Z respectively such that

$$\Sigma_{k=1}^{\infty} p^*(f_k)r(z_k) < ||T||_{(n)} + \delta \quad \text{and} \quad T(x) = \Sigma_{k=1}^{\infty} f_k(x)z_k \quad (x \in X) .$$

For each z_k there exists a sequence $\{y_{k,m}\}$ in Y such that

$$z_k = \Sigma_{m=1}^{\infty} y_{k,m} \quad \text{and} \quad \Sigma_{m=1}^{\infty} r(y_{k,m}) \leq (1 + \delta)r(z_k) .$$

Let

$$f_{k,m} = f_k \quad \text{for all} \quad m = 1, 2, \ldots .$$

Then we have

$$T(x) = \Sigma_{k=1}^{\infty} \Sigma_{m=1}^{\infty} f_{k,m}(x)y_{k,m} \quad \text{for all} \quad x \in X ,$$

and

$$||T||_{(n)} \leq \Sigma_{k=1}^{\infty} \Sigma_{m=1}^{\infty} p^*(f_{k,m})r(y_{k,m}) \leq \Sigma_{k=1}^{\infty} p^*(f_k)(1 + \delta)r(z_k)$$
$$< (1 + \delta)(||T||_{(n)} + \delta) .$$

Therefore $T \in N_{\ell^1}(X, Y)$ and the proof is complete.

We denote by $\widetilde{X}$ the completion of a normed space X. In view of the preceding result and the fact that nuclear maps are precompact, it follows that if $T \in L(X, Y)$, then $T \in N_{\ell^1}(X, Y)$ if and only if $\widetilde{T} \in N_{\ell^1}(\widetilde{X}, \widetilde{Y})$, where $\widetilde{T} : \widetilde{X} \to \widetilde{Y}$ is the unique extension of T .

(3.1.3) Proposition. <u>If</u> $T \in N_{\ell^1}(X, Y)$ <u>then</u> $T' \in N_{\ell^1}(Y', X')$ <u>and</u>

$$\|T'\|_{(n)} \leq \|T\|_{(n)} .$$

<u>If, in addition,</u> (Y, q) <u>is reflexive, then</u> $T \in N_{\ell^1}(X, Y)$ <u>if and only if</u> $T' \in N_{\ell^1}(Y', X')$; <u>in this case,</u>

$$\|T'\|_{(n)} = \|T\|_{(n)} .$$

<u>Proof</u>. For any $\delta > 0$, there are sequences $\{f_n\}$ and $\{y_n\}$ in X' and Y respectively such that

$$\Sigma_{k=1}^{\infty} p^*(f_k)q(y_k) < \|T\|_{(n)} + \delta \quad \text{and} \quad T(x) = \Sigma_{k=1}^{\infty} f_k(x)y_k \quad (x \in X) .$$

In view of the definition of adjoint maps, we have

$$T'(y') = \Sigma_{k=1}^{\infty} \langle y_k, y'\rangle f_k \quad \text{for all} \quad y' \in Y' .$$

As $q(y_k) = q^{**}(y_k)$, it follows that

$$\Sigma_{k=1}^{\infty} q^{**}(y_k)p^*(f_k) = \Sigma_{k=1}^{\infty} q(y_k)p^*(f_k) < \|T\|_{(n)} + \delta ,$$

and hence that $T' \in N_{\ell^1}(Y', X')$ and $\|T'\|_{(n)} \leq \|T\|_{(n)}$ because δ was arbitrary.

Suppose now that Y is reflexive and that $T' \in N_{\ell^1}(Y', X')$. Then $T'' \in N_{\ell^1}(X'', Y)$ and

$$\|T''\|_{(n)} \leq \|T'\|_{(n)}$$

by the first part. Denote by $e_X : X \to X''$ the evaluation map. Then $T = T'' e_X$, hence $T \in N_{\ell^1}(X, Y)$ and

$$\|T\|_{(n)} \leq \|T''\|_{(n)} \, \|| e_X \|| = \|T''\|_{(n)} \leq \|T'\|_{(n)} . \tag{3.5}$$

Formulae (3.4) and (3.5) show that $\|T\|_{(n)} = \|T'\|_{(n)}$.

It is not known whether $T' \in N_{\ell^1}(Y', X')$ implies $T \in N_{\ell^1}(X, Y)$. But we shall show (see (3.4.7)) that if the Banach space Y has the extension property (for definition, see §3.4), then the answer of this question is affirmative.

(3.1.4) Proposition. Let X and Y be normed spaces and $T \in L(X, Y)$. Then $T \in N_{\ell^1}(X, Y)$ if and only if T permits a factoring

$$X \xrightarrow{T_1} c_o \xrightarrow{T_2} \ell^1 \xrightarrow{T_3} Y ,$$

where T_1 and T_3 are continuous linear and $T_2 \in N_{\ell^1}(c_o, \ell^1)$.

Proof. The sufficiency is obvious. To prove the necessity, we can choose, in view of (2.3.14)(a) and the definition of nuclear maps, an $(\lambda_n) \in \ell^1$ and two null sequences $\{f_n\}$ and $\{y_n\}$ in X' and Y respectively such that

$$Tx = \sum_{n=1}^{\infty} \lambda_n f_n(x) y_n \quad \text{for all} \quad x \in X .$$

Let $T_1 : X \to c_o$ be defined by

$$T_1(x) = (f_n(x))_{n \in \mathbb{N}} \quad \text{for all} \quad x \in X ,$$

let $T_2 : c_o \to \ell^1$ be defined by

$$T_2((\eta_n)) = (\lambda_n \eta_n)_{n\in\mathbb{N}} \quad \text{for all} \quad (\eta_n) \in c_o ,$$

and let $T_3 : \ell^1 \to Y$ be defined by

$$T_3((\zeta_n)) = \Sigma_{j=1}^{\infty} \zeta_j y_j \quad \text{for all} \quad (\zeta_n) \in \ell^1 .$$

Then T_j $(j = 1, 2, 3)$ are continuous linear maps and $T = T_3 T_2 T_1$. Since ℓ^1 is the Banach dual of c_o , it follows that

$$T_2((\eta_n)) = \Sigma_{j=1}^{\infty} \lambda_j <(\eta_n), e_j> e_j ,$$

where e_j are the unit elements in ℓ^1 . Therefore, $T_2 \in N_{\ell^1}(c_o, \ell^1)$.

(3.1.5) Theorem. For two Hilbert spaces H_1 and H_2 and $T \in L(H_1, H_2)$, the following statements are equivalent.

(a) $T \in N_{\ell^1}(H_1, H_2)$.

(b) T is compact and $\Sigma_{n=1}^{\infty} \zeta_n < +\infty$, where ζ_n are all eigenvalues of the absolute value $[T]$ of T .

(c) T is compact and admits a representation of the form

$$Tx = \Sigma_n \zeta_n [x, e_n] d_n \quad \text{with} \quad \Sigma_n \zeta_n < +\infty ,$$

where $\zeta_n \geq 0$, $\{e_n\}$ and $\{d_n\}$ are orthonormal sequences in H_1 and H_2 respectively.

(d) $[T]^{\frac{1}{2}} \in \pi_{\ell^1}(H_1)$.

(e) T is the composition of the following two Hilbert-Schmidt operators

$$H_1 \xrightarrow{A} H \xrightarrow{B} H_2 ,$$

where H is a Hilbert space.

(f) $[T] \in N_{\ell^1}(H_1)$.

(g) <u>There is an orthonormal basis</u> $\{u_\iota : \iota \in \Lambda\}$ <u>in</u> H_1 <u>such that</u> $\Sigma_\Lambda \|Tu_\iota\| < +\infty$.

<u>Proof</u>. (a) ⇒ (b) : Nuclear operators are compact, hence we have by the polar representation of compact operators that

$$T(x) = W([T]x) = \Sigma \zeta_n [x, e_n] d_n \quad ,$$

where $\{e_n\}$ and $\{d_n\}$ are orthonormal sequences in H_1 and H_2 respectively, $\zeta_n > 0$, $[T]e_n = \zeta_n e_n$, W is a partial isometry whose initial set is $\overline{\operatorname{Im}[T]}$ and $d_n = We_n$. On the other hand, the nuclearity of T ensures that there exist two sequences $\{x_n\}$ and $\{y_n\}$ in H_1 and H_2 respectively such that

$$\Sigma_{n=1}^{\infty} \|x_n\| \, \|y_n\| < +\infty \quad \text{and} \quad T(x) = \Sigma_{n=1}^{\infty} [x, x_n] y_n \quad \text{for all} \quad x \in H_1 .$$

By Cauchy-Schwarz's inequality and Bessel's inequality we obtain

$$\begin{aligned}\Sigma_{n=1}^{\infty} \zeta_n &= \Sigma_{k=1}^{\infty} [[T]e_k, e_k] = \Sigma_{k=1}^{\infty} [(W[T])e_k, We_k] = \Sigma_{k=1}^{\infty} [Te_k, d_k] \\ &= \Sigma_{k=1}^{\infty} \Sigma_{n=1}^{\infty} [e_k, x_n][y_n, d_k] \leq \Sigma_{n=1}^{\infty} ((\Sigma_k |[e_k, x_n]|^2)^{\frac{1}{2}} (\Sigma_k |[y_n, d_k]|^2)^{\frac{1}{2}} \\ &\leq \Sigma_{n=1}^{\infty} \|x_n\| \, \|y_n\| < +\infty \quad .\end{aligned}$$

(b) ⇒ (c) : Follows from the polar representations of compact operators.

(c) ⇒ (d) : By a well-known result (see Schatten [2]), we have

$$[T]x = \Sigma \zeta_n [x, e_n] e_n \quad \text{and} \quad [T]^{\frac{1}{2}}(x) = \Sigma_n \sqrt{\zeta_n}\, [x, e_n] e_n \quad \text{for all} \quad x \in H_1 .$$

As $\Sigma_n (\sqrt{\zeta_n})^2 = \Sigma_n \zeta_n < +\infty$, it follows from (2.2.8) that $[T]^{\frac{1}{2}} \in \pi_{\ell^1}(H_1)$.

(d) ⇒ (e) : In view of (2.2.8) and the assumption, $T \in L^c(H_1, H_2)$, hence the polar representation of compact operators shows that $T = W[T]$, where W is a partial isometry whose initial set is $\overline{\text{Im}\,[T]}$. Clearly, $W[T]^{\frac{1}{2}}$ is a Hilbert-Schmidt operator from H_1 into H_2, hence

$$H = H_1,\ A = [T]^{\frac{1}{2}} \quad \text{and} \quad B = W[T]^{\frac{1}{2}}$$

have the required properties.

(e) ⇒ (a) : Let $\{d_n\}$ be an orthonormal sequence in $A(H_1)$ which is clearly separable, and let $x \in H_1$. Then we have

$$BA(x) = B(\Sigma_{n=1}^{\infty}[Ax, d_n]d_n) = \Sigma_{n=1}^{\infty}[Ax, d_n]Bd_n = \Sigma_{n=1}^{\infty}[x, A'd_n]Bd_n .$$

Clearly, A' is a Hilbert-Schmidt operator, it then follows from Cauchy-Schwarz's inequality that

$$\Sigma_{n=1}^{\infty}||A'd_n||\,||Bd_n|| \leq (\Sigma_{n=1}^{\infty}||A'd_n||^2)^{\frac{1}{2}}(\Sigma_{n=1}^{\infty}||Bd_n||^2)^{\frac{1}{2}} < +\infty .$$

Therefore $T \in N_{\ell^1}(H_1, H_2)$.

(a) ⇔ (f) : Follows from the polar representation:

$$T = W[T] \quad \text{and} \quad [T] = W'T ,$$

where W is a partial isometry whose initial set is $\overline{\text{Im}\,[T]}$.

(c) ⇒ (g) : We first notice that

$$\Sigma_n||Te_n|| = \Sigma_n||[T]e_n|| = \Sigma_n \zeta_n < +\infty .$$

We then extend $\{e_n\}$ to an orthonormal basis $\{u_\iota : \iota \in \Lambda\}$ in H_1 by adding an orthonormal set $\{w_r\}$ for which $[T]w_r = 0$, hence $Tw_r = 0$ and thus $\{u_\iota : \iota \in \Lambda\}$ has the required property.

(g) ⇒ (d) : Let $T = W[T]$ be the polar decomposition of T. Then

$$[[T]u_\iota , u_\iota] \leq \|[T]u_\iota\| = \|Tu_\iota\| \quad ,$$

hence

$$\Sigma_\iota \|[T]^{\frac{1}{2}}u_\iota\|^2 = \Sigma_\iota [[T]u_\iota , u_\iota] \leq \Sigma_\iota \|Tu_\iota\| < +\infty \quad ,$$

and thus $[T]^{\frac{1}{2}} \in \pi_{\ell^1}(H_1)$ by (2.2.8).

An $T \in L(H_1, H_2)$ satisfying one of the preceding equivalent properties is called an <u>operator of trace class</u>.

In view of (3.1.5), we define for any $T \in N_{\ell^1}(H_1, H_2)$ that

$$\|T\|_{(TR)} = \Sigma_{n=1}^{\infty} \zeta_n \quad ,$$

where ζ_n are all eigenvalues of $[T]$, and called the <u>trace norm</u>. It can be shown that

$$\|T\|_{(n)} = \|T\|_{(TR)} \quad \text{for all} \quad T \in N_{\ell^1}(H_1, H_2) \ .$$

Furthermore, we shall see in §4.3 that the topological dual of $(N_{\ell^1}(H_1, H_2), \|\cdot\|_{(n)})$ can be identified with the Banach space $B(H_1', H_2)$ of all continuous bilinear form on $H_1' \times H_2$ equipped with the bilinear norm or, equivalently, $(L(H_1', H_2'), |||\cdot|||)$ where $|||\cdot|||$ is the operator norm.

Let X, Y and Z be Banach spaces, let $S : Y \to Z$ be a surjective continuous linear map and $T \in L(X, Z)$. If there exists an $\widetilde{T} \in L(X, Y)$ such that $T = S \circ \widetilde{T}$, then $\widetilde{T}$ is called a <u>lifting</u> of T.

It can be shown that nuclear maps enjoy the nuclear extension property, as well as the nuclear lifting property.

Let E and F be locally convex spaces. An $T \in L(E, F)$ is called a <u>nuclear map</u> if there exist an absolutely convex 0-neighbourhood V in E and an infracomplete subset B of F such that

$$T(V) \subset B \quad \text{and} \quad T_{(V,B)} \in N_{\ell^1}(E_V, F(B)) ,$$

where $T_{(V,B)}$ is the induced map of T (for definition, see Chapter 0).

The set consisting of all nuclear maps from E into F, denoted by $N_{\ell^1}(E, F)$, is a vector subspace of $L^{\ell b}(E, F)$. An $T \in L(E, F)$ is nuclear if and only if T is the composition of the following three continuous linear maps

$$E \xrightarrow{T_1} X \xrightarrow{T_{(n)}} Y \xrightarrow{T_2} F ,$$

where X and Y are Banach spaces and $T_{(n)} \in N_{\ell^1}(X, Y)$, and this is the case if and only if there exist an $(\zeta_n) \in \ell^1$, an equicontinuous sequence $\{f_n\}$ in E' and a sequence $\{y_n\}$ contained in an infracomplete subset B of F such that for any $x \in E$, the series $\sum_{n=1}^{\infty} \zeta_n f_n(x) y_n$ is absolutely convergent to Tx.

From this it can be easily verified that if $T \in N_{\ell^1}(E, F)$ then $T' \in N_{\ell^1}(F'_\beta, E'_\beta)$. It can be shown that the composition of two continuous linear maps, in which one of them is nuclear, must be nuclear.

(3.1.6) Theorem (Pietsch). <u>Let</u> E <u>be a locally convex space</u>, <u>let</u> X_1, X_2 <u>be normed spaces and let</u> X_3 <u>be a Banach space</u>. <u>If</u> $T_1 \in \pi_{\ell^1}(E, X_1)$ <u>and</u> $T_i \in \pi_{\ell^1}(X_{i-1}, X_i)$ $(i = 2, 3)$, <u>then</u> $T_3T_2T_1 \in N_{\ell^1}(E, X_3)$.

<u>Proof</u>. By (2.2.5), each T_j can be decomposed in the following way:

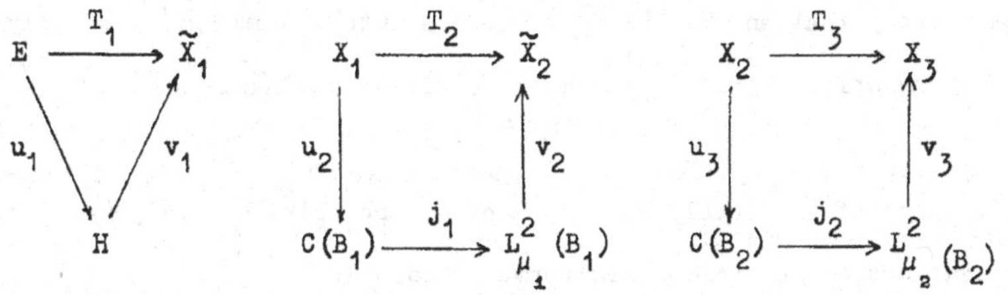

,

where H is a Hilbert space, j_i $(i = 1, 2)$ are the canonical embedding maps, μ_i are positive Radon measures on the compact Hausdorff spaces B_i. Denote by $\tilde{u}_i$ the unique continuous extension of u_i on $\tilde{X}_{i-1}$ $(i = 2, 3)$. Then we have

$$T_3T_2T_1 = v_3(j_2\tilde{u}_3v_2)(j_1\tilde{u}_2v_1)u_1 .$$

In view of (2.2.9) and (2.2.8), $j_1\tilde{u}_2v_1 : H \to L^2_{\mu_2}(B_2)$ is a Hilbert-Schmidt linear map and $j_2\tilde{u}_3v_2 : L^2_{\mu_2}(B_1) \to L^2_{\mu_2}(B_2)$ is a Hilbert-Schmidt linear map, it follows from (3.1.5) that the map

$$(j_2\tilde{u}_3v_2)(j_1\tilde{u}_2v_1) : H \to L^2_{\mu_2}(B_2)$$

is nuclear, and hence that $T_3T_2T_1 \in N_{\ell^1}(E, X_3)$.

3.2 Prenuclear seminorms and nuclear spaces

A seminorm p on E is said to be *prenuclear* if there exist a $\sigma(E', E)$-closed equicontinuous subset B of E' and a positive Radon measure μ on B such that

$$p(x) \leq \int_B |\langle x, x'\rangle| d\mu(x') \quad \text{for all} \quad x \in E .$$

Clearly prenuclear seminorms are continuous, and every $\sigma(E, E')$-continuous seminorm on E is prenuclear. On the other hand, it is easily seen from (2.2.2) that an $T \in L(E, F)$ is absolutely summing if and only if $q \circ T$ is a prenuclear seminorm whenever q is a continuous seminorm on F.

In terms of absolutely summing maps we are able to give some alternative characterizations of prenuclear seminorms as follows.

(3.2.1) Lemma. *Let* p *be a seminorm on* E *and* $V_p = \{x \in E : p(x) \leq 1\}$. *Then the following statements are equivalent.*

(a) p *is a prenuclear seminorm.*

(b) *The quotient map* $Q_p : E \to E_p$ *is absolutely summing.*

(c) V_p^o *is a prenuclear subset of* E'.

(d) *There is a continuous seminorm* r *on* E *with* $p \leq r$ *such that the canonical map* $Q_{p,r} : E_r \to E_p$ *is absolutely summing.*

(e) *There is an absolutely convex* 0-*neighbourhood* W *in* E *such that the inequality*

$$\sum_{k=1}^{n} p(x_k) \leq \sup\{\sum_{k=1}^{n} |\langle x_k, x'\rangle| : x' \in W^o\}$$

holds for any finite subset $\{x_1, \ldots, x_n\}$ of E.

Proof. It is well-known that $p = \hat{p} \circ Q_p$, where $\hat{p}$ is the quotient norm of p, it then follows from (2.2.2) that the statements (a), (b) and (e) are equivalent. Since $V^o = Q_p'((Q_p(V_p))^o)$, it follows from (2.2.2) that (b) and (c) are equivalent. The implication (d) $\Rightarrow$ (b) is obviously a consequence of (2.2.1). Therefore we complete the proof by showing that (e) implies (d).

Let r be the gauge of W. Then r is a continuous seminorm on E such that $p \leq r$. As $W^o = Q_r'((Q_r(W))^o)$ and $(Q_r(W))^o$ is the closed unit ball in the Banach dual of E_r, we have for any finite subset $\{Q_r(x_1), \ldots, Q_r(x_n)\}$ of E_r that

$$\begin{aligned}\Sigma_{k=1}^{n} \hat{p}(Q_{p,r}(Q_r(x_k))) &= \Sigma_{k=1}^{n} p(x_k) \leq \sup\{\Sigma_{k=1}^{n} |\langle x_k, x'\rangle| : x' \in W^o\} \\ &= \sup\{\Sigma_{k=1}^{n} |\langle Q_r(x_k), f\rangle| : f \in (Q_r(W))^o\} .\end{aligned}$$

Therefore $Q_{p,r}$ is an absolutely summing map.

Remark. As a consequence of the preceding result, we see that if $T \in L(E, F)$ and q is a prenuclear seminorm on F, then $q \circ T$ is a prenuclear seminorm on E.

(3.2.2) Lemma. For any prenuclear seminorm p on E there is an absolutely convex o-neighbourhood V in E with $V \subset \{x \in E : p(x) \leq 1\}$ such that $\tilde{E}_V$ is isometrically isomorphic to a subspace of $L^2_\mu(B)$, where μ is a positive Radon measure on a suitable $\sigma(E', E)$-closed equicontinuous

<u>subset</u> B <u>of</u> E' .

<u>Proof</u>. There exist a $\sigma(E', E)$-closed equicontinuous subset B of E' and a positive Radon measure μ on B such that

$$p(x) \leq \int_B |\langle x, x'\rangle| d\mu(x') \quad \text{for all} \quad x \in E .$$

We may assume that $\mu(B) = 1$. Let

$$V = \{x \in E : \int_B |\langle x, x'\rangle|^2 d\mu(x') \leq 1\} .$$

Then V is circled and $B^o \subset V$. Furthermore, the Cauchy-Schwarz inequality ensures that V is convex and $V \subset \{x \in E : p(x) \leq 1\}$ because of

$$\int_B |\langle x, x'\rangle| d\mu(x') \leq (\int_B |\langle x, x'\rangle|^2 d\mu(x'))^{\frac{1}{2}}$$

on account of $\mu(B) = 1$. It is easily seen that the gauge p_V of V has the expression

$$p_V(x) = (\int_B |\langle x, x'\rangle|^2 d\mu(x'))^{\frac{1}{2}} \quad \text{for all} \quad x \in E .$$

As $p_V = \hat{p}_V \circ Q_V$, it follows that $\tilde{E}_V$ is isometrically isomorphic to a subspace of $L^2_\mu(B)$.

(3.2.3) Lemma. <u>If</u> p <u>and</u> q <u>are prenuclear seminorms, then so do</u> p + q <u>and</u> ζp <u>for any</u> $\zeta \geq 0$.

<u>Proof</u>. The prenuclearity of ζp is obvious. To see the prenuclearity of p + q , we define $T : E \to E_p \times E_q$ by setting

$$T(x) = (Q_p(x), Q_q(x)) \quad \text{for all} \quad x \in E .$$

Then T is linear and $\operatorname{Ker} T = \operatorname{Ker}(p + q)$, hence the injection $\hat{T}$ associated with T is an isometric linear map from the normed space E_{p+q} into the normed space $E_p \times E_q$ because of

$$\|\hat{T}(Q_{p+q}(x))\| = \|Tx\| = \|(Q_p(x), Q_q(x))\| = p(x) + q(x) = (\widehat{p + q})(Q_{p+q}(x)) .$$

As $T = \hat{T} \circ Q_{p+q}$, we have only to show that T is absolutely summing in view of (2.2.1) and (3.2.1). To do this we first notice from (3.2.1) that there are absolutely convex o-neighbourhoods V and W in E such that the inequalities

$$\Sigma_{k=1}^{n} p(x_k) \leq \sup\{\Sigma_{k=1}^{n} |\langle x_k, x'\rangle| : x' \in V^o\} \quad \text{and}$$
$$\Sigma_{j=1}^{m} q(z_j) \leq \sup\{\Sigma_{j=1}^{m} |\langle z_j, z'\rangle| : z' \in W^o\}$$

hold for any finite subsets $\{x_1, \ldots, x_n\}$ and $\{z_1, \ldots, z_m\}$ of E. Setting $U = 2^{-1}(V \cap W)$. Then we have for any finite subset $\{x_1, \ldots, x_n\}$ of E that

$$\Sigma_{k=1}^{n} \|Tx_k\| = \Sigma_{k=1}^{n} (p(x_k) + q(x_k)) \leq \sup\{\Sigma_{k=1}^{n} |\langle x_k, 2f\rangle| : f \in (V \cap W)^o\}$$
$$= \sup\{\Sigma_{k=1}^{n} |\langle x_k, g\rangle| : g \in U^o\} .$$

Therefore T is an absolutely summing map.

A locally convex space $(E, \mathscr{P})$ is said to be <u>nuclear</u> if every $\mathscr{P}$-continuous seminorm on E is prenuclear.

(3.2.4) Theorem. <u>The following statements are equivalent.</u>

(a) $(E, \mathscr{P})$ <u>is a nuclear space</u>.

(b) <u>Every $\mathcal{P}$-equicontinuous subset of</u> E' <u>is prenuclear</u>.

(c) $Q_p \in \pi_{\ell^1}(E, E_p)$ <u>for any continuous seminorm</u> p <u>on</u> E .

(d) <u>For any continuous seminorm</u> p <u>on</u> E <u>there is a continuous seminorm</u> r <u>on</u> E <u>with</u> $p \leq r$ <u>such that</u> $Q_{p,r} \in \pi_{\ell^1}(E_r, E_p)$.

(e) <u>The identity map</u> I <u>on</u> E <u>is absolutely summing</u>.

(f) <u>The canonical embedding map from</u> $(\ell^1[E], \mathcal{P}_\pi)$ <u>into</u> $(\ell^1(E), \mathcal{P}_\varepsilon)$ <u>is a topological isomorphism from the first space onto the second</u>.

(g) $\pi_{\ell^1}(E, F) = L(E, F)$ <u>for any locally convex space</u> F .

(h) $\pi_{\ell^1}(E, Y) = L(E, Y)$ <u>for any normed space</u> Y .

(i) $Q_p \in N_{\ell^1}(E, \widetilde{E}_p)$ <u>for any continuous seminorm</u> p <u>on</u> E .

(j) $N_{\ell^1}(E, Y) = L(E, Y)$ <u>for any Banach space</u> Y .

(k) <u>For any continuous seminorm</u> p <u>on</u> E <u>there is a continuous seminorm</u> r <u>on</u> E <u>with</u> $p \leq r$ <u>such that</u> $Q_{p,r} \in N_{\ell^1}(\widetilde{E}_r, \widetilde{E}_p)$.

(ℓ) <u>For any continuous seminorm</u> p <u>on</u> E <u>there is a continuous seminorm</u> r <u>on</u> E <u>with</u> $p \leq r$ <u>such that the canonical embedding map from</u> $E'(V_p^o)$ <u>into</u> $E'(V_r^o)$ <u>is nuclear</u>.

<u>Moreover, if in addition,</u> $(E, \mathcal{P})$ <u>is metrizable, then</u> (a) <u>is equivalent to</u>

(m) $\ell^1[E] = \ell^1(E)$.

<u>Proof</u>. In view of (3.2.1) and (2.2.3), the statements (a) through (f) are mutually equivalent. The implication (g) ⇒ (e) is obvious, and (e) ⇒ (g) follows from (2.2.1). The implication (g) ⇒ (h) ⇒ (c) and (i) ⇒ (a) are obvious, while the implication (d) ⇒ (i) follows from a result of Grothendieck. The equivalence of (a) and (m) follows from (2.2.4). It is also clear that (k)

implies (i), and that (k) implies (ℓ) since the adjoint map of a nuclear map is nuclear. Therefore we complete the proof by showing the implications (i) ⇒ (j) ⇒ (k) and (ℓ) ⇒ (k) .

(i) ⇒ (j) : Let $T \in L(E, Y)$ and let p be the gauge of $T^{-1}(\Sigma)$, where Σ is the closed unit ball in Y . Then $\operatorname{Ker} p \subset \operatorname{Ker} T$ and the normability of Y ensures that $L^{\ell b}(E, Y) = L(E, Y)$, thus there is an $\widetilde{T} \in L(E_p, Y)$ such that $T = \widetilde{T} Q_p$; consequently $T \in N_{\ell^1}(E, Y)$.

(j) ⇒ (k) : For any continuous seminorm p on E , the nuclearity of Q_p shows that Q_p is of the form

$$Q_p(x) = \Sigma_{n=1}^{\infty} \zeta_n \langle x, x'_n \rangle \widetilde{y}_n ,$$

where $(\zeta_n) \in \ell^1$, $\{\widetilde{y}_n\}$ is a bounded sequence in $\widetilde{E}_p$ and $\{x'_n\}$ is an equicontinuous sequence in E' . Let $V_p = \{x \in E : p(x) \leq 1\}$, let

$$W = V_p \cap \{x \in E : |\langle x, x'_n \rangle| \leq 1 \text{ for all } n \geq 1\} ,$$

and let r be the gauge of W . Then r is continuous and $p \leq r$. It is known that the adjoint map Q'_r of Q_r is an isometry from the Banach space $E'(W^o)$ onto $(\widetilde{E}_r)'$, hence there exists $h_n \in (\widetilde{E}_r)'$ such that

$$x'_n = h_n Q_r \text{ and } \|h_n\| \leq 1 \text{ for all } n \geq 1 .$$

Furthermore, the canonical map $Q_{p,r} : \widetilde{E}_r \to \widetilde{E}_p$ is of the form

$$Q_{p,r}(Q_r(x)) = \Sigma_{n=1}^{\infty} \zeta_n \langle x, x'_n \rangle \widetilde{y}_n = \Sigma_{n=1}^{\infty} \zeta_n \langle Q_r(x), h_n \rangle \widetilde{y}_n ,$$

which shows that $Q_{p,r} \in N_{\ell^1}(\widetilde{E}_r, \widetilde{E}_p)$.

$(\ell) \Rightarrow (k)$: This can be deduced from (3.4.7) and (3.4.2)(c) below; but we present here a somewhat more direct and elementary proof. For any continuous seminorm p on E, there exist continuous seminorms q and r on E with $p \leqslant q \leqslant r$ such that the canonical embedding maps

$$E'(V_p^o) \xrightarrow{\;j_{q,p}\;} E'(V_q^o) \xrightarrow{\;j_{r,q}\;} E'(V_r^o)$$

are all nuclear. It is clear that $(E_p)''$ and $(E'(V_p^o))'$ are isometrically isomorphic, and that the following diagrams

$$\begin{array}{ccccc}
(E_p)' & \xrightarrow{\;Q'_{p,q}\;} & (E_q)' & \xrightarrow{\;Q'_{q,r}\;} & (E_r)' \\
\downarrow{\scriptstyle Q'_p} & & \downarrow{\scriptstyle Q'_q} & & \downarrow{\scriptstyle Q'_r} \\
E(V_p^o) & \xrightarrow[\;j_{q,p}\;]{} & E(V_q^o) & \xrightarrow[\;j_{r,q}\;]{} & E(V_r^o)
\end{array}$$

are commutative. Since Q'_p and Q'_q are isometries, it follows that $j_{q,p}$ can be regarded as the adjoint map of $Q_{p,q}$, hence the following mappings

$$(E_r)'' \xrightarrow{\;Q''_{q,r}\;} (E_q)'' \xrightarrow{\;Q''_{p,q}\;} (E_p)''$$

are nuclear. Since $Q''_{p,q}$ is a compact map and $Q_{p,q}$ is the restriction to $\widetilde{E}_q$ of $Q''_{p,q}$, it follows that $Q_{p,q}$ sends the closed unit ball in $\widetilde{E}_q$ into a relatively $\sigma(\widetilde{E}_p, E'(V^o))$-compact subset of $\widetilde{E}_p$, and hence that $Q_{p,q}$ is a weakly compact map. By Schaefer [1, p.169], $Q''_{p,q}((E_q)'') \subset \widetilde{E}_p$. On the other hand, the nuclearity of $Q''_{q,r}$ shows that

$$Q''_{q,r} = \sum_{n=1}^{\infty} \zeta_n f_n \otimes y_n ,$$

where $(\zeta_n) \in \ell^1$, $\{y_n\}$ is a bounded sequence in $(E_q)''$ and $\{f_n\}$ is an equicontinuous sequence in $(E_r)'''$. Let

$$z_n = Q''_{p,q}(y_n) \text{ and } g_n \text{ the restriction to } E_r \text{ of } f_n .$$

Then $z_n \in \widetilde{E}_p$ (since $Q''_{p,q}((E_q)'') \subset \widetilde{E}_p$) , and $\{g_n\}$ is an equicontinuous sequence in $(E_r)'$. Furthermore, for any $\widetilde{x} \in \widetilde{E}_r$

$$Q_{p,r}(\widetilde{x}) = (Q_{p,q} \circ Q_{q,r})(\widetilde{x}) = Q''_{p,r}(\Sigma_{n=1}^{\infty} \zeta_n <\widetilde{x}, f_n> y_n)$$
$$= \Sigma_{n=1}^{\infty} \zeta_n <\widetilde{x}, f_n> z_n = \Sigma_{n=1}^{\infty} \zeta_n <\widetilde{x}, g_n> z_n ;$$

thus $Q_{p,r}$ is a nuclear mapping.

As a consequence of (3.2.4) and (3.2.2), every nuclear space is topological isomorphic to a dense subspace of the projective limit of a suitable family of Hilbert spaces. This result has been generalized by Saxon [1] as follows: If E is a nuclear space and Y is an arbitrary infinite-dimensional Banach space, then there exists a neighbourhood basis $\mathcal{U}_E$ at 0 such that each normed space E_V $(V \in \mathcal{U}_E)$ is isomorphic to a subspace of Y .

As nuclear mappings are precompact, we obtain an immediate consequence of (3.2.4) and (1.2.4).

(3.2.5) Proposition. Nuclear spaces are Schwartz spaces.

We shall see from (3.5.4) that the converse of (3.2.5) is, in general, not true. In view of (3.2.5), all conclusions of (1.2.6), (1.2.7) and (1.2.8) hold for nuclear spaces. Furthermore, we have the following result.

(3.2.6) Proposition. For a nuclear space $(E, \mathcal{P})$, the following assertions hold.

(a) Every $\sigma(E, E')$-convergent sequence in E is also $\mathcal{P}$-convergent to the same limit.

(b) For any $V \in \mathcal{U}_E$ and any absolutely convex bounded subset A of E, the canonical map $K_{V,A} : E(A) \to E_V$ is nuclear, hence $K_{V,A}(A)$ is precompact and separable in E_V.

(c) A normed space is nuclear if and only if its dimension is finite.

Proof. It is clear that E is nuclear if and only if its completion $\tilde{E}$ of E is nuclear; thus (a) follows from (3.2.5) and (1.2.7). The conclusion (c) is obvious, while (b) follows from $K_{V,A} = Q_V j_A$, where $Q_V : E \to E_V$ is a nuclear map.

Example. Let $\Delta = [a, b]$ and let $\mathcal{D}(\Delta)$ be the vector space consisting of all infinitely differentiable complex-valued functions on the real line $\mathbb{R}$ which vanish outside of Δ. For any $m = 1, 2, \ldots$, we define

$$[\phi, \psi]_m = \Sigma_{j=0}^{m} \int_\Delta \phi^{(j)}(t) \overline{\psi^{(j)}(t)}\, dt \quad \text{and} \quad \|\psi\|_m = ([\psi, \psi]_m)^{\frac{1}{2}}, \qquad (2.1)$$

where $\psi^{(0)} = \psi$. Then $\mathcal{D}(\Delta)$, equipped with the sequence $\{\|\cdot\|_m : m \geq 0\}$ of seminorms, is a Fréchet space (in fact, it is a countably Hilbert space in the sense of Gelfand and Vilenkin [1]). We claim that $\mathcal{D}(\Delta)$ is a nuclear space.

In fact, for each $m \geq 1$, let $\mathcal{D}_m(\Delta)$ be the vector space consisting of all complex-valued functions which vanish outside of Δ , have continuous derivatives up to order $m - 1$ and the m-th derivatives belong to $L^2(\Delta)$ a.e.. Then $\mathcal{D}_m(\Delta)$ is a Hilbert space equipped with the inner product definied by (2.1). In order to verify the nuclearity of $\mathcal{D}(\Delta)$, it suffices, by (3.2.4) and (2.2.8), to show that the canonical embedding $J_m^{m+1} : \mathcal{D}_{m+1}(\Delta) \to \mathcal{D}_m(\Delta)$ is a Hilbert-Schmidt operator. To simplify the notation, we shall suppose $\Delta = [0, 2\pi]$. Let $M_m(\Delta)$ be the vector space of all complex-valued functions defined on Δ which have continuous derivatives of order $m - 1$ and the m-th derivatives belong to $L^2(\Delta)$ a.e.. Then $M_m(\Delta)$, equipped with the inner product defined by (2.1), is a Hilbert space, and $\mathcal{D}_m(\Delta)$ is a closed subspace of $M_m(\Delta)$. Let $\pi_m : M_m(\Delta) \to \mathcal{D}_m(\Delta)$ be the projector, let $u_m : M_{m+1}(\Delta) \to M_m(\Delta)$ and $v_m : \mathcal{D}_{m+1}(\Delta) \to M_{m+1}(\Delta)$ be the canonical embedding. Then $J_m^{m+1} = \pi_m u_m v_m$. Clearly, π_m and v_m are continuous, hence it is required to show that u_m is a Hilbert-Schmidt operator. Let us consider the orthonormal sequence

$$e_k^{m+1}(t) = e^{ikt}/(2\pi \Sigma_{j=0}^{m+1} k^{2j})^{\frac{1}{2}} \qquad k = 0, \pm 1, \pm 2, \ldots$$

in $M_{m+1}(\Delta)$, and the orthonormal sequence

$$g_k^m(t) = e^{ikt}/(2\pi \Sigma_{j=0}^{m} k^{2j})^{\frac{1}{2}} \qquad k = 0, \pm 1, \pm 2, \ldots$$

in $M_m(\Delta)$. Suppose $\psi \in M_{m+1}(\Delta)$. Then we have by the Fourier expansion of ψ that

$$u_m \psi = \Sigma_k \lambda_{k,m} [\psi, e_k^{m+1}]_{m+1} \, g_k^m ,$$

where

$$\lambda_{k,m} = \left(\frac{\Sigma_{j=0}^{m} k^{2j}}{\Sigma_{j=0}^{m+1} k^{2j}} \right)^{\frac{1}{2}} . \tag{2.2}$$

In view of (2.2), it is not hard to show that $\Sigma_{k=1}^{\infty} \lambda_{k,m}^{2} < +\infty$, as asserted by (2.2.8).

3.3 Prenuclear-bounded linear mappings

An $T \in L^{*}(E, F)$ is called a prenuclear-bounded map (or prenuclear map in the terminology of Wong [1]) if there is a prenuclear seminorm p on E such that $\{Tx \in F : p(x) \leq 1\}$ is a bounded subset of F or, equivalently, for any continuous seminorm q on F there is an $\lambda_q \geq 0$ such that

$$q(Tx) \leq \lambda_q p(x) \quad \text{for all} \quad x \in E .$$

Clearly, prenuclear-bounded maps are absolutely summing, but the converse need not be true. Lemma (3.2.3) shows that the set consisting of all prenuclear-bounded maps from E into F , denoted by $N_{\ell^1}^{(p)}(E, F)$, is a vector subspace of $L^{\ell b}(E, F)$, therefore

$$N_{\ell^1}^{(p)}(E, F) \subset \pi_{\ell^1}(E, F) \cap L^{\ell b}(E, F) .$$

If, in addition, F is a normed space, then

$$N_{\ell^1}^{(p)}(E, F) = \pi_{\ell^1}(E, F) .$$

Theorem (3.2.4) shows that the absolutely summing property of the identity map on E characterizes the nuclearity of E . The prenuclear-bounded property of the identity map on E characterizes the nuclearity and normability of E as shown by the following result.

(3.3.1) Theorem. The identity map I on E is a prenuclear-bounded map if and only if E is normable and a finite-dimensional space.

Proof. Necessity. There is a prenuclear seminorm p on E such that for any continuous seminorm q on E it is possible to find an $\lambda_q \geq 0$ such that

$$q(x) \leq \lambda_q p(x) \quad \text{for all} \quad x \in E .$$

Hence the topology on E is determined by the single prenuclear seminorm p, thus E is normable. As $N^{(p)}_{\ell^1}(E) \subset \pi_{\ell^1}(E)$, it follows from (3.2.4) that E is nuclear, and hence that the dimension of E is finite.

Sufficiency. Under the hypotheses, E is nuclear space, hence $I \in \pi_{\ell^1}(E)$, and thus $I \in N^{(p)}_{\ell^1}(E)$ since E is normable.

As an application of Theorems (3.2.4) and (3.3.1), we obtain the following famous theorem of Dvoretzky and Rogers [1] .

(3.3.2) Corollary (Dvoretzky and Rogers). A normed space X is finite dimensional if and only if every summable sequence in X is absolutely summable.

In terms of prenuclear-bounded maps we are able to give other criteria for nuclear spaces as follows.

(3.3.3) Theorem. The following statements are equivalent.

(a) E is a nuclear space.

(b) $N^{(p)}_{\ell^1}(E, F) = L^{\ell b}(E, F)$ <u>for any locally convex space</u> F.

(c) $L^{\ell b}(E, F) \subseteq \pi_{\ell^1}(E, F)$ <u>for any locally convex space</u> F.

(d) $L^{p}(E, F) \subseteq N^{(p)}_{\ell^1}(E, F)$ <u>for any locally convex space</u> F.

<u>Proof</u>. As $N^{(p)}_{\ell^1}(E, F) \subseteq \pi_{\ell^1}(E, F)$ and $L^p(E, F) \subseteq L^{\ell b}(E, F)$, the implication (b) ⇒ (c) and (b) ⇒ (d) follow. The implication (a) ⇒ (b) is clear by making use of the definitions of nuclearity and of prenuclear-bounded maps. Therefore we complete the proof by showing the implications (c) ⇒ (a) and (d) ⇒ (a).

(c) ⇒ (a) : Suppose that the statement (c) holds. Then

$$L^{\ell b}(E, Y) \subseteq \pi_{\ell^1}(E, Y) \quad \text{for any normed space } Y.$$

Therefore $L(E, Y) = L^{\ell b}(E, Y) \subseteq \pi_{\ell^1}(E, Y)$ for any normed space Y, and thus E is nuclear by (3.1.4).

(d) ⇒ (a) : As $N^{(p)}_{\ell^1}(E, F) \subseteq \pi_{\ell^1}(E, F)$, it suffices to show, in view of (3.1.4), that

$$L(E, Y) \subseteq N^{(p)}_{\ell^1}(E, F) \quad \text{for any normed space } Y.$$

In fact, for any normed space Y, we have

$$L(E, Y) = L^{\ell b}(E, Y) = L^p(E, Y_\sigma) \quad \text{and} \quad N^{(p)}_{\ell^1}(E, Y) = N^{(p)}_{\ell^1}(E, Y_\sigma)$$

because bounded subsets of Y are $\sigma(Y, Y')$-bounded. It then follows that

$$L(E, Y) = L^p(E, Y_\sigma) \subseteq N^{(p)}_{\ell^1}(E, Y_\sigma) = N^{(p)}_{\ell^1}(E, Y).$$

In order to give some characterizations of prenuclear-bounded maps, we need the following result whose proof is straightforward and hence will be omitted.

(3.3.4) Lemma. *Let* E, F *and* G *be locally convex spaces, let* $T \in L(E, F)$ *and* $S \in L(F, G)$. *If one of* T *and* S *is prenuclear-bounded, then* $S \circ T \in N^{(p)}_{\ell^1}(E, G)$.

We are now able to give some criteria of prenuclear-bounded linear maps as follows.

(3.3.5) Proposition. *For an* $T \in L(E, F)$, *the following statements are equivalent.*

(a) $T \in N^{(p)}_{\ell^1}(E, F)$.

(b) *There exist an* $\sigma(E', E)$-*closed equicontinuous subset* B *of* E' *and a positive Radon measure* μ *on* B *such that for any continuous seminorm* q *on* F , *it is possible to find an* $\zeta_q \geq 0$ *for which the following inequality holds*:

$$q(Tx) \leq \zeta_q \int_B |\langle x, x'\rangle| d\mu(x') \quad \textit{for all} \quad x \in E .$$

(c) *There exists an absolutely convex* o-*neighbourhood* V *in* E *such that for any continuous seminorm* q *on* F *it is possible to find an* $\beta_q \geq 0$ *for which the inequality*

$$\Sigma_{k=1}^{n} q(Tx_k) \leq \beta_q \sup\{\Sigma_{k=1}^{n} |\langle x_k, x'\rangle| : x' \in V^o\}$$

holds for any finite subset $\{x_1, \ldots, x_n\}$ *of* E .

(d) T <u>is the composition of the following continuous linear maps</u>

$$E \xrightarrow{Q} X \xrightarrow{\tilde{T}} Y \xrightarrow{J} F \ .$$

<u>where</u> X <u>and</u> Y <u>are normed spaces and</u> $\tilde{T} \in \pi_{\ell^1}(X, Y)$.

<u>Proof</u>. In view of (3.2.1), the statements (a), (b) and (c) are mutually equivalent. The implication (d) ⇒ (a) follows from (3.3.4) in view of $\pi_{\ell^1}(X, Y) = N^{(p)}_{\ell^1}(X, Y)$. To prove the implication (a) ⇒ (d), let p be a prenuclear seminorm on E such that $C = \{Tx \in F : p(x) \leq 1\}$ is a bounded subset of F . Then $\operatorname{Ker} p \subseteq \operatorname{Ker} T$, hence there is an $S \in L(E_p, F(C))$ such that $T = j_C \, S \, Q_p$, where $j_C : F(C) \to F$ is the canonical embedding map and $Q_p : E \to E_p$ is the quotient map. On the other hand, the prenuclearity of p implies that there is a continuous seminorm r on E with $p \leq r$ such that $Q_{p,r} \in \pi_{\ell^1}(E_r, E_p)$ by (3.2.1). The map $\tilde{T}$, defined by

$$\tilde{T} = S \, Q_{p,r} \ ,$$

belongs to $\pi_{\ell^1}(E_r, F(C))$, and

$$T = j_C \, \tilde{T} \, Q_r \ .$$

(3.3.6) Corollary. $N_{\ell^1}(E, F) \subset N^{(p)}_{\ell^1}(E, F)$ <u>and the composition of two prenuclear-bounded maps is nuclear</u>.

<u>Proof</u>. Since nuclear maps are absolutely summing, the inclusion $N_{\ell^1}(E, F) \subset N^{(p)}_{\ell^1}(E, F)$ follows from (3.3.5)(d) and the fact that an $T \in L(E, F)$ is nuclear if and only T permits a factoring

$$E \xrightarrow{T_1} X \xrightarrow{S} Y \xrightarrow{T_2} F ,$$

where X and Y are Banach spaces and $S \in N_{\ell^1}(X, Y)$. The second assertion is a consequence of (3.3.5)(d) and a result of Grothendieck (see Pietsch [1, (3.3.5)]).

Example. Absolutely summing maps are neither prenuclear-bounded nor precompact nor nuclear.

For any infinite-dimensional locally convex space E , it is trivial that $E_\sigma = (E, \sigma(E, E'))$ is a nuclear space. By (3.2.4) and (3.2.1), we have

$$I \in \pi_{\ell^1}(E_\sigma) \quad \text{and} \quad I \notin N^{(p)}_{\ell^1}(E_\sigma) ,$$

where I is the identity map on E . It is known from (3.3.3) and the inclusion $N_{\ell^1}(E, F) \subset N^{(p)}_{\ell^1}(E, F)$ that

$$I \notin L^p(E_\sigma) \quad \text{and} \quad I \notin N_{\ell^1}(E_\sigma) .$$

3.4 Quasi-nuclear-bounded linear mappings

Nuclear representations of nuclear maps suggest the following definition: A seminorm p on E is said to be quasi-nuclear if there exist an $(\zeta_n) \in \ell^1$ and an equicontinuous sequence $\{f_n\}$ in E' such that

$$p(x) \leq \sum_{n=1}^{\infty} |\zeta_n \langle x, f_n \rangle| \quad \text{for all } x \in E .$$

Clearly quasi-nuclear seminorms on E are continuous, and $\sigma(E, E')$-

continuous seminorms on E are quasi-nuclear. If $T \in L(E, F)$ and q is a quasi-nuclear seminorm on F, then $q \circ T$ is clearly a quasi-nuclear seminorm on E.

(3.4.1) Lemma. Quasi-nuclear seminorms on E are prenuclear; consequently, a locally convex space $(E, \mathcal{F})$ is nuclear if and only if every continuous seminorm on E is quasi-nuclear.

Proof. The second assertion is a consequence of the first one. Let p be a quasi-nuclear seminorm on E. There exist an $(\zeta_n) \in \ell^1$ and an equicontinuous sequence $\{f_n\}$ in E' such that

$$p(x) \leq \sum_{n=1}^{\infty} |\zeta_n \langle x, f_n \rangle| \quad \text{for all} \quad x \in E .$$

Let $\zeta = \sum_{n=1}^{\infty} |\zeta_n|$ and let V be an absolutely convex o-neighbourhood in E such that $f_n \in V^o$ for all n. Then $B = \zeta V^o$ is a $\sigma(E', E)$-closed equicontinuous subset of E' such that $\zeta_n f_n \in B$ for all n. Denote by δ_n the point mass at $\zeta_n f_n$. Then

$$\zeta_n \langle x, f_n \rangle = \int_B \langle x, x' \rangle d\delta_n(x') \quad \text{for all} \quad x \in E \quad \text{and} \quad n \geq 1 .$$

The Hahn-Banach extension theorem shows that the linear map

$$g \mapsto \sum_{n=1}^{\infty} |\zeta_n| g(f_n) \quad \text{for all} \quad g \in C(B)$$

defines a positive Radon measure μ on B such that

$$p(x) \leq \sum_{n=1}^{\infty} |\zeta_n \langle x, f_n \rangle| \leq \int_B |\langle x, x' \rangle| d\mu(x') \quad \text{for all} \quad x \in E ,$$

thus p is prenuclear.

An $T \in L^*(E, F)$ is called a quasi-nuclear-bounded map if there is a quasi-nuclear seminorm p on E such that $\{Tx \in F : p(x) \leq 1\}$ is a bounded subset of F.

The set consisting of all quasi-nuclear-bounded maps from E into F, denoted by $N^{(q)}_{\ell^1}(E, F)$, is a vector subspace of $L^{\ell b}(E, F)$. Furthermore, we have

$$N_{\ell^1}(E, F) \subseteq N^{(q)}_{\ell^1}(E, F) \subseteq N^{(p)}_{\ell^1}(E, F) \subseteq \pi_{\ell^1}(E, F) \cap L^{\ell b}(E, F)$$

by (3.4.1) and nuclear representations of nuclear maps. (3.3.6) ensures that the composition of two quasi-nuclear-bounded maps is nuclear. It is easily seen that the composition of two continuous linear maps, in which one of them is quasi-nuclear-bounded, is quasi-nuclear-bounded. A similar argument given in the proof of (a) $\Rightarrow$ (d) of (3.3.5) shows that an $T \in L(E, F)$ is quasi-nuclear-bounded if and only if T permits a factoring

$$E \xrightarrow{\ Q\ } X \xrightarrow{\ \tilde{T}\ } Y \xrightarrow{\ J\ } F ,$$

where X and Y are normed spaces and $\tilde{T} \in N^{(q)}_{\ell^1}(X, Y)$. We shall study later some important properties of quasi-nuclear-bounded maps, but we shall present below, by using the notion of quasi-nuclear-bounded maps, some characterization of nuclear spaces.

Clearly, p is a quasi-nuclear seminorm on E if and only if $Q_p \in N^{(q)}_{\ell^1}(E, E_p)$. On the other hand, if p is a quasi-nuclear seminorm on E, then there exist an $(\zeta_n) \in \ell^1$ and an equicontinuous sequence $\{f_n\}$ in E' such that

$$p(x) \leq \Sigma_{n=1}^{\infty} |\zeta_n f_n(x)| \quad \text{for all} \quad x \in E . \qquad (4.1)$$

Without loss of the generality one can assume that $\zeta = \Sigma_{n=1}^{\infty} |\zeta_n| \neq 0$. Then there is an absolutely convex o-neighbourhood W in E such that $\zeta f_n \in W^o$ for all n . The gauge of W , denoted by r , satisfies

$$p(x) \leq (\Sigma_{n=1}^{\infty} |\zeta_n|) \sup_n |f_n(x)| \leq r(x) \quad \text{for all} \quad x \in E$$

by (4.1). As Q_r' being an isometry from $(E_r)'$ onto $E'(W^o)$, there exists $h_n \in (Q_r(W))^o$ such that $h_n \circ Q_r = \zeta f_n$ for all n . Therefore,

$$\begin{aligned}\hat{p}(Q_{p,r}(Q_r(x))) = p(x) &\leq \Sigma_{n=1}^{\infty} |\zeta_n f_n(x)| \\ &= \Sigma_{n=1}^{\infty} |\zeta_n \zeta^{-1} \langle Q_r(x), h_n \rangle| \quad \text{for all} \quad x \in E\end{aligned}$$

by (4.1), and thus $Q_{p,r} \in N_{\ell^1}^{(q)}(E_r, E_p)$. This proves that p is a quasi-nuclear seminorm on E if and only if there exists a continuous seminorm r on E with $p \leq r$ such that $Q_{p,r} \in N_{\ell^1}^{(q)}(E_r, E_p)$. Therefore, we have proved the equivalence of (a), (b) and (c) of the following results.

(3.4.2) Theorem. <u>The following statements are equivalent.</u>

(a) E <u>is a nuclear space.</u>

(b) $Q_p \in N_{\ell^1}^{(q)}(E, E_p)$ <u>for any continuous seminorm</u> p <u>on</u> E .

(c) <u>For any continuous seminorm</u> p <u>on</u> E <u>there is a continuous seminorm</u> r <u>on</u> E <u>with</u> $p \leq r$ <u>such that</u> $Q_{p,r} \in N_{\ell^1}^{(q)}(E_r, E_p)$.

(d) $L^{\ell b}(E, F) = N_{\ell^1}^{(q)}(E, F)$ <u>for any locally convex space</u> F .

(e) $L(E, Y) = N_{\ell^1}^{(q)}(E, Y)$ <u>for any normed space</u> Y .

(f) $L^p(E, F) \subset N_{\ell^1}^{(q)}(E, F)$ <u>for any locally convex space</u> F .

<u>Proof</u>. The implications (d) $\Rightarrow$ (e) $\Rightarrow$ (b) and (d) $\Rightarrow$ (f) are obvious, and the implication (f) $\Rightarrow$ (a) follows from (3.3.3)(d). Therefore, we complete the proof by showing that (c) implies (d). For any $T \in L^{\ell b}(E, F)$, there exist an absolutely convex o-neighbourhood V in E and an absolutely convex bounded subset B of F such that

$$T = j_B T_{(V,B)} Q_V , \tag{4.2}$$

where $T_{(V,B)} \in L(E_V, F(B))$ is the induced map of T. Let p be the gauge of V. By the statement (c), there exists a continuous seminorm r on E with $p \leq r$ such that $Q_{p,r} \in N^{(q)}_{\ell^1}(E_r, E_p)$. In view of (4.2), we have

$$T = j_B T_{(V,B)} Q_{p,r} Q_r ,$$

hence $T \in N^{(q)}_{\ell^1}(E, F)$.

Suppose now that (X, p) and (Y, q) are normed spaces. Then an $T \in L(X, Y)$ is quasi-nuclear-bounded if and only if there exists a sequence $\{f_n\}$ in X' such that

$$\Sigma_{n=1}^{\infty} p^*(f_n) < +\infty \quad \text{and} \quad q(Tx) \leq \Sigma_{n=1}^{\infty} |f_n(x)| \quad \text{for all} \quad x \in X . \tag{4.3}$$

Therefore our definition of quasi-nuclear-bounded maps from one normed space into another coincides with that of quasi-nuclear maps defined by Pietsch [1], hence the example constructed by Pietsch [1, pp. 44 and 60] shows that $N^{(q)}_{\ell^1}(E, F)$ is, in general, a proper vector subspace of $N^{(p)}_{\ell^1}(E, F)$. But we shall show in §4.3 that if X and Y are Hilbert spaces, then $N^{(q)}_{\ell^1}(X, Y) = \pi_{\ell^1}(X, Y)$ and $\|\cdot\|_{(qn)} = \|\cdot\|_{(s)}$ (see (4.3.10)).

As suggested by (4.3), we define naturally

$$\|T\|_{(qn)} = \inf\{\sum_{n=1}^{\infty} p^*(f_n) : \{f_n\} \subset X' \text{ is such that (4.3) holds}\}.$$

It is not hard to show that $\|\cdot\|_{(qn)}$ is a norm on $N^{(q)}_{\ell^1}(X, Y)$, which is called the <u>quasi-nuclear norm</u>, and that

$$\|T\|_{(s)} \leq \|T\|_{(qn)} \leq \|T\|_{(n)} \quad \text{for all } T \in N_{\ell^1}(X, Y) . \tag{4.4}$$

We shall show in §4.3 that $\|\cdot\|_{(qn)}$ is the restriction of $\|\cdot\|_{(s)}$ to $N^{(q)}_{\ell^1}(X, Y)$ (see (4.3.7)). If (Y, q) is complete, then so does $(N^{(q)}_{\ell^1}(X, Y), \|\cdot\|_{(qn)})$ (see the remark after (3.4.6)), and we shall show (see (4.3.8)) that

$$\text{the } \|\cdot\|_{(s)}\text{-closure of } L^f(X, Y) \subset N^{(q)}_{\ell^1}(X, Y) . \tag{4.5}$$

It is not known whether $L^f(X, Y)$ is dense in $(N^{(q)}_{\ell^1}(X, Y), \|\cdot\|_{(qn)})$. But, if either X' or Y has the approximation property, then the answer is affirmative (see (3.4.5)). Before proving this result, we need the following

(3.4.3) Lemma. <u>Let</u> (X, p) <u>and</u> (Y, q) <u>be Banach spaces and</u> $T \in N^{(q)}_{\ell^1}(X, Y)$. <u>Then for any</u> $\delta > 0$ <u>there exist</u> $(\mu_n) \in \ell^1$, <u>closed subspaces</u> M <u>and</u> N <u>of</u> ℓ^∞ <u>and</u> ℓ^1 <u>respectively such that</u> $\sum_{n=1}^{\infty} |\mu_n| < \|T\|_{(qn)} + \delta$ <u>and</u> T <u>permits a factoring</u>

$$X \xrightarrow{S_1} M \xrightarrow{S_2} N \xrightarrow{S_3} Y ,$$

<u>where</u> $S_1 \in L^c(X, M)$ <u>and</u> $S_3 \in L^c(N, Y)$ <u>with norms</u> ≤ 1, <u>and</u> $S_2 \in N^{(q)}_{\ell^1}(M, N)$ <u>satisfies</u>

$$S_2((\eta_n)) = (\mu_n \eta_n)_{n\in \mathbb{N}} \quad \underline{\text{for all}} \quad (\eta_n) \in M \ . \qquad (4.6)$$

<u>Proof</u>. There exist an $(\zeta_n) \in \ell^1$ and an equicontinuous sequence $\{f_n\}$ in X' such that

$$\Sigma_{n=1}^{\infty} |\zeta_n| < \|T\|_{(qn)} + \frac{\delta}{2} \quad \text{and} \quad q(Tx) \leq \Sigma_{n=1}^{\infty} |\zeta_n f_n(x)| \quad \text{for all} \quad x \in X. \qquad (4.7)$$

As $(\zeta_n) \in \ell^1$, we can choose an sequence $\{n_k\}$ of positive numbers with $n_1 < n_2 < \dots$ such that

$$\Sigma_{j=n_k}^{n_{k+1}} |\zeta_j| < 4^{-k}\delta \ .$$

Let us define

$$\zeta_j = \begin{cases} 1 & \text{if} \quad 1 \leq j < n_2 \\ 2^{-k} & \text{if} \quad n_{k+1} \leq j < n_{k+2} \end{cases} \ .$$

Then $\lim_j \zeta_j = 0$ and $\Sigma_{j=1}^{\infty} |\zeta_j| \zeta_j^{-1} \leq \Sigma_{j=1}^{\infty} |\zeta_j| + \frac{\delta}{2}$, thus the sequence (μ_j), defined by

$$\mu_j = |\zeta_j| \zeta_j^{-1} \quad (j \geq 1) \ ,$$

belongs to ℓ^1. Let us define $S_1 : X \to \ell^{\infty}$ by

$$S_1 x = (\zeta_n^{\frac{1}{2}} f_n(x))_{n\geq 1} \quad \text{for all} \quad x \in X \ ,$$

and $S : \ell^{\infty} \to \ell^1$ by

$$S((\eta_n)) = (\mu_n \eta_n)_{n\geq 1} \quad \text{for all} \quad (\eta_n) \in \ell^{\infty} \ .$$

Then $S_1 \in L^c(X, \ell^{\infty})$ and $S \in N_{\rho 1}(\ell^{\infty}, \ell^1)$. Suppose further that

$$M = \overline{S_1(X)} \text{ and } S_2 = S|_M \text{ and } N = \overline{S_2(S_1X)} \subset \ell^1 .$$

Then M and N are closed vector subspaces of ℓ^∞ and ℓ^1 respectively, $S_2 \in N^{(q)}_{\ell^1}(M, N)$ and (4.6) holds. Define $A : \ell^1 \to \ell^1$ by setting

$$A((\alpha_n)) = (\zeta_n^{\frac{1}{2}} \alpha_n) \quad \text{for all} \quad (\alpha_n) \in \ell^1 .$$

Then $A \in L^c(\ell^1)$ with $|||A||| \leq 1$, hence the restriction to N of A, denoted by $A|_N$, is a compact operator from N into $\overline{A\, S_2 S_1(X)}$ with $|||A|_N||| \leq 1$. In view of (4.7), there exists an $B \in L(\overline{A\, S_2 S_1(X)}, Y)$ with $|||B||| \leq 1$ such that

$$T = B \circ (A|_N) \circ S_2 \circ S_1 . \tag{4.8}$$

Clearly, $S_3 = B \circ (A|_N) \in L^c(N, Y)$ with $|||S_3||| \leq 1$, the result then follows from (4.8).

A locally convex space E is said to have the <u>approximation property</u> if the identity map on E belongs to the closure of $L^f(E)$ in $L_p(E)$, where $L_p(E)$ is the locally convex space of $L(E)$ equipped with the topology of precompact convergence.

As each $T \in L^f(E)$ can be represented in the form

$$T(x) = \Sigma_{k=1}^{n} \langle x, f_k \rangle u_k \quad \text{for all} \quad x \in E ,$$

for some $\{f_1, \ldots, f_n\} \subset E'$ and $\{u_1, \ldots, u_n\} \subset E$, it follows that E has the approximation property if and only if for any precompact subset K of E and any absolutely convex o-neighbourhood V in E, there exist finite subsets $\{f_1, \ldots, f_n\}$ of E' and $\{u_1, \ldots, u_n\}$ of E such that

$\Sigma_{k=1}^{n} <x, f_k> u_k - x \in V$ for all $x \in K$.

Each Hilbert space has the approximation property; the Banach space $C(K)$ and its Banach dual, where K is a compact Hausdorff space, have the approximation property. If (Ω, μ) is a finite measure space, then the Banach space $L_{\mu}^{p}(\Omega)$ $(1 \le p \le +\infty)$ has the approximation property. However, in view of a famous example of Enflo [1], there are reflexive, separable Banach spaces which do not have the approximation property.

Grothendieck has shown that a Banach space Y has the approximation property if and only if for any Banach space X, $L^f(X, Y)$ is dense in $(L^c(X, Y), |||\cdot|||)$, and that the Banach dual X' of a Banach space X has the approximation property if and only if for any Banach space Y, $L^f(X, Y)$ is dense in $(L^c(X, Y), |||\cdot|||)$ (for the proofs of these results, see Schaefer [1, p.113]).

In view of (3.4.3) and Grothendieck's results mentioned above, we obtain the following

(3.4.4) Theorem. <u>Let</u> X <u>and</u> Y <u>be Banach spaces</u>. <u>If either</u> X' <u>or</u> Y <u>has the approximation property, then</u> $L^f(X, Y)$ <u>is dense in</u> $(N_{\ell^1}^{(q)}(X, Y), \|\cdot\|_{(qn)})$.

We shall show in §4.3 that under the assumptions of (3.4.4), the completion of $(L^f(X, Y), \|\cdot\|_{(s)})$ is isometrically isomorphic to $(N_{\ell^1}^{(q)}(X, Y), \|\cdot\|_{(qn)})$. Moreover, if X and Y are <u>Hilbert spaces</u>, then $L^f(X, Y)$ is dense in $(\pi_{\ell^1}(X, Y), \|\cdot\|_{(s)})$ (see (4.3.10)).

It is known that $N_{\ell^1}(E, Y) \subset N^{(q)}_{\ell^1}(E, Y)$, where Y is a Banach space. It is naturally to ask under what conditions on E (or Y) , quasi-nuclear-bounded maps are nuclear. To answer this question, we require the following terminology.

A Banach space Y is said to have the extension property if, whenever X is a normed space, N a subspace of X and $T \in L(N, Y)$, then there exists an $\overline{T} \in L(X, Y)$ such that

$$|||T||| = |||\overline{T}||| \quad \text{and} \quad T(x) = \overline{T}(x) \quad \text{for all} \quad x \in N .$$

By making use of the Hahn-Banach extension theorem, it is not hard to show that the Banach space $\ell^{\infty}_{\wedge}$ has the extension property. Kelley [1] and Hasumi [1] have shown that a Banach space Y has the extension property if and only if there exists a compact Hausdorff extremally disconnected space K such that Y and $C(K)$ are isometrically isomorphic.

(34.5) Theorem. Let E be a locally convex space and (Y, q) a Banach space. If Y has the extension property, then $N^{(q)}_{\ell^1}(E, Y) = N_{\ell^1}(E, Y)$.

Proof. Let $T \in N^{(q)}_{\ell^1}(E, Y)$. There exist an $(\zeta_n) \in \ell^1$ and an equicontinuous sequence $\{f_n\}$ in E' such that

$$q(Tx) \leq \Sigma_{n=1}^{\infty} |\zeta_n f_n(x)| \quad \text{for all} \quad x \in E .$$

Let us define $T_1 : E \to \ell^{\infty}$ by

$$T_1 x = (f_n(x))_{n\geq 1} \quad \text{for all} \quad x \in E ,$$

and $S : \ell^{\infty} \to \ell^{1}$ by

$$S((\eta_n)) = (\zeta_n \eta_n)_{n\geq 1} \quad \text{for all} \quad (\eta_n) \in \ell^{\infty} .$$

Then T_1 is continuous, $S \in N_{\ell^1}(\ell^{\infty}, \ell^{1})$ and

$$q(Tx) \leq \|ST_1(x)\|_1 \quad \text{for all} \quad x \in E ,$$

where $\|\cdot\|_1$ is the norm on ℓ^1 , hence there is an $T_3 \in L(ST_1(E), Y)$ such that

$$|||T_3||| \leq 1 \quad \text{and} \quad T_3 \circ S \circ T_1 = T \tag{4.9}$$

because of $\text{Ker}\ (ST_1) \subset \text{Ker}\ T$. Since Y has the extension property and $ST_1(E) \subset \ell^1$, there exists an $\overline{T}_3 \in L(\ell^1, Y)$ such that

$$|||\overline{T}_3||| = |||T_3||| \leq 1 \quad \text{and} \quad \overline{T}|_{ST_1(E)} = T_3 .$$

In view of (4.9), we have $T = \overline{T}_3\, ST$, where $S \in N_{\ell^1}(\ell^{\infty}, \ell^{1})$, therefore $T \in N_{\ell^1}(E, Y)$.

By using the definition of $\|\cdot\|_{(qn)}$, a similar argument given in the proof of (3.4.5) shows that if X is a normed space and if the Banach space Y has the extension property, then

$$N^{(q)}_{\ell^1}(X, Y) = N_{\ell^1}(X, Y) \quad \text{and} \quad \|\cdot\|_{(qn)} = \|\cdot\|_{(n)} .$$

We shall show in §4.3 that if Y is a Banach space and K is a compact Hausdorff space, then

$$N^{(q)}_{\ell^1}(C(K), Y) = N_{\ell^1}(C(K), Y) \quad \text{and} \quad \|\cdot\|_{(qn)} = \|\cdot\|_{(n)} .$$

Let Y be a normed space and let Σ^{o} be the closed unit ball in Y' . We determine an index set Λ so that a map $\iota \mapsto g_{\iota}$ exists from Λ onto Σ^{o} . Then the map j_Y , defined by

$$j_Y(y) = [\langle y, g_{\iota}\rangle, \Lambda] \quad \text{for all} \quad y \in Y ,$$

is an isometry from Y into the Banach space ℓ^{∞}_{Λ} . Therefore, every normed space can be regarded as a subspace of ℓ^{∞}_{Λ} for some index set Λ .

(3.4.6) Corollary (Pietsch). Let E be a locally convex space, let Y be a normed space and $T \in L(E, Y)$. Then $T \in N^{(q)}_{\ell^1}(E, Y)$ if and only if there exist a normed space Z and an isometry j_Y from Y into Z such that $j_Y T \in N_{\ell^1}(E, Z)$.

Proof. The necessity follows from (3.4.5) and the preceding remark by virtue of the fact that the composition of two continuous linear maps, in which one of them is quasi-nuclear-bounded, must be quasi-nuclear-bounded. To prove the sufficiency, we first notice that $j_Y T \in N^{(q)}_{\ell^1}(E, Z)$, hence there exists a quasi-nuclear seminorm p on E such that $\{j_Y(Tx) : p(x) \leq 1\}$ is a bounded subset of Z . As j_Y is an isometry from Y into Z , it follows that $\{Tx : p(x) \leq 1\}$ is a bounded subset of Y , and hence that $T \in N^{(q)}_{\ell^1}(E, Y)$.

It should be noted that if E is a normed space and if $T \in N^{(q)}_{\ell^1}(E, Y)$, then $\|T\|_{(qn)} = \|j_Y T\|_{(n)}$.

From the preceding result, it is easily seen that $N^{(q)}_{\ell^1}(E, Y) \subset L^{p}(E, Y)$, and that if X is a normed space and Y is a Banach space, then

$(N^{(q)}_{\ell^1}(E, Y), \|\cdot\|_{(qn)})$ is complete because $(N^{(q)}_{\ell^1}(X, Y), \|\cdot\|_{(qn)})$ can be regarded as a closed subspace of the Banach space $(N_{\ell^1}(X, \ell^\infty_\Lambda), \|\cdot\|_{(n)})$, where Λ is the closed unit ball in Y'.

(3.4.7) Proposition. Let E and F be infrabarrelled spaces and $T \in L(E, F)$. If $T' \in N_{\ell^1}(F'_\beta, E'_\beta)$, then $T \in N^{(q)}_{\ell^1}(E, F)$. Moreover, if F is a Banach space which has the extension property, then $T \in N_{\ell^1}(E, F)$ if and only if $T' \in N_{\ell^1}(F', E'_\beta)$.

Proof. The infrabarrelledness of E and F ensures that the evaluation maps $e_E : E \to E''_\beta$ and $e_F : F \to F''_\beta$ are (into) topological isomorphisms. As $T' \in N_{\ell^1}(F'_\beta, E'_\beta)$, it follows that $T'' \in N_{\ell^1}(E''_\beta, F''_\beta)$, and hence that

$$e_F T = T'' e_E \in N_{\ell^1}(E, F''_\beta) \subset N^{(q)}_{\ell^1}(E, F''_\beta) .$$

Consequently, there exists a quasi-nuclear seminorm p on E such that $\{e_F(Tx) \in F''_\beta : p(x) \leq 1\}$ is a bounded subset of F''_β, and thus $\{Tx \in F : p(x) \leq 1\}$ is a bounded subset of F because of

$$\{Tx \in F : p(x) \leq 1\} = F \cap \{e_F(Tx) \in F''_\beta : p(x) \leq 1\} .$$

This shows that $T \in N^{(q)}_{\ell^1}(E, F)$.

Clearly, the second assertion is a consequence of the first one and (3.4.5).

(3.4.8) Example. The adjoint map of an absolutely summing (resp.

quasi-nuclear-bounded) map need not be absolutely summing (resp. quasi-nuclear-bounded).

For a fixed $(a_n) = (n^{-\frac{1}{3}}) \in c_o$, we define $T : \ell^1 \to \ell^2$ by the equation

$$T(\Sigma_{n=1}^{\infty} \zeta_n e_n) = \Sigma_{n=1}^{\infty} \zeta_n \, n^{-\frac{1}{3}} e_n \quad \text{for all} \quad (\zeta_n) \in \ell^1,$$

where e_n are unit vectors in ℓ^1. Then we claim that T is quasi-nuclear-bounded. To do this, it suffices to show, in view of (4.5), that there exists a sequence $\{T_n\}$ in $L^f(\ell^1, \ell^2)$ such that $\|T_n - T\|_{(s)} \to 0$ as $n \to \infty$.

In fact, let $S : \ell^1 \to \ell^1$ be defined by

$$S(\Sigma_{n=1}^{\infty} \zeta_n e_n) = \Sigma_{n=1}^{\infty} \zeta_n a_n e_n \quad \text{for all} \quad (\zeta_n) \in \ell^1,$$

and let $S_k : \ell^1 \to \ell^1$ be defined by

$$S_k(\Sigma_{n=1}^{\infty} \zeta_n e_n) = \Sigma_{k=1}^{k} \zeta_n a_n e_n \quad \text{for all} \quad (\zeta_n) \in \ell^1 .$$

Then $|||S_k - S||| \to 0$ as $k \to \infty$, and $T = J S$, where $J : \ell^1 \to \ell^2$ is the canonical embedding map. Clearly, J is absolutely summing, hence

$$\|T - J S_k\|_{(s)} = \|J S - J S_k\|_{(s)} \leq \|J\|_{(s)} |||S - S_k||| \to 0 \quad \text{as} \quad k \to \infty .$$

As $J S_k \in L^f(\ell^1, \ell^2)$, it follows that $T \in N^{(q)}_{\ell^1}(\ell^1, \ell^2)$.

Clearly, the adjoint map $T' : \ell^2 \to \ell^\infty$ of T is given by the equation

$$T'(\Sigma_{n=1}^{\infty} \mu_n e_n) = \Sigma_{n=1}^{\infty} \mu_n \, n^{-\frac{1}{3}} e_n \quad \text{for all} \quad (\mu_n) \in \ell^2 .$$

Moreover, we claim that T' is not absolutely summing, hence it is not quasi-nuclear-bounded.

Let $x^{(m)} = (m^{-\frac{2}{3}}\delta_n^{(m)}) \in \ell^2$ $(m = 1, 2, \ldots)$. Then $[x^{(m)}, \mathbb{N}]$ is a weakly summable sequence in ℓ^2 because of

$$\Sigma_{m=1}^{\infty}|<(\mu_n), x^{(m)}>| = \Sigma_{m=1}^{\infty}|\mu_m m^{-\frac{2}{3}}| \leq (\Sigma_{m=1}^{\infty}|\mu_m|^2)^{\frac{1}{2}}(\Sigma_{m=1}^{\infty} m^{-\frac{4}{3}})^{\frac{1}{2}} < +\infty .$$

But

$$T'x^{(m)} = T'(\Sigma_{n=1}^{\infty} m^{-\frac{2}{3}}\delta_n^{(m)}e_n) = \Sigma_{n=1}^{\infty} m^{-\frac{2}{3}}\delta_n^{(m)} n^{-\frac{1}{3}} e_n = m^{-1}e_m$$

and

$$||T'x^{(m)}||_{\infty} = \frac{1}{m} \quad \text{for all} \quad m \geq 1 ,$$

it then follows that $\Sigma_{m=1}^{\infty}||T'x^{(m)}||_{\infty} = +\infty$, and hence that T' is not absolutely summing.

3.5 Properties of nuclear spaces

In view of (2.2.2) and the definition of nuclear spaces, it is easily seen that the completion of every nuclear space is nuclear, that subspaces of a nuclear space are nuclear and that the product of an arbitrary family of nuclear spaces is nuclear; consequently, the topological projective limit of an arbitrary family of nuclear spaces is nuclear. Furthermore, by using (3.4.1), it is not hard to show that every separated quotient space of a nuclear space by a closed vector subspace is nuclear, and that the locally convex direct sum of a <u>countable family</u> of nuclear spaces is nuclear; consequently, the topological inductive limit of a <u>countable family</u> of nuclear spaces is nuclear. The following example shows that the locally convex direct sum of an uncountable family of nuclear space is, in general, not nuclear.

<u>Example</u> (a). <u>The locally convex direct sum of a non-countable family of nuclear spaces need not be nuclear</u>.

Let Λ be a non-countable set. By a well-known result, we identify the topological dual of the locally convex direct sum $\mathbb{C}^{(\Lambda)}$ with the product space $\mathbb{C}^{\Lambda}$, hence closed equicontinuous subsets of $\mathbb{C}^{\Lambda}$ are compact subsets of $\mathbb{C}^{\Lambda}$. If $\mathbb{C}^{(\Lambda)}$ is nuclear, then these compact subsets should be metrizable by (3.2.5) and (1.2.6)(a). But a compact subset such as $[0, 1]^{\Lambda}$ is metrizable if and only if Λ is countable. This contradiction shows that $\mathbb{C}^{(\Lambda)}$ is not a nuclear space.

(b) <u>The separated quotient space of a complete nuclear space by a closed vector subspace need not be complete</u>.

It is well-known that the locally convex spaces $\mathbb{C}^{\mathbb{N}}$ and $\mathbb{C}^{(\mathbb{N})}$ are nuclear spaces, hence the locally convex spaces

$$E = (\mathbb{C}^{\mathbb{N}})^{(\mathbb{N})} \quad \text{and} \quad F = (\mathbb{C}^{(\mathbb{N})})^{\mathbb{N}}$$

are nuclear; consequently, the locally convex direct sum $E \oplus F$ is a nuclear space too. But it is known from Köthe [1] that there is a closed subspace M of $E \oplus F$ such that the quotient space $(E \oplus F)/_M$ is not complete, therefore $E \oplus F$ is such an example what we want.

For any non-zero $y \in F$, the map $x' \mapsto T_{x'}$, defined by

$$T_{x'}(x) = \langle x, x' \rangle y \quad \text{for all} \quad x \in E ,$$

is a topological isomorphism from the locally convex space E' equipped with the $\mathfrak{Z}$-topology, denoted by $(E', \mathcal{I}(\mathfrak{Z}))$, onto a subspace of $L_{\mathfrak{Z}}(E, F)$. Also for any non-zero $x' \in E'$, the map $y \mapsto S_y$, defined by

$$S_y(x) = \langle x, x' \rangle y \quad \text{for all} \quad x \in E ,$$

is a topological isomorphism from F onto a subspace of $L_{\mathfrak{Z}}(E, F)$. If $L_{\mathfrak{Z}}(E, F)$ is nuclear, then so do $(E', \mathcal{I}(\mathfrak{Z}))$ and F. The converse also holds as shown by the following result, due to Randtke [1].

(3.5.1) Theorem. <u>The locally convex space</u> $L_{\mathfrak{Z}}(E, F)$ <u>is nuclear if and only if both</u> $(E', \mathcal{I}(\mathfrak{Z}))$ <u>and</u> F <u>are nuclear</u>. <u>In particular</u>, $L_\beta(E, F)$ <u>is nuclear if and only if</u> $(E', \beta(E', E))$ <u>and</u> F <u>are nuclear</u>.

<u>Proof</u>. To prove the sufficiency, it is sufficient to show, by (3.4.1), that each gauge $p_{(A,U)}$ of $P(A, U)$ is a quasi-nuclear seminorm on $L(E, F)$. In fact, let p_{A^o} be the gauge of A^o defined on E' and q_U the gauge of U defined on F. Then the assumptions imply that there are (ζ_n), (μ_n) in ℓ^1 and two equicontinuous sequences $\{x''_n\}$ and $\{y'_n\}$ in $(E', \mathcal{I}(\mathfrak{Z}))'$ and F' respectively such that

$$p_{A^o}(u') \leq \Sigma_{n=1}^{\infty} |\zeta_n \langle u', x''_n \rangle| \quad \text{for all} \quad u' \in E' \tag{5.1}$$

and

$$q_U(v) \leqslant \sum_{k=1}^{\infty} |\mu_k \langle v, y_k'\rangle| \quad \text{for all} \quad v \in F . \qquad (5.2)$$

As $T'(y_k') \in E'$ for all $T \in L(E, F)$ and k, it follows that the maps $f_{k,n}$, defined by

$$f_{k,n}(T) = \langle T'(y_k'), x_n''\rangle \quad \text{for all} \quad T \in L(E, F) ,$$

are linear functionals on $L(E, F)$. We claim that $\{f_{k,n} : (k, n) \in \mathbb{N}^2\}$ is an equicontinuous sequence in $(L_{\mathcal{B}}(E, F))'$. To do this, let $W \in \mathcal{U}_F$ and let $B \in \mathcal{B}$ be such that

$$y_k' \in W^{\circ} \quad \text{and} \quad x_n'' \in B^{\circ\circ} \quad \text{for all} \quad k \quad \text{and} \quad n ,$$

where $B^{\circ\circ}$ is the polar of B° taken in $(E', \mathcal{I}(\mathcal{B}))'$. For any $T \in P(B, W)$, we have $T'(W^{\circ}) \subset B^{\circ}$, thus

$$|f_{k,n}(T)| = |\langle T'(y_k'), x_n''\rangle| \leqslant 1 \quad \text{for all} \quad T \in P(B, W) ,$$

which obtains our assertion.

Now for any $T \in L(E, F)$, we have by (5.1) and (5.2) that

$$\begin{aligned} p_{(A,U)}(T) &= \sup\{|\langle Tx, y'\rangle| : x \in A \text{ and } y' \in U^{\circ}\} \\ &= \sup\{q_U(Tx) : x \in A\} \\ &\leqslant \sum_{k=1}^{\infty} |\mu_k| \sup\{|\langle Tx, y_k'\rangle| : x \in A\} \\ &= \sum_{k=1}^{\infty} |\mu_k| p_{A^{\circ}}(T'y_k') \\ &\leqslant \sum_{k=1}^{\infty} |\mu_k| \sum_{n=1}^{\infty} |\zeta_n \langle T'y_k', x_n''\rangle| \\ &= \sum_{k=1}^{\infty} \sum_{n=1}^{\infty} |\mu_k| |\zeta_n| |f_{k,n}(T)| , \end{aligned}$$

thus $p_{(A,U)}$ is a quasi-nuclear seminorm.

A similar argument given in the proof of (1.4.5) yields the following result.

(3.5.2) Proposition. <u>The locally convex space</u> E' <u>is nuclear if and only if for any</u> $B \in \mathcal{B}$ <u>there is an</u> $A \in \mathcal{B}$ <u>with</u> $B \subset A$ <u>such that</u> $j_{A,B} \in N_{\ell^1}(E(B), E(A))$.

If E is an infrabarrelled space, then $\{V^o : V \in \mathcal{U}_E\}$ is a fundamental family of $\beta(E', E)$-bounded subsets, hence an infrabarrelled space E is nuclear if and only if its strong bidual $(E'', \beta(E'', E'))$ is nuclear by (3.2.4) and (3.5.2). A slight generalization of this result is the following

(3.5.3) Corollary. <u>Let</u> E <u>be an</u> σ-<u>infrabarrelled space</u>. <u>If</u> $(E'', \beta(E'', E'))$ <u>is nuclear then so does</u> E .

<u>Proof</u>. For any closed, absolutely convex o-neighbourhood U in E , there is a closed absolutely convex $\beta(E', E)$-bounded subset N of E' with $U^o \subset N$ such that $j_{N,U^o} \in N_{\ell^1}(E'(U^o), E'(N))$ by (3.5.2), hence we have

$$x' = \sum_{n=1}^{\infty} \langle x', \psi_n \rangle u'_n \quad \text{in } E'(N) \quad (\text{for any } x' \in E'(U^o)) ,$$

where we can assume that $u'_n \in N$ and $\psi_n \in (E'(U^o))'$ satisfy

$$\sum_{n=1}^{\infty} (p_{U^o})^*(\psi_n) < +\infty .$$

Since the countable set $\{u'_n\}$ is $\beta(E', E)$-bounded and since E is σ-infrabarrelled, it follows that $\{u'_n\}$ is equicontinuous, and hence that there is a closed, absolutely convex o-neighbourhood V in E such that $u'_n \in V^o$ for all n. Since

$$\sum_{n=1}^{\infty} (p_{U^o})^*(\psi_n) p_{V^o}(u'_n) < +\infty \quad \text{and} \quad x' = \sum_{n=1}^{\infty} \langle x', \psi_n \rangle u'_n \quad \text{for all} \quad x' \in E'(U^o),$$

it follows that the canonical embedding map $E'(U^o) \to E'(V^o)$ is nuclear, and hence from (3.2.4) that E is a nuclear space.

A locally convex space E is called a <u>co-nuclear</u> (or <u>dual nuclear</u>) <u>space</u> if its strong dual $(E', \beta(E', E))$ is a nuclear space.

It is known that the product space $\mathbb{C}^{\Lambda}$ is nuclear, where Λ is an infinitely non-countable index set, and that the locally convex direct sum $\mathbb{C}^{(\Lambda)}$ is not nuclear, hence $\mathbb{C}^{\Lambda}$ is a nuclear space which is not a co-nuclear space. But for metrizable case the situation never appears as shown by Grothendieck, namely, a metrizable locally convex space is nuclear if and only if it is a co-nuclear space (see (4.6.9)). As the strong dual of a (DF)-space (resp. Fréchet space) is a Fréchet space (resp. (DF)-space), and every (DF)-space, which is either nuclear or co-nuclear, is infrabarrelled (see (1.4.10) since co-nuclear spaces are co-Schwartz), it follows that a (DF)-space is nuclear if and only if it is a co-nuclear space. In view of this remark, we are able to give an example showing that Schwartz spaces are, in general, not nuclear as follows.

(3.5.4) Examples. (a) <u>Schwartz spaces need not be nuclear spaces</u>.
(b) <u>Co-Schwartz spaces need not be co-nuclear</u>.

<u>Proof</u>. It is known from Köthe [1, §31.5] that there exists a

Fréchet-Montel space G in which there is a closed subspace N such that $G/_N$ is <u>not</u> a Montel space. In view of (1.4.13), G is a Fréchet co-Schwartz space which is <u>not</u> a Schwartz space, hence G'_β is a Schwartz space which is not nuclear since every nuclear spaces is a Schwartz space. Also G is a co-Schwartz space which is not co-nuclear.

Since co-nuclear spaces are co-Schwartz spaces, all conclusions of (1.4.8) hold for co-nuclear spaces. Furthermore, we have the following result.

(3.5.5) Proposition. <u>For a co-nuclear space</u> $(E, \mathcal{P})$, <u>the following assertions hold.</u>

(a) <u>For any bounded subset</u> B <u>of</u> E <u>there exists a closed, absolutely convex bounded subset</u> A <u>of</u> E <u>with</u> $B \subset A$ <u>such that</u> B <u>is precompact in the normed space</u> $E(A)$. <u>In particular, bounded subsets of</u> E <u>are precompact and separable</u>.

(b) <u>For any</u> $V \in \mathcal{U}_E$ <u>and absolutely convex bounded subset</u> B <u>of</u> E , <u>the canonical map</u> $K_{V,B} : E(B) \to E_V$ <u>is nuclear</u>.

<u>Proof</u>. We first notice from (3.5.2) that there exists a closed, absolutely convex bounded subset A of E with $B \subset A$ such that $j_{A,B} : E(B) \to E(A)$ is nuclear, hence part (a) holds, and

$$K_{V,B} = K_{V,A}\, j_{A,B} \ .$$

Therefore $K_{V,B}$ is nuclear.

To prove that every co-nuclear space has the approximation property, we require the following result.

(3.5.6) Lemma. Let $(E, \mathcal{P})$ be a locally convex space, and let G be a vector subspace of E' which separates points of E. Then for any $\sigma(E, G)$-closed, absolutely convex $\mathcal{P}$-neighbourhood V of 0 in E, $G \cap V^o$ is dense in V^o for the topology $\mathcal{P}^o$ of uniform convergence on $\mathcal{P}$-precompact subsets of E, where V^o is the polar of V taken in E'.

Proof. Since V^o is an $\mathcal{P}$-equicontinuous subset of E', it suffices to show that $G \cap V^o$ is $\sigma(E', E)$-dense in V^o or, equivalently, $V^o = (G \cap V^o)^{oo}$. Indeed, the identity map $I : (E, \mathcal{P}) \to (E, \sigma(E, G))$ is continuous, it follows that the adjoint map $I' : G \to E'$ is injective, and hence that

$$V^{\pi} = V^o \cap G = (I')^{-1}(V^o) = (I(V))^{\pi},$$

where V^{π} is the polar of V taken in G. As V is $\sigma(E, G)$-closed, the bipolar theorem ensures that $V = V^{\pi o} = V^{\pi\pi}$, hence

$$V^o = (V^{\pi o})^o = (V^o \cap G)^{oo},$$

as asserted.

(3.5.7) Proposition (Terzioğlu [1]). A locally convex space $(E, \mathcal{P})$ has the approximation property if it satisfies the following two conditions:

(i) for any $\mathcal{P}$-precompact subset C of E, there exists a closed, absolutely convex, bounded subset B of E with $C \subset B$ such that C is a precompact subset of the normed space $E(B)$;

(ii) <u>for any closed, absolutely convex, bounded subset</u> B <u>of</u> E <u>and any absolutely convex</u> o-<u>neighbourhood</u> V <u>in</u> E, <u>the canonical map</u> $K_{V,B} : E(B) \to E_V$ <u>is the limit of a sequence of continuous linear maps with finite rank for the operator norm</u> $|||\cdot|||$.

<u>In particular, every co-nuclear space has the approximation property</u>.

<u>Proof</u>. Let C be a $\mathcal{P}$-precompact subset of E and V an absolutely convex o-neighbourhood in E. In view of the definition of the approximation property, it is required to show that there exist finite subsets $\{f_1, \ldots, f_n\}$ and $\{u_1, \ldots, u_n\}$ of E' and E respectively such that

$$p_V(\Sigma_{i=1}^n \langle x, f_i \rangle u_i - x) \leq 1 \quad \text{for all} \quad x \in C, \tag{5.3}$$

where p_V is the gauge of V.

In fact, by (i) there exists a closed, absolutely convex, bounded subset B of E with $C \subset B$ such that C is a precompact subset of $E(B)$. Since $K_{V,B} = Q_V j_B$, it follows from (ii) that there exist finite subsets $\{h_1, \ldots, h_n\}$ and $\{u_1, \ldots, u_n\}$ of the Banach dual $E(B)'$ of $E(B)$ and E respectively such that

$$\Sigma_{i=1}^n \langle x, h_i \rangle u_i - x \in \tfrac{1}{2} V \quad \text{for all} \quad x \in B. \tag{5.4}$$

Without loss of generality one can assume that $\|h_i\| \leq 1$ and $p_V(u_i) \neq 0$ $(i = 1, 2, \ldots, n)$. Let $\mathcal{P}|_{E(B)}$ be the relative topology on $E(B)$ induced by $\mathcal{P}$ and $G = (E(B), \mathcal{P}|_{E(B)})'$. Then G is a vector subspace of $E(B)'$ which separates points of $E(B)$ and B is $\sigma(E(B), G)$-closed, absolutely convex o-neighbourhood in the normed space $E(B)$. By (3.5.6), $G \cap B^{o}$ is dense in B^{o} for the topology on $E(B)'$ of uniform convergence on precompact subsets of the normed space $E(B)$, where B^{o} is the polar of B taken in $E(B)'$. Since $h_i \in B^{o}$, there exists $g_i \in G \cap B^{o}$ such that

$$|\langle x, h_i\rangle - \langle x, g_i\rangle| \leq (2n\, p_V(u_i))^{-1} \quad \text{for all } x \in C \text{ and } i = 1, \ldots, n . \qquad (5.5)$$

By the Hahn-Banach extension theorem, each g_i can be extended to a $\mathcal{P}$-continuous linear functional f_i on E . For any $x \in C$, we have by (5.4) and (5.5) that

$$p_V(\Sigma_{i=1}^{n}\langle x, f_i\rangle u_i - x) \leq p_V(\Sigma_{i=1}^{n}\langle x, h_i\rangle u_i - x) + p_V(\Sigma_{i=1}^{n}(\langle x, f_i\rangle - \langle x, h_i\rangle)u_i)$$
$$\leq \tfrac{1}{2} + \Sigma_{i=1}^{n}|\langle x, f_i\rangle - \langle x, h_i\rangle| p_V(u_i) \leq 1 ,$$

which obtains our required inequality (5.3).

Finally, if E is a co-nuclear space, then E satisfies conditions (i) and (ii) by Proposition (3.5.5) and (3.1.1).

It can be shown (see Schaefer [1, p.109]) that if E has a neighbourhood basis $\mathcal{U}_E$ at 0 consisting of absolutely convex sets such that the Banach space $\widetilde{E}_V$ has the approximation property for every $V \in \mathcal{U}_E$, then E has the approximation property. In view of this result and (3.2.4) and (3.2.2), each nuclear space has the approximation property because each Hilbert space has the approximation property.

We conclude this section with another characterization of the nuclearity in terms of the abstract kernel theorem.

Following Pietsch [1], an $\psi \in B(E, F)$ is called a <u>nuclear bilinear form</u> if there exist an $(\zeta_n) \in \ell^1$ and two equicontinuous sequences $\{x'_n\}$ and $\{y'_n\}$ in E' and F' respectively such that

$$\psi(x, y) = \sum_{n=1}^{\infty} \zeta_n \langle x, x'_n\rangle\langle y, y'_n\rangle \quad \text{for all} \quad (x, y) \in E \times F .$$

Denote by $B^n(E, F)$ the set consisting of all nuclear bilinear forms on $E \times F$. If $T \in L^{eq}(E, F(\mathcal{J})')$ is such that $b_T \in B^n(E, F)$, then

$$b_T(x, y) = \sum_{n=1}^{\infty} \zeta_n \langle x, x'_n\rangle\langle y, y'_n\rangle \quad \text{for all} \quad (x, y) \in E \times F , \tag{5.6}$$

where $(\zeta_n) \in \ell^1$, $\{x'_n\} \subset V_2^o$ and $\{y'_n\} \subset U_2^o$ for some $V_2 \in \mathcal{U}_E$ and $U_2 \in \mathcal{U}_F$. As $T \in L^{eq}(E, F(\mathcal{J})')$, there exist $V_1 \in \mathcal{U}_E$ and $U_1 \in \mathcal{U}_F$ such that $T(V_1) \subset U_1^o$, hence $V = V_1 \cap V_2$ and $U = U_1 \cap U_2$ are o-neighbourhoods in E and F respectively such that

$$T(V) \subset U^o, \ x'_n \in V^o \ \text{and} \ y'_n \in U^o \ \text{for all} \ n .$$

Since U^o is $\sigma(F', F)$-compact, U^o must be an infracomplete subset of F'_β. Furthermore, we claim that the induced map $T_{(V,U^o)}$ of T is a nuclear map from E_V into $F'(U^o)$.

In fact, since $(Q_V)'$ is an isometry from $(E_V)'$ onto $E'(V^o)$, there exists an equicontinuous sequence $\{f_n\}$ in $(E_V)'$ such that $f_n \circ Q_V = x'_n$ for all n. By (5.6), we have

$$\begin{aligned} &\lim_n q_{U^o}(T_{(V,U^o)}(Q_V(x)) - \sum_{k=1}^{n} \zeta_k \langle Q_V(x), f_k\rangle y'_k) \\ &= \lim_n q_{U^o}(Tx - \sum_{k=1}^{n} \zeta_k \langle x, x'_k\rangle y'_k) \\ &= \lim_n \sup\{|\langle y, Tx\rangle - \sum_{k=1}^{n} \zeta_k \langle x, x'_k\rangle\langle y, y'_k\rangle| \ : y \in U\} = 0 \quad \text{for all} \quad x \in E , \end{aligned}$$

as asserted. Therefore $T \in N_{\ell^1}(E, F'_\beta)$.

(3.5.8) Lemma. <u>Let</u> $T \in L^{eq}(E, F(\mathfrak{J})')$. <u>If</u> $b_T \in B^n(E, F)$, <u>then</u> $T \in N_{\ell^1}(E, F'_\beta)$. <u>Conversely, if</u> $(F, \mathfrak{J})$ <u>is</u> σ-<u>infrabarrelled and</u> $T \in N_{\ell^1}(E, F'_\beta)$, <u>then</u> $b_T \in B^n(E, F)$.

<u>Proof</u>. We have only to prove the second part. There exist an $(\zeta_n) \in \ell^1$, an equicontinuous sequence $\{x'_n\}$ in E' and a sequence $\{y'_n\}$ contained in some infracomplete subset B of F'_β such that

$$Tx = \sum_{n=1}^{\infty} \zeta_n \langle x, x'_n\rangle y'_n \quad \text{for all} \quad x \in E .$$

Clearly, each $y \in F$ can be regarded as a continuous linear functional on $F'(B)$, hence

$$b_T(x, y) = \langle y, Tx\rangle = \sum_{n=1}^{\infty} \zeta_n \langle x, x'_n\rangle\langle y, y'_n\rangle \quad \text{for all} \quad (x, y) \in E \times F .$$

Now the σ-infrabarrelledness of F ensures that $\{y'_n\}$ is equicontinuous, hence $b_T \in B^n(E, F)$.

(3.5.9) Theorem (Abstract kernel theorem). <u>The following statements are equivalent</u>.

(a) E <u>is nuclear</u>.

(b) $B^n(E, F) = B(E, F)$ <u>for any locally convex space</u> F .

(c) $L^{eq}(E, F(\mathfrak{J})') \subset N_{\ell^1}(E, F'_\beta)$ <u>for any locally convex space</u> F .

(d) $L^{eq}(E, F(\mathfrak{J})') \subset N^{(p)}_{\ell^1}(E, F'_\beta)$ <u>for any locally convex space</u> F.

(e) <u>For any locally convex space</u> F <u>and any</u> $\psi \in B(E, F)$, <u>there</u>

exist a prenuclear seminorm p on E and a continuous seminorm q on F such that

$$|\psi(x, y)| \leq p(x)q(y) \quad \text{for all} \quad x \in E \quad \text{and} \quad y \in F .$$

In particular, if E is a nuclear space, then for any $\psi \in B(E, F)$ there exists a unique $T \in N_{\ell^1}(E, F'_\beta)$ such that

$$\psi(x, y) = \langle y, Tx\rangle \quad \text{for all} \quad (x, y) \in E \times F . \tag{5.7}$$

Proof. The implication (b) ⇒ (c) and (5.7) follow from (3.5.8), and the implications (c) ⇒ (d) and (a) ⇒ (e) are obvious. Therefore we complete the proof by showing the implications (e) ⇒ (d) ⇒ (a) ⇒ (b).

(e) ⇒ (d) : Let $T \in L^{eq}(E, F(\mathfrak{J})')$. Then $b_T \in B(E, F)$, hence

$$|\langle y, Tx\rangle| = |b_T(x, y)| \leq p(x)q(y) , \tag{5.8}$$

where p is a prenuclear seminorm on E and q is a continuous seminorm on F. Let $U = \{y \in F : q(y) \leq 1\}$ and $N = \{Tx \in F : p(x) \leq 1\}$. Then U is a o-neighbourhood in F, hence for any absolutely convex bounded subset B of F, there is an $\zeta > 0$ such that $B \subset \zeta U$. It then follows from (5.8) that $N \subset \zeta B^{o}$, and hence that $T \in N^{(p)}_{\ell^1}(E, F'_\beta)$.

(d) ⇒ (a) : In view of (3.2.4)(h), it suffices to show that $L(E, Y) \subseteq \pi_{\ell^1}(E, Y)$ for any normed space Y. To this end, let $T \in L(E, Y)$ and let $e_Y : Y \to Y''$ be the evaluation map. Then $S = e_Y T \in L(E, Y'') = L^{eq}(E, (Y'_\beta)')$, hence there exist a prenuclear seminorm p on E and $\zeta > 0$ such that

$$\{Sx \in Y'' : p(x) \leq 1\} \subset \zeta\,\Sigma^{oo} ,$$

where Σ is the closed unit ball in Y. Since $T(E) \subset Y$ and e_Y is an isometry, it follows that $\{Tx \in Y : p(x) \leq 1\} \subset \zeta\Sigma$, and hence that $T \in N^{(p)}_{\ell^1}(E, Y) = \pi_{\ell^1}(E, Y)$ because Y is a normed space.

(a) $\Rightarrow$ (b) : Let $\psi \in B(E, F)$. Then there exist $V \in \mathcal{U}_E$ and $U \in \mathcal{U}_F$ such that

$$|\psi(x, y)| \leq p_V(x)q_U(y) \text{ for all } (x, y) \in E \times F . \tag{5.9}$$

The nuclearity of E ensures that there exists an $W \in \mathcal{U}_E$ with $W \subset V$ such that $Q_{V,W} \in N_{\ell^1}(E_W, E_V)$, hence there exist two sequences $\{h_n\}$ and $\{x_n\}$ in $(E_W)'$ and E respectively such that

$$\Sigma_{n=1}^{\infty} (\hat{p}_W)^*(h_n)\hat{p}_V(Q_V(x_n)) < +\infty \text{ and } Q_{V,W}(Q_W(x)) = \Sigma_{n=1}^{\infty} \langle Q_W(x), h_n\rangle Q_V(x) \quad (x \in E). \tag{5.10}$$

As $(Q_W)'$ is an isometry from $(E_W)'$ onto $E'(W^o)$, we have

$$u_n' = (Q_W)'(h_n) \in E'(W^o) \text{ and } p_{W^o}(u_n') = (\hat{p}_W)^*(h_n) \text{ for all } n ,$$

hence we obtain from (5.10) that $\Sigma_{n=1}^{\infty} p_{W^o}(u_n')p_V(x_n) < +\infty$ and

$$\begin{aligned} &\lim_n \hat{p}_V(Q_{V,W}(Q_W(x)) - \Sigma_{k=1}^{n} \langle Q_W(x), h_k\rangle Q_V(x_k)) \\ &= \lim_n \hat{p}_V(Q_V(x - \Sigma_{k=1}^{n} \langle x, u_k'\rangle x_k) \\ &= \lim_n p_V(x - \Sigma_{k=1}^{n} \langle x, u_k'\rangle) = 0 \text{ for all } x \in E . \end{aligned}$$

It then follows from (5.9) that

$$\psi(x, y) = \Sigma_{k=1}^{\infty} \langle x, u_k' \rangle \psi(x_k, y) \quad \text{for all} \quad (x, y) \in E \times F . \tag{5.11}$$

On the other hand, the continuity of $\psi^{(x_k)}$ on F ensures that there is $\nu_k' \in F'$ such that

$$\psi(x_k, y) = \langle y, \nu_k' \rangle \quad \text{for all} \quad y \in F .$$

Replace $\psi(x_k, y)$ by $\langle y, \nu_k' \rangle$ in (5.11), we obtain

$$\psi(x, y) = \Sigma_{k=1}^{\infty} \langle x, u_k' \rangle \langle y, \nu_k' \rangle \quad \text{for all} \quad (x, y) \in E \times F . \tag{5.12}$$

Since

$$|\langle y, \nu_k' \rangle| = |\psi(x_k, y)| \leq p_V(x_k) q(y) \quad ,$$

it follows that

$$q_{U^o}(\nu_k') = \sup\{|\langle y, \nu_k' \rangle| \ : \ y \in U\} \leq p_V(x_k) \quad ,$$

and hence that

$$\Sigma_{n=1}^{\infty} p_{W^o}(u_n') q_{U^o}(\nu_n') \leq \Sigma_{n=1}^{\infty} p_{W^o}(u_n') p_V(x_n) < +\infty . \tag{5.13}$$

Now formulae (5.12) and (5.13) show that $\psi \in B^n(E, F)$.

3.6 Nuclearity of Köthe spaces

By a Köthe set we mean a set P consisting of sequences $a = [a_n]$ of positive numbers satisfying the following two conditions:

(i) for any $m \geq 1$ there exists an $a = [a_n] \in P$ such that $a_m > 0$,

(ii) for any finite elements $a^{(k)} = [a_n^{(k)}] \in P$ $(k = 1, 2, \ldots, m)$ there exists an $a = [a_n] \in P$ such that

$$\max\{a_n^{(k)} : 1 \leq k \leq m\} \leq a_n \quad \text{for all} \quad n \geq 1 .$$

For any Köthe set P, the set defined by

$$\lambda(P) = \{x = [x_n] \in \mathbb{C}^{\mathbb{N}} : p_a(x) = \sum_{n=1}^{\infty} a_n |x_n| < \infty \quad \text{for all} \quad a = [a_n] \in P\}$$

is clearly a vector subspace of $\mathbb{C}^{\mathbb{N}}$, which is called the Köthe space (or the Köthe space determined by P).

Clearly each Köthe space $\lambda(P)$ is a normal sequence space containing $\mathbb{K}^{(\mathbb{N})}$, thus we might consider its α-dual $\lambda(P)^{\times}$. To this end, we first notice that $e^{(j)} = [\delta_n^{(j)}] \in \lambda(P)$, where

$$\delta_n^{(n)} = 1 \quad \text{and} \quad \delta_n^{(j)} = 0 \quad \text{for} \quad j \neq n .$$

It then follows that

$$K_P = \{\zeta = [\zeta_n] \in \lambda(P) : \zeta_n \geq 0 \quad \text{for all} \quad n \geq 1\}$$

is a Köthe set. Moreover, the Köthe space determined by K_P is the α-dual of $\lambda(P)$ as shown by the following result.

(3.6.1) Proposition. <u>Let</u> P <u>be a Köthe set and</u>

$$K_P = \{\zeta = [\zeta_n] \in \lambda(P) : \zeta_n \geq 0 \ \underline{\text{for all}} \ n \geq 1\} . \qquad (6.1)$$

<u>Then</u> $\lambda(P)^\times = \lambda(K_P)$.

Proof. Since $K_P \subset \lambda(P)$ and $|\zeta| = [|\zeta_n|] \in K_P$ for any $\zeta = [\zeta_n] \in \lambda(P)$, it follows that $\lambda(P)^\times \subseteq \lambda(K_P)$. Conversely, let $v = [v_n] \in \lambda(K_P)$. For any $\zeta = [\zeta_n] \in \lambda(P)$, we have $|\zeta| \in K_P$, hence

$$\sum_{n=1}^{\infty} |v_n \zeta_n| < +\infty$$

which shows that $v \in \lambda(P)^\times$ by the definition of α-duals.

Let P be a Köthe set. For any $a = [a_n] \in P$,

$$p_a(x) = \sum_{n=1}^{\infty} a_n |x_n| \quad \text{for all} \quad x = [x_n] \in \lambda(P)$$

is clearly a regular seminorm on $\lambda(P)$. The family $\{p_a : a \in P\}$ determines a Hausdorff locally convex topology on $\lambda(P)$, which is denoted by $\mathcal{J}_P$ and is called the <u>natural topology determined by</u> P .

In view of (3.6.1) and the definition of the Köthe topology, it is clear that $\sigma_K(\lambda(P)^\times, \lambda(P))$ coincides with the natural topology $\mathcal{J}_{K_P}$ on $\lambda(K_P)$ determined by K_P .

It is not hard to show that $(\lambda(P), \mathcal{J}_P)$ is complete. Furthermore, $\mathcal{J}_P$ is coarser than $\sigma_K(\lambda(P), \lambda(P)^\times)$ as a part of the following result shows.

(3.6.2) Proposition. For a Köthe set P, the following assertions hold:

(a) The topological dual of $(\lambda(P), \mathcal{T}_P)$, denoted by $\lambda(P)'$, is isomorphic to the vector subspace of $\mathbb{C}^{\mathbb{N}}$ consisting of sequences $\eta = [\eta_n]$ with the following property

$$|\eta_n| \leq \beta a_n \quad \text{for all} \quad n \geq 1 \tag{6.2}$$

for some $\beta \geq 0$ and $a = [a_n] \in P$. Consequently, $\lambda(P)'$ is the normal hull in $\mathbb{C}^{\mathbb{N}}$ of P.

(b) $\lambda(P)' \subset \lambda(P)^{\times}$ and $\lambda(P)'$ is $\sigma_K(\lambda(P)^{\times}, \lambda(P))$-dense in $\lambda(P)^{\times}$.

(c) The relative topology $\mathcal{T}_{K_P}|_{\lambda(P)'}$ on $\lambda(P)'$ induced by $\mathcal{T}_{K_P}$ is coarser than $\beta(\lambda(P)', \lambda(P))$.

Proof. (a) If $\eta = [\eta_n] \in \mathbb{C}^{\mathbb{N}}$ satisfies (6.2), then η defines a $\mathcal{T}_P$-continuous linear functional f on $\lambda(P)$ by the following equation

$$f(x) = \Sigma_{n=1}^{\infty} \eta_n x_n \quad \text{for all} \quad x = [x_n] \in \lambda(P). \tag{6.3}$$

Conversely, if $f \in \lambda(P)'$ then there exist an $\beta \geq 0$ and $a = [a_n] \in P$ such that

$$|f(x)| \leq \beta\, p_a(x) = \beta\, \Sigma_{n=1}^{\infty} a_n |x_n| \quad \text{for all} \quad x = [x_n] \in \lambda(P). \tag{6.4}$$

Since $e^{(j)} = [\delta_n^{(j)}] \in \lambda(P)$, it follows from (6.3) that

$$|\eta_j| = |f(e^{(j)})| \leq \beta a_j \quad \text{for all} \quad j \geq 1.$$

For any $x = [x_n] \in \lambda(P)$, the sequence $\{\Sigma_{j=1}^{n} x_j e^{(j)}, n \geq 1\}$, in $\lambda(P)$, is $\mathcal{T}_P$-convergent to x, thus the continuity of f ensures that

$$f(x) = \lim_{n\to\infty} f(\Sigma_{j=1}^{n} x_j e^{(j)}) = \Sigma_{j=1}^{\infty} x_j \eta_j .$$

(b) In view of part (a) and the definition of α-dual, we have

$$\lambda(P)' \subset \lambda(P)^{\times} .$$

To prove the density of $\lambda(P)'$ in $(\lambda(P)^{\times}, \sigma_K(\lambda(P)^{\times}, \lambda(P)))$, let $v = [v_n] \in \lambda(P)^{\times}$. For any $\zeta = [\zeta_n] \in \lambda(P)$ and $\delta > 0$, there exists an integer $N = N(\delta) \geq 1$ such that

$$\Sigma_{j=1}^{p} |\zeta_{n+j} v_{n+j}| < \delta \quad \text{for all } n \geq N \text{ and } p = 1, 2, \ldots . \tag{6.4}$$

Since P is a Köthe set, for any $m \geq 1$ there exists an $a = [a_n] \in P$ such that $a_m > 0$, hence $e^{(m)} = [\delta_n^{(m)}] \in \lambda(P)'$ by part (a) because

$$\delta_n^m \leq a_m^{-1} a_n \quad \text{for all } n \geq 1 .$$

Therefore $\Sigma_{j=1}^{m} v_j e^{(j)} \in \lambda(P)'$ for all $m \geq 1$. As the seminorm

$$q_\zeta(y) = \Sigma_{n=1}^{\infty} |\zeta_n y_n| \qquad \text{for all } y = [y_n] \in \lambda(P)^{\times}$$

is a $\sigma_K(\lambda(P)^{\times}, \lambda(P))$-continuous seminorm on $\lambda(P)^{\times}$, it follows from (6.4) that

$$q_\zeta(v - \Sigma_{j=1}^{n} v_j e^{(n)}) < \delta \quad \text{for all } n \geq N ,$$

and hence that $\lambda(P)'$ is $\sigma_K(\lambda(P)^{\times}, \lambda(P))$-dense in $\lambda(P)^{\times}$.

(c) We first notice that

$$\mathcal{T}_{K_P}|_{\lambda(P)'} = \sigma_K|\lambda(P)^\times, \lambda(P))|_{\lambda(P)'} .$$

For any $\zeta = [\zeta_n] \in \lambda(P)$, the singleton $\{|\zeta|\}$ is $\sigma(\lambda(P), \lambda(P)')$-bounded, hence the conclusion follows.

It is known from (3.6.2) that the natural topology $\mathcal{T}_P$ on $\lambda(P)$ is coarser than the Köthe topology $\sigma_K(\lambda(P), \lambda(P)^\times)$. It is natural to ask under what condition $\mathcal{T}_P$ coincides with $\sigma_K(\lambda(P), \lambda(P)^\times)$. They are equal whenever $(\lambda(P), \mathcal{T}_P)$ is barrelled as shown by the following result.

(3.6.3) Proposition. <u>Let</u> P <u>be a Köthe set and let</u> $\mathcal{T}_P$ <u>be the natural topology on</u> $\lambda(P)$ <u>determined by</u> P . <u>If</u> $(\lambda(P), \mathcal{T}_P)$ <u>is barrelled</u>, <u>then</u>

$$\lambda(P)' = \lambda(P)^\times \quad \underline{\text{and}} \quad \mathcal{T}_P = \sigma_K(\lambda(P), \lambda(P)^\times) .$$

<u>Proof</u>. For any $v = [v_n] \in \lambda(P)^\times$, (3.6.2) shows that there exists a sequence $\{f^{(n)}, n \geq 1\}$ in $\lambda(P)'$ which is $\sigma_K(\lambda(P)^\times, \lambda(P))$-convergent to v . The barrelledness of $(\lambda(P), \mathcal{T}_P)$ implies that $(\lambda(P)', \sigma(\lambda(P)', \lambda(P)))$ is quasi-complete, and surely sequentlally complete, thus $\{f^{(n)}, n \geq 1\}$ is $\sigma(\lambda(P)', \lambda(P))$-convergent to some $g \in \lambda(P)'$. Therefore $g = f$ and thus $f \in \lambda(P)'$.

In view of (2.3.12)(g), $\sigma_K(\lambda(P), \lambda(P)^\times)$ is the coarsest normal topology on $\lambda(P)$ which is consistent with $\langle\lambda(P), \lambda(P)^\times\rangle$. As $\lambda(P)' = \lambda(P)^\times$ and $\mathcal{T}_P \leq \sigma_K(\lambda(P), \lambda(P)^\times)$, it follows that $\mathcal{T}_P = \sigma_K(\lambda(P), \lambda(P)^\times)$.

The remainder of this section is devoted to a study of the nuclearity of Köthe spaces equipped with the natural topology.

A Köthe set P is said to satisfy the <u>Grothendieck-Pietsch condition</u> if for any $a = [a_n] \in P$ there exists an $b = [b_n] \in P$ such that

$$\left[\frac{a_n}{b_n} \right] \in \ell^1$$

here we use the convention that $\frac{a_n}{b_n} = 0$ if $b_n = 0$.

It is clear that a Köthe set P satisfies the Grothendieck-Pietsch condition if and only if for any $a = [a_n] \in P$ there exist $b = [b_n] \in P$ and $\eta = [\eta_n] \in \ell^1$ such that

$$a_n \leq \eta_n b_n \quad \text{for all} \quad n \geq 1 .$$

(3.6.4) Theorem (Grothendieck and Pietsch). <u>Let</u> P <u>be a Köthe set and let</u> $\mathcal{J}_P$ <u>be the natural topology determined by</u> P. <u>Then</u> $(\lambda(P), \mathcal{J}_P)$ <u>is nuclear if and only if</u> P <u>satisfies the Grothendieck-Pietsch condition</u>.

<u>Proof</u>. The sufficiency is clear by making use of (3.4.1). We employ (3.2.4)(j) to verify the necessity. Suppose that $(\lambda(P), \mathcal{J}_P)$ is nuclear and that $[a_n] \in P$. Then the map T, defined by

$$T([x_n]) = (a_n x_n)_{n\geq 1} \quad \text{for all} \quad [x_n] \in \lambda(P) \tag{6.5}$$

is clearly a continuous linear map from $(\lambda(P), \mathcal{J}_P)$ into ℓ^1, hence nuclear. There exist an equicontinuous sequence $\{f_k : k \geq 1\}$ in $\lambda(P)'$

and a sequence $\{y^{(k)}\}$ in ℓ^1 such that

$$\Sigma_{k=1}^{\infty} \|y^{(k)}\|_1 < +\infty \quad \text{and} \quad T(x) = \Sigma_{k=1}^{\infty} \langle x, f_k \rangle y^{(k)} \quad \text{for all} \quad x = [x_n] \in \lambda(P), \tag{6.6}$$

where $\|\cdot\|_1$ is the usual norm on ℓ^1. As $e^{(j)} = [\delta_n^{(j)}] \in \lambda(P)$, it follows from (6.5) and (6.6) that

$$[a_n \delta_n^{(j)}] = T(e^{(j)}) = \Sigma_{k=1}^{\infty} \langle e^{(j)}, f_k \rangle y^{(k)} .$$

In particular, we have by taking the j-th coordinate that

$$a_j = \Sigma_{k=1}^{\infty} \langle e^{(j)}, f_k \rangle y_j^{(k)} . \tag{6.7}$$

On the other hand, the equicontinuity of $\{f_k\}$ ensures that there exists an $b = [b_n] \in P$ such that

$$|\langle x, f_k \rangle| \leq p_b(x) = \Sigma_{n=1}^{\infty} b_n |x_n| \quad \text{for all} \quad k \geq 1 \quad \text{and} \quad x = [x_n] \in \lambda(P).$$

It then follows from (6.7) that

$$a_j \leq \Sigma_{k=1}^{\infty} |\langle e^{(j)}, f_k \rangle| \, |y_j^{(k)}| \leq b_j \Sigma_{k=1}^{\infty} |y_j^{(k)}| \quad \text{for all} \quad j \geq 1 .$$

For each $j \geq 1$, we let

$$\eta_j = \Sigma_{k=1}^{\infty} |y_j^{(k)}| .$$

Then $a_j \leq \eta_j b_j$, and

$$\Sigma_{j=1}^{\infty} \eta_j = \Sigma_{j=1}^{\infty} \Sigma_{k=1}^{\infty} |y_j^{(k)}| = \Sigma_{k=1}^{\infty} \|y^{(k)}\|_1 < +\infty ,$$

which shows that $(\eta_j) \in \ell^1$. Therefore P satisfies the Grothendieck-Pietsch condition.

The set P, defined by

$$P = \{[n^k] : k = 1, 2, \ldots\}$$

is obviously a (countable) Köthe set, the natural topology on $\lambda(P)$ is determined by the family $\{p_k : k = 1, 2, \ldots\}$ of regular seminorms, where

$$p_k(x) = \Sigma_{n=1}^{\infty} n^k |x_n| \quad \text{for any } x = [x_n] \in \lambda(P).$$

$(\lambda(P), \mathcal{T}_P)$ is called the <u>Fréchet space of rapidly decreasing sequences</u>, and denoted by s. For any $k \geq 1$, since

$$\left[\frac{n^k}{n^{k+2}} \right] \in \ell^1$$

it follows that P satisfies the Grothendieck-Pietsch condition; thus we obtain the following result.

(3.6.5) Corollary. <u>The Fréchet space</u> s <u>of rapidly decreasing sequences is nuclear</u>.

It is known from (3.6.2)(c) that $\sigma_K(\lambda(P)^{\times}, \lambda(P))$ is coarser than $\beta(\lambda(P)', \lambda(P))$ on $\lambda(P)'$. The nuclearity of $(\lambda(P), \mathcal{T}_P)$ ensures that they are equal as the following result, due to Köthe [3], shows.

(3.6.6) Theorem. <u>Let</u> P <u>be a Köthe set and let</u> $\mathcal{T}_P$ <u>be the natural topology on</u> $\lambda(P)$. <u>If</u> $(\lambda(P), \mathcal{T}_P)$ <u>is nuclear, then the following assertions hold.</u>

(a) <u>For any</u> $\mathcal{T}_P$<u>-bounded subset</u> B <u>of</u> $\lambda(P)$ <u>there exists an</u> $\zeta = [\zeta_n] \in \lambda(P)$ <u>such that</u>

$\zeta_n \geq 0$ <u>and</u> $\sup\{|x_n| : x = [x_n] \in B\} \leq \zeta_n$ <u>for all</u> $n \geq 1$. (6.8)

(b) $\beta(\lambda(P)', \lambda(P))$ <u>coincides with</u> $\sigma_K(\lambda(P)^\times, \lambda(P))$ <u>on</u> $\lambda(P)'$, <u>hence</u> $(\lambda(P)', \beta(\lambda(P)', \lambda(P))$ <u>is dense in</u> $(\lambda(P)^\times, \mathcal{T}_{K_P})$, <u>where</u> $\mathcal{T}_{K_P}$ <u>is the natural topology on</u> $\lambda(K_P) = \lambda(P)^\times$ <u>determined by the Köthe set</u>

$$K_P = \{\zeta = [\zeta_n] \in \lambda(P) : \zeta_n \geq 0 \text{ for all } n \geq 1\} .$$

<u>Proof</u>. (a) For each $n \geq 1$, let

$$\zeta_n = \sup\{|x_n| : x = [x_n] \in B\} .$$

We claim that $\zeta = [\zeta_n] \in \lambda(P)$. Indeed, the nuclearity of $\lambda(P)$ ensures that P satisfies the Grothendieck-Pietsch condition by (3.6.4), hence for any $a = [a_n] \in P$ there exists an $b = [b_n]$ such that $[\frac{a_n}{b_n}] \in \ell^1$. On the other hand, the boundedness of B shows that there is $C \geq 0$ such that

$$p_b(x) = \Sigma_{n=1}^{\infty} |x_n| b_n \leq C \text{ for all } x = [x_n] \in B ;$$

thus

$$0 \leq \zeta_n \leq b_n^{-1} C \quad \text{for all } n \geq 1$$

and

$$\Sigma_{n=1}^{\infty} \zeta_n a_n \leq C \Sigma_{n=1}^{\infty} \frac{a_n}{b_n} < +\infty .$$

This proves part (a).

(b) In view of (3.6.2)(b) and (c), it suffices to show that

$$\beta(\lambda(P)', \lambda(P)) \leq \sigma_K(\lambda(P)^\times, \lambda(P))|_{\lambda(P)'} . \qquad (6.9)$$

For any bounded subset B of $\lambda(P)$, there exists an $\zeta = [\zeta_n] \in \lambda(P)$ such that (6.8) holds, hence

$$\{\eta = [\eta_n] \in \lambda(P)^{\times} : \sum_{n=1}^{\infty} \zeta_n |\eta_n| \leq 1\} \cap \lambda(P)' \subset (\Gamma B)^{o}$$

which obtains our assertion (6.9).

In the sequel $\mathfrak{I}_P$ will be assumed to be the natural topology on $\lambda(P)$ determined by the Köthe set P, and $\mathfrak{I}_{K_P}$ will be the natural topology on $\lambda(K_P) = \lambda(P)^{\times}$ determined by the Köthe set

$$K_P = \{\zeta = [\zeta_n] \in \lambda(P) : \zeta_n \geq 0 \text{ for all } n \geq 1\} .$$

(3.6.7) Corollary (Köthe). Let $(\lambda(P), \mathfrak{I}_P)$ be a nuclear space. Then it is co-nuclear if and only if K_P satisfies the Grothendieck-Pietsch condition.

Proof. By (3.6.6)(b), on $\lambda(P)'$, $\beta(\lambda(P)', \lambda(P))$ coincides with $\mathfrak{I}_{K_P}$, and $(\lambda(P)^{\times}, \sigma_K(\lambda(P)^{\times}, \lambda(P)))$ is the completion of $(\lambda(P)', \beta(\lambda(P)', \lambda(P)))$. In view of (3.6.4), K_P satisfies the Grothendieck-Pietsch condition if and only if $(\lambda(P)^{\times}, \sigma_K(\lambda(P)^{\times}, \lambda(P)))$ is nuclear, and this is the case if and only if $(\lambda(P)', \beta(\lambda(P)', \lambda(P)))$ is nuclear.

(3.6.8) Proposition. Let $(\lambda(P), \mathfrak{I}_P)$ be a nuclear space and $\zeta = [\zeta_n] \in \mathbb{C}^{\mathbb{N}}$. Then $\zeta \in \lambda(P)$ if and only if for any $a = [a_n] \in P$, the set

$$\{a_n \zeta_n : n \geq 1\}$$

is bounded.

Proof. The necessity follows from

$$|a_n \zeta_n| \leq \Sigma_{j=1}^{\infty} |a_j \zeta_j| < \infty \quad \text{for all} \quad n \geq 1 .$$

To prove the sufficiency, let $b = [b_n] \in P$ be such that $[\frac{a_n}{b_n}] \in \ell^1$, and let $M \geq 0$ be such that

$$|\zeta_n b_n| \leq M \quad \text{for all} \quad n \geq 1 .$$

As

$$\Sigma_{n=1}^{\infty} a_n |\zeta_n| = \Sigma_{n=1}^{\infty} |\zeta_n| \frac{a_n b_n}{b_n} \leq M \Sigma_{n=1}^{\infty} \frac{a_n}{b_n} < \infty$$

we conclude that $\zeta \in \lambda(P)$.

(3.6.9) Corollary. Let $(\lambda(P), \mathcal{T}_P)$ be a nuclear space. For each $a \in P$, let

$$\tilde{p}_a(\zeta) = \sup_n |\zeta_n a_n| \quad \text{for all} \quad \zeta = [\zeta_n] \in \lambda(P) .$$

Then $\mathcal{T}_P$ is determined by the family $\{\tilde{p}_a : a \in P\}$ of seminorms.

Proof. Clearly each $\tilde{p}_a$ is a seminorm on $\lambda(P)$ such that

$$\tilde{p}_a(\zeta) = \sup_n |\zeta_n a_n| \leq p_a(\zeta) = \Sigma_{n=1}^{\infty} |\zeta_n a_n| \text{ for all } \zeta = [\zeta_n] \in \lambda(P). \quad (6.10)$$

Conversely, for any $a = [a_n] \in P$, the nuclearity of $\lambda(P)$ shows that there exists an $b = [b_n] \in P$ such that $[\frac{a_n}{b_n}] \in \ell^1$, hence

$$\tilde{p}_a(\zeta) = \Sigma_n |\zeta_n| a_n = \Sigma_n |\zeta_n| b_n \frac{a_n}{b_n} \leq \tilde{p}_b(\zeta) (\Sigma_{n=1}^{\infty} \frac{a_n}{b_n}) . \quad (6.11)$$

The result then follows from (6.10) and (6.11).

As the Fréchet space s of rapidly decreasing sequences is nuclear, we obtain an immediate consequence of (3.6.8) as follows.

(3.6.10) Corollary. Let $\zeta = [\zeta_n] \in \mathbb{C}^{\mathbb{N}}$. Then $\zeta \in s$ if and only if for any $k \geq 1$, the set $\{n^k \zeta_n : n \geq 1\}$ is bounded, and this is the case if and only if for any $k \geq 1$, $\lim_n n^k |\zeta_n| = 0$. Consequently, the natural topology on s is determined by $\{\tilde{p}_k : k \geq 1\}$, where

$$\tilde{p}_k|\zeta| = \sup_n |n^k \zeta_n| \quad \text{for all } \zeta = [\zeta_n] \in s .$$

The following result is due to Terzioğlu [4].

(3.6.11) Corollary. Let $t = [t_n]$ be a decreasing sequence of positive numbers. Then $t \in s$ if and only if $\Sigma_n t_n^{\mu} < \infty$ for any $\mu > 0$.

Proof. Let $t \in s$. For any $\mu > 0$ we choose a natural number $k \geq 1$ such that $k\mu > 2$. As the set $\{n^k t_n : n \geq 1\}$ is bounded, there is an $M > 0$ such that

$$n^k t_n \leq M \quad \text{for all} \quad n \geq 1 ,$$

hence

$$\Sigma_n t_n^{\mu} \leq \Sigma_n M^{\mu} n^{-k\mu} \leq M^{\mu} \Sigma_n \frac{1}{n^2} < \infty .$$

Conversely, for any $k \geq 1$ there exists an $M > 0$ such that

$$\Sigma_{j=1}^{\infty} \sqrt[k]{t_j} = M .$$

As $[t_n]$ is decreasing, it follows that

$$n \sqrt[k]{t_n} \leq \Sigma_{j=1}^{n} \sqrt[k]{t_j} \leq M \quad \text{for all} \quad n \geq 1 ,$$

and hence that

$$n^k t_n \leq M^k \quad \text{for all} \quad n \geq 1 .$$

Therefore the set $\{n^k t_n : n \geq 1\}$ is bounded, thus $t \in s$ by (3.6.10).

A sequence $\{e_n\}$ in a topological vector space E is called a <u>basis</u> for E if for each $x \in E$ there exists a unique sequence $\{f_n(x)\}$ of scalars such that

$$x = \Sigma_{n=1}^{\infty} f_n(x) e_n .$$

Then $\{f_n\}$ is the well-defined sequence of linear functionals on E, and is called the <u>biorthogonal sequence</u> to $\{e_n\}$.

A basis $\{e_n\}$ of a locally convex space E is said to be <u>equicontinuous</u> if for any $U \in \mathcal{U}_E$ there exists an $V \in \mathcal{U}_E$ such that

$$\sup_n |\langle x, f_n \rangle| \, p_U(e_n) \leq p_V(x) \quad \text{for all} \quad x \in E .$$

One of the deepest results for the theory of nuclear spaces is the 'basis theorem': A complete nuclear space which has an equicontinuous basis, is isomorphic to a Köthe space (see Pietsch [1, pp.172-173] or Mitiagin [1]). By making use of the basis theorem, it can be easily shown that every nuclear Fréchet space with a basis is isomorphic to a Köthe space; because of this result, we shall often formulate our problem for a nuclear Köthe space instead of a nuclear Fréchet space with a basis. From this it is natural to ask whether every nuclear Fréchet space has a basis. The answer is negative as shown by Mitiagin and Zobin [1].

A basis $\{v_n\}$ of another locally convex space F is said to be quasisimilar to the basis $\{e_n\}$ in E if there exist a permutation σ of $\mathbb{N}$ and a sequence $\{\zeta_n\}$ of non-zero scalars such that the linear map

$$e_n \mapsto \zeta_n v_{\sigma(n)}$$

is a well-defined topological isomorphism from E onto F.

It is an open problem whether every equicontinuous basis of a nuclear Köthe space is quasisimilar to the canonical basis of the space. Dragilev has shown that this is indeed so for a particular class of nuclear Fréchet Köthe spaces which includes all nuclear power series spaces (for definition, see below). A generalization of Dragilev's theorem has been obtained by Alpseymen.

A Köthe set P is called a power set of infinite type if it satisfies the following two conditions:

(Ii) for any $a = [a_n] \in P$,

$$0 < a_n \leq a_{n+1} \quad \text{for all } n \geq 1,$$

(Iii) for each $a = [a_n] \in P$ there exists an $b = [b_n] \in P$ such that

$$a_n^2 \leq b_n \quad \text{for all } n \geq 1.$$

As P is a Köthe set, the condition (Iii) is equivalent to

(Iii)* for any $a = [a_n] \in P$ and $c = [c_n] \in P$ there exists an $b = [b_n] \in P$ such that

$$a_n c_n \leq b_n \quad \text{for all} \quad n \geq 1 .$$

A Köthe set Q is called a <u>power set of finite type</u> if it satisfies the following two conditions:

(Fi) for any $a = [a_n] \in Q$,

$$0 < a_{n+1} \leq a_n \quad \text{for all} \quad n \geq 1 ,$$

(Fii) for any $a = [a_n] \in Q$ there exists an $b = [b_n] \in Q$ such that

$$\sqrt{a_n} \leq b_n \quad \text{for all} \quad n \geq 1 .$$

The Köthe space determined by a power set of infinite type P is called an G_∞<u>-space</u> (or a <u>smooth sequence space of infinite type</u>) and denoted by $\lambda_\infty(P)$, while the Köthe space determined by a power set of finite type Q is called an G_1<u>-space</u> (or a <u>smooth sequence space of finite type</u>) and denoted by $\lambda_1(Q)$. By a <u>smooth sequence space</u> we mean either an G_∞-space or an G_1-space.

It is easy to check that for an increasing sequence $Q = [\theta_n]$ of positive numbers, the set P , defined by

$$P = \{[k^{\theta_n}]_{n\geq 1} : k = 1, 2, \ldots\}$$

is a power set of infinite type, and the set Q defined by

$$Q = \{[r^{\theta_n}]_{n\geq 1} : 0 < r < 1\}$$

is a power set of finite type. The G_∞-space determined by P is called a <u>power series space of infinite type</u> and denoted by $\lambda_\infty(\theta)$ or $\lambda_\infty([\theta_n])$, while the G_1-space determined by Q is called a <u>power series space of finite type</u> and denoted by $\lambda_1(\theta)$ or $\lambda_1([\theta_n])$.

It is clear that

$$\lambda_\infty([\theta_n]) \subset \ell^1$$

whenever $[\theta_n]$ is an increasing sequence of positive numbers.

<u>Examples</u>. (a) s is a power series space of infinite type determined by $[\log n]_{n\geq 1}$; hence $s \subset \ell^1$.

In fact, it is obvious that

$$\log k \leq k \quad \text{for all} \quad k \geq 1 .$$

If $\zeta = [\zeta_n] \in s$, then we have for any $k \geq 1$ that

$$\begin{aligned}\Sigma_{n=1}^{\infty} |\zeta_n| k^{\log n} &= \Sigma_n |\zeta_n| e^{\log n \log k} \leq \Sigma_n |\zeta_n| e^{k \log n} \\ &= \Sigma_n |\zeta_n| n^k < \infty ,\end{aligned}$$

which shows that $s \subseteq \lambda_\infty([\log n])$. Conversely, if $\zeta = [\zeta_n] \in \lambda_\infty([\log n])$, for any $k \geq 1$ we choose a natural number $m \geq 1$ such that $e^k \leq m$, hence

$$|\zeta_n| n^k = |\zeta_n| e^{k \log n} \leq |\zeta_n| m^{\log n} \quad \text{for all} \quad n \geq 1 .$$

Therefore

$$\Sigma_n |\zeta_n| n^k \leq \Sigma_n |\zeta_n| m^{\log n} < \infty ,$$

which shows that $\lambda_\infty([\log n]) \subseteq s$.

(b) The sequence $[n]_{n\geq 0}$ is increasing, the natural topology on the power series space $\lambda_\infty([n])$ of infinite type is determined by the sequence $\{p_k\}$ of seminorms, where

$$p_k(\zeta) = \Sigma_{n=0}^{\infty} |\zeta_n| k^n \quad \text{for all} \quad \zeta = [\zeta_n] \in \lambda_\infty([n]),$$

and $\lambda_\infty([n])$ is complete. It is called the <u>space of coefficients of power series expansions of entire functions</u>, and denoted by Γ .

Let H be the vector space of all entire functions $x(z)$ defined on the complex plane. It is well known that H is a Fréchet space for the compact-open topology which is determined by a sequence $\{q_k\}$ of seminorms, where

$$q_k(x) = \sup\{|x(z)| : |z| \leq k\} \quad \text{for all} \quad x \in H .$$

For any $\zeta = [\zeta_n] \in \Gamma$, the map $\zeta \mapsto x$, defined by

$$x(z) = \Sigma_{n=0}^{\infty} \zeta_n z^n \quad \text{for any} \quad z \in \mathbb{C} ,$$

is an algebraic isomorphic from Γ onto H . Furthermore,

$$q_k(x) \leq \Sigma_{n=0}^{\infty} |\zeta_n| k^n = p_k(\zeta) ,$$

hence Banach's open mapping theorem shows that this map is a topological isomorphism from Γ onto the Fréchet space H of all entire functions equipped with the compact-open topology.

The following result, due to Terzioğlu [4, 5], characterizes the nuclearity of G_∞-spaces.

(3.6.12) Proposition. <u>For an</u> G_∞<u>-space</u> $\lambda_\infty(P)$, <u>the following statements are equivalent</u>.

(a) $\lambda_\infty(P)$ <u>is nuclear</u>.

(b) <u>There exists an</u> $[a_n] \in P$ <u>such that</u> $[\frac{1}{a_n}] \in \ell^1$.

(c) <u>For any</u> $k \geq 1$ <u>there exist an</u> $[b_n] \in P$ <u>and</u> $M > 0$ <u>such that</u>

$$(n+1)^{2^k} \leq Mb_n \quad \text{for all} \quad n \geq 1 .$$

(d) <u>For each</u> $[a_n] \in P$ <u>and</u> $k \geq 1$, <u>there exist an</u> $[c_n] \in P$ <u>and</u> $M > 0$ <u>such that</u>

$$(n+1)^{2^k} a_n \leq Mc_n \quad \text{for all} \quad n \geq 1 .$$

<u>Proof</u>. (a) ⇒ (b) : For any $[c_n] \in P$, Grothendieck-Pietsch's theorem (3.6.4) shows that there exists an $[a_n] \in P$ such that $[\frac{c_n}{a_n}] \in \ell^1$. Since P is a power set of infinite type, it follows that

$$\frac{c_1}{a_n} \leq \frac{c_n}{a_n} \quad \text{for all} \quad n \geq 1 ,$$

and hence that $[\frac{1}{a_n}] \in \ell^1$.

(b) ⇒ (a) : For any $[b_n] \in P$, there exists an $[c_n] \in P$ such that

$$\max\{a_n, b_n\} \leq c_n \quad \text{for all} \quad n \geq 1 . \qquad (6.12)$$

Since $[\frac{1}{a_n}] \in \ell^1$, it follows from (6.12) that $[\frac{1}{c_n}] \in \ell^1$. For this $[c_n]$, the condition (Iii) shows that there is an $[d_n] \in P$ such that

$$c_n^2 \leq d_n \quad \text{for all} \quad n \geq 1 . \tag{6.13}$$

It then follows from (6.12) and (6.13) that

$$\frac{b_n}{d_n} \leq \frac{b_n}{c_n^2} \leq \frac{1}{c_n} \quad ,$$

and hence that $[\frac{b_n}{d_n}] \in \ell^1$ since $[\frac{1}{c_n}] \in \ell^1$. Therefore $\lambda_\infty(P)$ is nuclear by (3.6.4).

(b) ⇒ (c) : Let $[a_n] \in P$ be such that $[\frac{1}{a_n}] \in \ell^1$. Since $0 < a_n \leq a_{n+1}$ for all $n \geq 1$, we have

$$\frac{n+1}{a_n} \leq \Sigma_{j=1}^{n+1} \frac{1}{a_j} \leq \Sigma_{j=1}^{\infty} \frac{1}{a_j} = N < \infty \quad \text{for all} \quad n \geq 1$$

hence

$$(n + 1) \leq Na_n \quad \text{for all} \quad n \geq 1 .$$

For each $k \geq 1$, the condition (Iii) shows that there exists an $[b_n] \in P$ such that

$$a_n^{2^k} \leq b_n \quad \text{for all} \quad n \geq 1 ,$$

hence

$$(n + 1)^{2^k} \leq Na_n^{2^k} \leq Nb_n \quad \text{for all} \quad n \geq 1 .$$

(c) ⇒ (d) : For any $k \geq 1$ there exist an $[b_n] \in P$ and $M > 0$ such that

$$(n+1)^{2^k} \leq Mb_n \quad \text{for all} \quad n \geq 1 .$$

For any $[a_n] \in P$, the condition (Iii)* shows that there exists an $[c_n] \in P$ such that $a_n b_n \leq c_n$ for all $n \geq 1$; thus

$$(n+1)^{2^k} a_n \leq Mc_n \quad \text{for all} \quad n \geq 1 .$$

(d) ⇒ (a) : For each $[a_n] \in P$ and $k \geq 1$ there exist an $[c_n] \in P$ and $M > 0$ such that

$$(n+1)^{2^k} a_n \leq Mc_n \quad \text{for all} \quad n \geq 1 .$$

As $\Sigma_n (n+1)^{-2^k} < \infty$, we conclude that $[\frac{a_n}{b_n}] \in \ell^1$, and hence that $\lambda_\infty(P)$ is nuclear by (3.6.4).

(3.6.13) Corollary. <u>If</u> $\lambda_\infty(P)$ <u>is a nuclear</u> G_∞<u>-space, then</u>

$$\lambda_\infty(P) \subset s \subset \ell^1 .$$

<u>Proof.</u> It suffices to show that $\lambda_\infty(P) \subset s$. Let $\zeta = [\zeta_n] \in \lambda_\infty(P)$. For any $k \geq 1$, there exist, by (3.6.12)(c), an $[b_n] \in P$ and $M > 0$ such that

$$n^k \leq (n+1)^{2^k} \leq Mb_n \quad \text{for all} \quad n \geq 1 ,$$

hence

$$\Sigma_n |\zeta_n| n^k \leq M \Sigma_n |\zeta_n| b_n < \infty .$$

Thus $\zeta \in s$.

(3.6.14) Corollary. <u>Let</u> $\theta = [\theta_n]$ <u>be an increasing sequence of</u>

<u>positive numbers</u>. <u>Then</u> $\lambda_\infty([\theta_n])$ <u>is nuclear if and only if there exists an</u> $t \in (0, 1)$ <u>such that</u>

$$\sum_n t^{\theta_n} < \infty .$$

<u>Proof</u>. (3.6.12) shows that $\lambda_\infty([\theta_n])$ is nuclear if and only if there exists a natural number $k > 1$ such that $[k^{-\theta_n}] \in \ell^1$, thus $t = k^{-1}$ has the required property.

Since $[2^{-n}] \in \ell^1$, we obtain an immediate consequence of (3.6.14) and (3.6.13).

(3.6.15) Corollary. <u>The space</u> Γ <u>of coefficients of power series expansions of entire functions is nuclear and</u> $\Gamma \subset s$.

The following result, due to Ramanujan [1], gives some interesting properties of power series spaces of infinite type which are nuclear.

(3.6.16) Corollary. <u>Let</u> $0 \leq \theta_n \leq \theta_{n+1}$ <u>for all</u> $n \geq 1$ <u>and</u> $\lim_n \theta_n = \infty$. <u>If the power series space</u> $\lambda_\infty([\theta_n])$ <u>of infinite type is nuclear, then the following assertions hold</u>.

(a) $\lambda_\infty([\theta_n]) = \{\zeta = [\zeta_n] \in \mathbb{C}^{\mathbb{N}} : \lim_n |\zeta_n|^{-\theta_n} = 0\}$.

(b) <u>The set</u> $\{\frac{\log n}{\theta_n} : n \geq 1\}$ <u>is bounded</u>.

(c) <u>If</u> $\zeta = [\zeta_n] \in \lambda_\infty([\theta_n])$, <u>then</u> $[n\,\zeta_n] \in \lambda_\infty([\theta_n])$.

<u>Proof</u>. (a) By (3.6.14), the nuclearity of $\lambda_\infty([\theta_n])$ ensures

that there exists an $t \in (0, 1)$ such that $\Sigma_n t^{\theta_n} < \infty$. If $[\zeta_n] \in \mathbb{C}^{\mathbb{N}}$ is such that $\lim_n |\zeta_n|^{-\theta_n} = 0$, then for any $k \geq 1$ there is a natural number $N > 0$ such that

$$|\zeta_n|^{-\theta_n} < k^{-1} t \quad \text{for all} \quad n \geq N ,$$

hence

$$\Sigma_{n=N}^{\infty} |\zeta_n| k^{\theta_n} \leq \Sigma_{n=N}^{\infty} t^{\theta_n} < \infty$$

which implies that $\zeta \in \lambda_\infty([\theta_n])$.

Conversely, if there exists an $\zeta = [\zeta_n] \in \lambda_\infty([\theta_n])$ such that $\lim_n |\zeta_n|^{-\theta_n} \neq 0$, then there exist an $\delta > 0$ and a sequence $\{n_j\}$ of natural numbers such that

$$|\zeta_{n_j}|^{-\theta_{n_j}} > \delta \quad \text{for all} \quad j = 1, 2, \ldots . \tag{6.14}$$

We choose an integer $k \geq 1$ satisfying $\delta k \geq 1$, then (6.14) shows that

$$\Sigma_{j=1}^{\infty} |\zeta_{n_j}| k^{\theta_{n_j}} > \Sigma_{j=1}^{\infty} (k\delta)^{\theta_{n_j}} = \infty$$

which contradicts the fact that $\zeta \in \lambda_\infty([\theta_n])$.

(b) By (3.6.14), there exists an $k > 1$ such that $\Sigma_{n=1}^{\infty} k^{-\theta_n} = M < \infty$. Since $k^{-\theta_n} \geq k^{-\theta_{n+1}}$ for all $n \geq 1$, we have

$$nk^{-\theta_n} \leq \Sigma_{j=1}^{n} k^{-\theta_j} \leq M \quad \text{for all} \quad n \geq 1 ,$$

hence

$$n^{-\theta_n} \leq kM^{-\theta_n} \quad \text{for all} \quad n \geq 1 ,$$

and thus

$$\frac{1}{\theta_n} \log n \leq \frac{1}{\theta_n} \log M + \log k \quad \text{for all} \quad n \geq 1 . \tag{6.15}$$

As $\lim_n \theta_n = \infty$, it follows that the sequence $\{\frac{1}{\theta_n}, n \geq 1\}$ is bounded, and hence from (6.15) that $\{\frac{\log n}{\theta_n} : n \geq 1\}$ is bounded.

(c) Let $M > 0$ be such that $\frac{\log n}{\theta_n} \leq M$ for all $n \geq 1$. Then

$$n \leq e^{M\theta_n} \quad \text{for all} \quad n \geq 1 .$$

For any $k \geq 1$, let $m \geq 1$ be such that $m \geq ke^M$. Then we obtain

$$|n \zeta_n| k^{\theta_n} \leq |\zeta_n| k^{\theta_n} e^{M\theta_n} = |\zeta_n| (ke^M)^{\theta_n} \leq |\zeta_n| m^{\theta_n} .$$

As $[\zeta_n] \in \lambda_\infty([\theta_n])$, we conclude that

$$\Sigma_n |n \zeta_n| k^{\theta_n} \leq \Sigma_n |\zeta_n| m^{\theta_n} < \infty .$$

Therefore $[n \zeta_n] \in \lambda_\infty([\theta_n])$.

The following result, due to Dubinsky and Ramanujan [1], is concerning with the decomposition property of nuclear G_∞-spaces.

(3.6.17) Proposition. _Let_ $\zeta = [\zeta_n] \in \mathbb{C}^{\mathbb{N}}$ _and let_ $\lambda_\infty(P)$ _be a nuclear_ G_∞_-space_. _Then_ $\zeta \in \lambda_\infty(P)$ _if and only if for any_ $t > 0$, $[|\zeta_n|^t] \in \lambda_\infty(P)$; _in this case_,

$$\lambda_\infty(P) \subset \lambda_\infty(P) \cdot \lambda_\infty(P) .$$

Proof. In view of (3.6.13) and the normality of $\lambda_\infty(P)$, we may assume that $0 < \zeta_n < 1$ for all $n \geq 1$ and $t \in \mathbb{N}$.

If $\zeta \in \lambda_\infty(P)$, then $\zeta_n^t \leq \zeta_n$ for all $n \geq 1$, hence $[\zeta_n^t] \in \lambda_\infty(P)$. Conversely, if for any $t > 0$, $[\zeta_n^t] \in \lambda_\infty(P)$, then $[\zeta_n^2] \in \lambda_\infty(P)$. For any $[a_n] \in P$, Grothendieck-Pietsch's theorem (3.6.4) shows that there exists an $[b_n] \in P$ such that $[\frac{a_n}{b_n}] \in \ell^1$. In view of (Iii), there exists an $[c_n] \in P$ such that $b_n^2 \leq c_n$ for all $n \geq 1$. By Cauchy-Schwarz's inequality,

$$\Sigma_n |\zeta_n| a_n = \Sigma_n |\zeta_n| b_n (\frac{a_n}{b_n}) \leq (\Sigma_n |\zeta_n|^2 b_n^2)^{\frac{1}{2}} (\Sigma_n (\frac{a_n}{b_n})^2)^{\frac{1}{2}}$$

$$\leq (\Sigma_n |\zeta_n^2| c_n)^{\frac{1}{2}} (\Sigma_n (\frac{a_n}{b_n})^2)^{\frac{1}{2}} < \infty ,$$

hence $[\zeta_n] \in \lambda(P)$.

The following result, due to Terzioğlu [4,5], characterizes the nuclearity of G_1-spaces.

(3.6.18) Proposition. <u>For an</u> G_1-<u>space</u> $\lambda_1(Q)$, <u>the following statements are equivalent</u>.

(a) $\lambda_1(Q)$ <u>is nuclear</u>.

(b) $Q \subset \ell^1$.

(c) $Q \subset s$.

(d) $\ell^\infty \subset \lambda_1(Q)$.

(e) <u>For any</u> $[a_n] \in Q$ <u>and</u> $k \geq 1$ <u>there exist an</u> $[c_n] \in Q$ <u>and</u> $M > 0$ <u>such that</u>

$$(n+1)^{2^k} a_n \leq M c_n \quad \underline{\text{for all}} \quad n \geq 1 .$$

<u>Proof</u>. In view of the definition of α-dual and (3.6.2)(a), the

equivalence of (b) and (d) are obvious.

(a) ⇒ (b) : For any $[a_n] \in Q$, (3.6.4) shows that there exists an $[b_n] \in Q$ such that $[\frac{a_n}{b_n}] \in \ell^1$. Since $b_n \geq b_{n+1}$ for all $n \geq 1$, it follows that

$$\frac{a_n}{b_1} \leq \frac{a_n}{b_n} \quad \text{for all} \quad n \geq 1 ,$$

and hence that $[a_n] \in \ell^1$.

(b) ⇒ (c) : As each $[a_n] \in Q$ is decreasing, it suffices by (3.6.11) to show that for any $\mu > 0$, $[a_n^{\mu}] \in \ell^1$. Indeed, let $k \geq 1$ be such that $2^{-k} \leq \mu$. In view of (Fii), there exists an $[b_n] \in Q$ such that

$$a_n^{-2^k} \leq b_n \quad \text{for all} \quad n \geq 1 ,$$

hence

$$a_n^{\mu} = (a_n^{-2^k})^{\mu 2^k} \leq b_n^{\mu 2^k} \quad \text{for all} \quad n \geq 1 . \tag{6.16}$$

Since $\mu 2^k \geq 1$ and $[b_n] \in \ell^1 \subset \ell^{\mu 2^k}$, it follows from (6.16) that $[a_n^{\mu}] \in \ell^1$.

(c) ⇒ (a) : For any $[a_n] \in Q$, by (Fii) there exists an $[b_n] \in Q$ such that

$$a_n \leq b_n^2 \quad \text{for all} \quad n \geq 1 . \tag{6.17}$$

As $s \subset \ell^1$, it follows from (6.17) that $[\frac{a_n}{b_n}] \in \ell^1$, and hence from (3.6.4) that $\lambda_1(Q)$ is nuclear.

(b) ⇒ (e) : For each $[a_n] \in Q$ there exists an $[c_n] \in Q$ such that

$$a_n \leq c_n^2 \quad \text{for all} \quad n \geq 1 .$$

For any $k \geq 1$ there exists an $[b_n] \in Q$ such that

$$c_n \leq b_n^{2^k} \quad \text{for all} \quad n \geq 1 .$$

Since $c_n \geq c_{n+1}$ for all $n \geq 1$, it follows from (b) that

$$(n + 1)b_n \leq \Sigma_{n=1}^{\infty} b_n = N < \infty \quad \text{for all} \quad n \geq 1 ,$$

and hence that

$$(n + 1)^{2^k} \frac{a_n}{c_n} \leq (n + 1)^{2^k} c_n \leq (n + 1)^{2^k} b_n^{2^k} \leq N^{2^k} \quad \text{for all} \quad n \geq 1 .$$

(3.6.19) Corollary. _Suppose that_ $0 \leq \theta_n \leq \theta_{n+1}$ _for all_ $n \geq 1$. _Then the power series space_ $\lambda_1([\theta_n])$ _of finite type is nuclear if and only if_

$$[r^{\theta_n}] \in \ell^1 \quad \text{for all} \quad 0 < r < 1 ,$$

and this is the case if and only if

$$[r^{\theta_n}] \in s \quad \text{for all} \quad 0 < r < 1 .$$

It can be shown that every nuclear smooth sequence space is co-nuclear; moreover, the strong dual of a nuclear G_1-space is a dense subspace of a nuclear G_∞-space, and the strong dual of a nuclear G_∞-space is a dense subspace of a nuclear G_1-space under some additional condition (see Terzioğlu [4] and [5]).

3.7 Universal nuclear spaces

It is known from (3.6.5) that the Fréchet space s of rapidly decreasing sequences is a nuclear space. It was shown by T. and Y. Kōmura [1] in 1966 that the space s is a universal nuclear space which answers the question posed by Grothendieck, namely they had verified the following:

(3.7.1) Theorem (T. and Y. Komura). _A locally convex space_ $(E, \mathcal{P})$ _is nuclear if and only if it is topologically isomorphic to a subspace of the product space_ s^Λ _for some index set_ Λ. _Consequently, every metrizable nuclear space is topologically isomorphic to a subspace of_ $s^{\mathbb{N}}$.

The sufficiency is obvious. The proof of the necessity, taken from Köthe [2], is based on the following several lemmas.

(3.7.2) Lemma. Let $(E, \mathcal{P})$ be a nuclear space, let U be an absolutely convex o-neighbourhood in E and $\mu > 0$. Then there exists an absolutely convex o-neighbourhood V in E such that

$$V \subset E_n + n^{-\mu} U \quad \text{for all} \quad n \geq 1 , \tag{7.1}$$

where E_n is some vector subspace of E whose dimension is less than or equal to n.

Proof. We first prove the result for $\mu \leq \frac{1}{2}$. In view of (3.2.2) and (3.2.4), we might assume that $\widetilde{E}_U$ is a Hilbert spaces, hence (3.1.5) shows that there is an absolutely convex o-neighbourhood W in E with $W \subset U$ such that $\widetilde{E}_W$ is a Hilbert space and the canonical map $Q_{U,W} : \widetilde{E}_W \to \widetilde{E}_U$ is of the form

$$Q_U(x) = Q_{U,W}(Q_W(x)) = \Sigma_{k=1}^{\infty} \lambda_k [Q_W(x), e_k] d_k \quad (x \in E) , \tag{7.2}$$

where $\{e_k\}$ and $\{d_k\}$ are orthonormal sequences in $\widetilde{E}_W$ and $\widetilde{E}_U$ respectively, and $\{\lambda_k\}$ is such that $\lambda_k \downarrow 0$ and $\lambda = \Sigma_{k=1}^{\infty} \lambda_k < \infty$. By the properties of $\{\lambda_k\}$, we have

$$\lambda_{n+1} \leq n^{-1} \Sigma_{k=1}^{n} \lambda_k \leq n^{-1} \lambda \quad \text{for all} \quad n \geq 1 .$$

It then follows from the Cauchy-Schwarz inequality that

$$\begin{aligned}
\hat{p}_U(\Sigma_{k=n+1}^{\infty} \lambda_k [Q_U(x), e_k] d_k) &\leq \Sigma_{k=n+1}^{\infty} |\lambda_k [Q_W(x), e_k]| \, p_U(d_k) \\
&\leq \sqrt{\lambda_{n+1}} \; \Sigma_{k=n+1}^{\infty} \sqrt{\lambda_k} \; |[Q_W(x), e_k]| \\
&\leq \sqrt{\lambda_{n+1}} \; (\Sigma_{k=n+1}^{\infty} \lambda_k)^{\frac{1}{2}} (\Sigma_{k=n+1}^{\infty} |[Q_W(x), e_k]|^2)^{\frac{1}{2}} \\
&\leq \sqrt{\frac{\lambda}{n}} \sqrt{\lambda} \; \hat{p}_W(Q_W(x)) \leq \frac{\lambda}{\sqrt{n}} \quad \text{for all} \quad x \in W .
\end{aligned} \tag{7.3}$$

If we denote by G_n the vector subspace of $\tilde{E}_U$ generated by $\{d_1, \ldots, d_n\}$, then formulae (7.2) and (7.3) show that

$$\lambda^{-1} Q_U(W) \subset G_n + \frac{1}{\sqrt{n}} \overline{Q_U(U)} \quad \text{for all} \quad n \geq 1 , \tag{7.4}$$

where $\overline{Q_U(U)}$ is the closure in $\tilde{E}_U$ of $Q_U(U)$.

It should be noted from (7.2) that the part of $Q_U(x)$ $(x \in W)$, contained in G_n, has coefficients which are uniformly bounded since $|[Q_W(x), e_k]| \leq 1$ for all $x \in W$ and $k = 1, 2, \ldots, n$. On the other hand, the density of E_U in $\tilde{E}_U$ shows that there exists $\{u_1, \ldots, u_n\}$ in E such that $\hat{p}_U(Q_U(u_k) - e_k) \leq 1$ for all $k = 1, \ldots, n$. From this it is not hard to see that if E_n is the vector subspace of E generated by $\{u_1, \ldots, u_n\}$, then

$$\frac{1}{2\lambda} W \subset E_n + \frac{1}{\sqrt{n}} U \quad \text{for all} \quad n \geq 1 . \tag{7.5}$$

If we take $V = \frac{1}{2\lambda} W$, then we obtain our assertion by making use of (7.5).

To prove the general case, it suffices to show that μ satisfies $\frac{1}{2} < \mu < \frac{m}{2}$, where m is some integer. In view of the first case, there exist absolutely convex o-neighbourhoods $U_1, \ldots, U_m$ such that

$$U_j \subset E_n^{(j)} + \frac{1}{\sqrt{n}} U_{j-1} , \quad j = 1, 2, \ldots, m \quad \text{and} \quad n \geq 1 , \tag{7.6}$$

where $U_o = U$ and $E_n^{(j)}$ are vector subspaces of E whose dimensions are less than or equal to n. It then follows from (7.6) that

$$U_m \subset E_n^{(1)} + \dots + E_n^{(m)} + n^{-(m/2)} U .$$

If we denote by E_{mn} the vector subspace of E generated by $E_n^{(1)} + \dots + E_n^{(m)}$, then $\dim E_{mn} \leq mn$ and

$$U_m \subset E_{mn} + n^{-(m/2)} U . \tag{7.7}$$

On the other hand, as $\mu < \frac{m}{2}$, we have

$$n^{m/2} > n^{\mu} = [m(n+1)]^{\mu} m^{-\mu} \left(\frac{n}{n+1} \right)^{\mu} \geq (m(n+1))^{\mu} (2m)^{-\mu} . \tag{7.8}$$

It then follows from (7.7) and (7.8) that

$$(2m)^{-\mu} U_m \subset E_{mn} + (m(n+1))^{-\mu} U \quad \text{for all } n \geq 1 ,$$

which obtains our assertion when we take $V = (2m)^{-\mu} U_m$.

(3.7.3) Lemma. Let $(E, \mathcal{P})$ be a nuclear space, and let B be a closed absolutely convex equicontinuous subset of E' such that $E'(B)$ is a Hilbert space. Then for any integer $k > 0$ there exists an orthonormal sequence $\{f_n^{(k)} : n \geq 1\}$ in $E'(B)$ such that

$$\{\Sigma_{n=1}^{\infty} \mu_n n^k f_n^{(k)} : \Sigma_{n=1}^{\infty} |\mu_n|^2 \leq 1\} \tag{7.9}$$

is an equicontinuous subset of E'.

Proof. For a given $k > 0$, the nuclearity of E together with (3.2.2) and (3.7.2) ensure that there exists an absolutely convex o-neighbourhood V in E with $V \subset B^{o}$ such that $\widetilde{E}_V$ is a Hilbert space and

$$V \subset E_n + n^{-k} B^o , \tag{7.10}$$

where E_n is a vector subspace of E generated by $\{x_1, \ldots, x_n\}$. The adjoint map of the canonical map $Q_{B^o,V} : E_V \to E_{B^o}$ is the canonical injection $j_{B,V^o} : E'(B) \to E'(V^o)$ and nuclear. As $E'(B)$ and $E'(V^o)$ are Hilbert spaces, it follows from (3.1.5) that j_{B,V^o} is of the form

$$j_{B,V^o}(f) = \Sigma_{i=1}^{\infty} \lambda_i [f, f_i^{(k)}] g_i \quad \text{for all} \quad f \in E'(B) ,$$

where $\{f_i^{(k)}\}$ and $\{g_i\}$ are orthonormal sequences in $E'(B)$ and $E'(V^o)$ respectively, and $\{\lambda_i\}$ is such that $\lambda_i \downarrow 0$ and $\Sigma_{i=1}^{\infty} \lambda_i < \infty$. As j_{B,V^o} being injection, we have

$$f_m^{(k)} = j_{B,V^o}(f_m^{(k)}) = \lambda_m g_m \quad \text{and} \quad p_{V^o}(f_m^{(k)}) = \lambda_m p_{V^o}(g_m) = \lambda_m \quad \text{for all} \quad m \geq 1. \tag{7.11}$$

Furthermore, we claim that

$$\lambda_{n+1} \leq n^{-k} \quad \text{for all} \quad n \geq 1 . \tag{7.12}$$

In fact, it is clear that B is the closed unit ball in $E'(B)$ consisting of all $\Sigma_{n=1}^{\infty} \mu_n f_n^{(k)}$ with $\Sigma_{n=1}^{\infty} |\mu_n|^2 \leq 1$. As E_n is a vector subspace of E generated by $\{x_1, \ldots, x_n\}$, it follows from the definition of the polarity that

$$B \cap E_n^o = \{f = \Sigma_{n=1}^{\infty} \mu_n f_n^{(k)} \in B : \langle x_i , f\rangle = 0 \quad \text{for all} \quad i = 1, 2, \ldots, n\}.$$

Clearly the element $\phi_o = \Sigma_{j=1}^{n+1} c_j f_j^{(k)}$, satisfying

$$\Sigma_{j=1}^{n+1} |c_j|^2 = 1 \quad \text{and} \quad \Sigma_{j=1}^{n+1} c_j \langle x_i , f_j^{(k)} \rangle = 0 \quad (i = 1, 2, \ldots, n) ,$$

belongs to $B \cap E_n^o$, hence (7.11) shows that

$$(p_{V^o}(\psi_o))^2 = (p_{V^o}(\Sigma_{j=1}^{n+1} c_j f_j^{(k)}))^2 = (p_{V^o}(\Sigma_{j=1}^{n+1} \lambda_j c_j g_j))^2$$

$$= \Sigma_{j=1}^{n+1} \lambda_j^2 |c_j|^2 \geq \lambda_{n+1}^2 (\Sigma_{j=1}^{n+1} |c_j|^2) = \lambda_{n+1}^2$$

because $(E'(V^o), p_{V^o})$ is a Hilbert space, and thus

$$\lambda_{n+1} \leq p_{V^o}(\psi_o) . \tag{7.13}$$

On the other hand, since E_n is a vector subspace of E , we conclude from (7.10) that $B \cap E_n^o \subset n^{-k} V^o$, and hence from (7.13) that

$$\lambda_{n+1} \leq p_{V^o}(\psi_o) \leq n^{-k}$$

because of $\psi_o \in B \cap E_n^o$. Therefore we obtain the required inequality (7.12).

Finally, by making use of (7.12), we are able to verify that the set (7.9) is contained in a multiple of V^o . In fact, let $\Sigma_{n=1}^{\infty} |\mu_n|^2 \leq 1$. In view of (7.11) and (7.12), we have

$$(p_{V^o}(\Sigma_{n=1}^{\infty} \mu_n n^k f_n^{(k)}))^2 = (p_{V^o}(\Sigma_{n=1}^{\infty} \mu_n n^k \lambda_n g_n))^2 = \Sigma_{n=1}^{\infty} |\mu_n|^2 (\lambda_n n^k)^2$$

$$= |\lambda_1 \mu_1|^2 + \Sigma_{n=2}^{\infty} |\mu_n|^2 (\lambda_n n^k)^2$$

$$\leq |\lambda_1 \mu_1|^2 + (\Sigma_{n=2}^{\infty} |\mu_n|^2 \lambda_n^2 (n-1)^{2k}) \sup_{m>1} \left(\frac{m}{m-1}\right)^{2k}$$

$$\leq |\lambda_1|^2 + 2^{2k} \Sigma_{n=2}^{\infty} |\mu_n|^2 \leq |\lambda_1|^2 + 2^{2k} ,$$

as asserted.

(3.7.4) Lemma. Let $(E, \mathscr{P})$ be a nuclear space, and let B be a closed absolutely convex equicontinuous subset of E' such that $E'(B)$ is a Hilbert space. Then there exists an orthonormal sequence $\{\psi_n : n \geq 1\}$ in $E'(B)$ such that every set $\{n^k \psi_n : n \geq 1\}$ is an equicontinuous subset of E' for any fixed integer $k \geq 1$.

Proof. For every $k \geq 1$, (3.7.3) ensures that there exists an orthonormal sequence $\{f_n^{(k)} : n \geq 1\}$ in $E'(B)$ such that

$$\{\Sigma_{n=1}^{\infty} \mu_n n^k f_n^{(k)} : \Sigma_{n=1}^{\infty} |\mu_n|^2 \leq 1\} \tag{7.14}$$

is an equicontinuous subset of E'. For any fixed $n \geq 1$, let us define

$$h_{(n-1)(n-1)+t} = f_t^{(n)} \quad \text{for} \quad 1 \leq t \leq n ,$$

$$h_{n^2-(t-1)} = f_n^{(t)} \quad \text{for} \quad 1 \leq t < n .$$

Orthogonalizing the sequence $\{h_n : n \geq 1\}$ we obtain a new orthonormal sequence $\{\psi_m : m \geq 1\}$ in $E'(B)$ such that

$$[\psi_m, f_n^{(k)}] = 0 \quad \text{for all } f_n^{(k)} \text{ with } k^2 < m \text{ and } n^2 < m . \tag{7.15}$$

On the other hand, for a fixed $k \geq 1$, in terms of the basis $\{f_n^{(k)} : n \geq 1\}$, each ψ_m can be expressed by

$$\psi_m = \Sigma_{n=1}^{\infty} \alpha_n^{(m,k)} f_n^{(k)} \quad \text{with} \quad \Sigma_{n=1}^{\infty} |\alpha_n^{(m,k)}|^2 = 1 .$$

In view of (7.15), we have

$$\alpha_n^{(m,k)} = 0 \quad \text{for} \quad k^2 < m \quad \text{and} \quad n^2 < m , \tag{7.16}$$

so that

$$\psi_m = \Sigma_{n^2 \geq m} \alpha_n^{(m,k)} f_n^{(k)} \quad \text{if} \quad m > k^2 . \tag{7.17}$$

For any m with $m > k^2$, since

$$m^{k/2}\psi_m = \Sigma_{n=1}^{\infty} m^{k/2}\alpha_n^{(m,k)} f_n^{(k)} = \Sigma_{n^2 \geq m} m^{k/2}\alpha_n^{(m,k)} f_n^{(k)} ,$$

it follows from (7.16) and (7.17) that

$$\Sigma_{n=1}^{\infty} |m^{k/2}\alpha_n^{(m,k)}|^2 \leq \Sigma_{n^2 \geq m} |n^k \alpha_n^{(m,k)}|^2 \quad \text{with} \quad \Sigma_{n=1}^{\infty} |\alpha_n^{(m,k)}|^2 = 1 ,$$

and hence from (7.14) that the set $\{m^{k/2}\psi_m : m \geq k^2 + 1\}$ is equicontinuous. Consequently, $\{m^{k/2}\psi_m : m \geq 1\}$ is an equicontinuous subset of E', and thus $\{m^k\psi_m : m \geq 1\}$ is an equicontinuous subset of E' for any fixed integer $k \geq 1$.

<u>Proof of the necessity of (3.7.1)</u>: We choose a fundamental system $\{B_\lambda : \lambda \in \Lambda\}$ of closed absolutely convex equicontinuous subsets of E' such that each $E'(B_\lambda)$ is a Hilbert space. For any $\lambda \in \Lambda$, (3.7.4) ensures that there exists an orthonormal sequence $\{\psi_n^{(\lambda)}, n \geq 1\}$ in $E'(B_\lambda)$ such that every set $\{n^k\psi_n^{(\lambda)} : n \geq 1\}$ is equicontinuous in E' $(k \geq 1)$. Let us define T_λ on E by setting

$$T_\lambda(x) = (\langle x, \psi_n^{(\lambda)}\rangle)_{n\geq 1} \quad \text{for all} \quad x \in E . \tag{7.18}$$

Let V be an absolutely convex o-neighbourhood in E such that $n^{k+2}\psi_n^{(\lambda)} \in V^o$ for all $n \geq 1$, and let p_V be the gauge of V. Then

$$\Sigma_{n=1}^{\infty} n^k |\langle x, \psi_n^{(\lambda)}\rangle| = \Sigma_{n=1}^{\infty} \frac{1}{n^2} n^{k+2} |\langle x, \psi_n^{(\lambda)}\rangle| \leq (\Sigma_{n=1}^{\infty} \frac{1}{n^2}) p_V(x) \quad \text{for all} \quad x \in E ,$$

hence T_λ is a continuous map from E into s. Clearly T is linear. Therefore, the map T, defined by

$$T(x) = (T_\lambda(x))_{\lambda\in\Lambda} \quad \text{for all} \quad x \in E ,$$

is a continuous linear map from E into s^Λ. Obviously, T is injective, thus we complete the proof by showing that T^{-1} is a continuous map from the subspace $T(E)$ of s^Λ onto E.

In fact, for any $\gamma \in \Lambda$, let $\{\psi_n^{(\gamma)} : n \geq 1\}$ be an orthonormal sequence in $E'(B_\gamma)$. Since any $f \in B_\gamma$ is of the form $\Sigma_{n=1}^{\infty} \mu_n^{(\gamma)} \psi_n^{(\gamma)}$ with $\Sigma_{n=1}^{\infty} |\mu_n^{(\gamma)}|^2 \leq 1$, it follows from the Cauchy-Schwarz inequality that

$$|\langle x, f\rangle| \leq \Sigma_{n=1}^{\infty} |\mu_n^{(\gamma)}| |\langle x, \psi_n^{(\gamma)}\rangle| \leq (\Sigma_{n=1}^{\infty} |\langle x, \psi_n^{(\gamma)}\rangle|^2)^{\frac{1}{2}} ,$$

and hence that

$$\{x \in E : \Sigma_{n=1}^{\infty} |\langle x, \psi_n^{(\gamma)}\rangle| \leq 1\} \subset B_\gamma^o . \tag{7.19}$$

Setting

$$U_\gamma = \{(a_n^{(\gamma)}) \in s : \Sigma_{n=1}^{\infty} n^2 |a_n^{(\gamma)}| \leq 1\} \quad \text{and} \quad U_\lambda = s \quad \text{for} \quad \lambda \in \Lambda \quad \text{with} \quad \lambda \neq \gamma . \tag{7.20}$$

Then $U = U_\gamma \times \prod_{\lambda \neq \gamma} U_\lambda$ is a o-neighbourhood in $s^\wedge$. For any

$T(x) = (T_\lambda(x))_{\lambda\in\wedge} \in T(E) \cap U$, we have by (7.18) and (7.20) that

$$\sum_{n=1}^{\infty} |\langle x, \psi_n^{(\gamma)}\rangle| \leq \sum_{n=1}^{\infty} n^2 |\langle x, \psi_n^{(\gamma)}\rangle| \leq 1 ,$$

and thus $T^{-1}(U) \subset B_\gamma^o$ by (7.19), as asserted.

3.8 Supplements - Nuclearity associated to a G_∞-space

The concept of nuclearity, studied in this chapter, depends upon on the Banach space ℓ^1. If ℓ^1 is replaced by an abstract sequence space λ equipped with a suitable locally convex topology, then this concept can be extended to that of λ-nuclearity in a natural way. One of the first applications of this idea is due to Martineau who replaced ℓ^1 by the nuclear Fréchet space s of rapidly decreasing sequences. After a few year, Brudovskii [1, 2] rediscovered the ideas of Martineau and along with many new results made an error that led Köthe [4] to introduce the idea of uniform s-nuclearity. In 1970, Ramanujan [1] was to begin the study of cases in which ℓ^1 is replaced by a nuclear power series space of infinite type. In extending this notion, Dubinsky and Ramanujan [1] replaced the power series space by a nuclear G_∞-space $\lambda_\infty(P)$ where the power set P of infinite type is assumed to be countable. In 1973, Terzioğlu [5] removed this restriction and considered this type of nuclearity for smooth sequence spaces.

Throughout this section $\lambda_\infty(P_0)$ will be assumed to be a fixed <u>nuclear</u> G_∞-space. In view of (3.6.13),

$$\lambda_\infty(P_0) \subset s \subset \ell^1 . \tag{8.1}$$

A seminorm p on E is said to be <u>quasi</u>-$\lambda_\infty(P_0)$-<u>nuclear</u> if there exist an $[\zeta_n] \in \lambda_\infty(P_0)$ and an equicontinuous sequence $\{f_n\}$ in E' such that

$$p(x) \leq \Sigma_{n=1}^{\infty} |\zeta_n \langle x, f_n \rangle| \quad \text{for all} \quad x \in E .$$

By (8.1), $\lambda_\infty(P_0)$-quasi-nuclear seminorms are quasi-nuclear, and surely continuous.

A locally convex space E is said to be $\lambda_\infty(P_0)$-<u>nuclear</u> if each continuous seminorm on E is $\lambda_\infty(P_0)$-quasi-nuclear.

As the Fréchet space s of rapidly decreasing sequences being a nuclear G_∞-space, it follows from (8.1) that $\lambda_\infty(P_0)$-nuclear spaces are s-nuclear, and <u>a fortiori</u> nuclear. There exist nuclear spaces which are not $\lambda_\infty(P_0)$-nuclear, but nuclear (DF)-spaces are s-nuclear as shown by Pietsch [1, p.179]. s-nuclear spaces have the same permanence properties as nuclear spaces (see Brudovskii [1, 2]).

In order to mention other characterizations of $\lambda_\infty(P_0)$-nuclear spaces, we required the following terminology: An $T \in L(E, F)$ is said to be <u>quasi</u>-$\lambda_\infty(P_0)$-<u>nuclear-bounded</u> if there exists a quasi-$\lambda_\infty(P_0)$-nuclear seminorm p on E such that the set $\{Tx \in F : p(x) \leq 1\}$ is bounded in F.

Quasi-s-nuclear-bounded maps are called <u>mappings of type</u>-s by Randtke [1]. It is easily shown that a continuous seminorm p on E is quasi-$\lambda_\infty(P_o)$-nuclear if and only if the quotient map $Q_p : E \to E_p$ is quasi-$\lambda_\infty(P_o)$-nuclear-bounded, that if $T \in L(E, F)$ and q is a quasi-$\lambda_\infty(P_o)$-nuclear seimnorm on F, then $q \circ T$ is a quasi-$\lambda_\infty(P_o)$-nuclear seminorm on E, and that the composition of two continuous linear maps, in which one of them is quasi-$\lambda_\infty(P_o)$-nuclear-bounded, is quasi-$\lambda_\infty(P_o)$-nuclear-bounded.

Denote by $N^{(q)}_{\lambda_\infty(P_o)}(E, F)$ the set of all quasi-$\lambda_\infty(P_o)$-nuclear-bounded maps from E into F, and by $s(E, F)$ the set of all mappings of type s. Then (8.1) shows that

$$N^{(q)}_{\lambda_\infty(P_o)}(E, F) \subset s(E, F) \subset N^{(q)}_{\ell^1}(E, F) .$$

A similar argument given in the proof of (3.4.2) yields the following result.

(3.8.1) Theorem. <u>The following statements are equivalent</u>.

(a) E <u>is</u> $\lambda_\infty(P_o)$-<u>nuclear</u>.

(b) $Q_p \in N^{(q)}_{\lambda_\infty(P_o)}(E, E_p)$ <u>for any continuous seminorm</u> p <u>on</u> E.

(c) <u>For any continuous seminorm</u> p <u>on</u> E <u>there is a continuous seminorm</u> r <u>on</u> E <u>with</u> $p \leq r$ <u>such that</u> $Q_{p,r} \in N^{(q)}_{\lambda_\infty(P_o)}(E_r, E_p)$.

(d) $L^{\ell b}(E, F) = N^{(q)}_{\lambda_\infty(P_o)}(E, F)$ <u>for any locally convex space</u> F.

(e) $L(E, Y) = N^{(q)}_{\lambda_\infty(P_o)}(E, Y)$ <u>for any normed space</u> Y.

(f) $L^p(E, F) \subset N^{(q)}_{\lambda_\infty(P_o)}(E, F)$ <u>for any locally convex space</u> F.

Let X and Y be Banach spaces. An $T \in L(X, Y)$ is said to be $\lambda_\infty(P_0)$-nuclear if there exists an $[\zeta_n] \in \lambda_\infty(P_0)$, an equicontinuous sequence $\{f_n\}$ in X' and a weakly $\lambda_\infty(P_0)^\times$-summable sequence $\{y_n\}$ in Y such that

$$Tx = \sum_{n=1}^{\infty} \zeta_n \langle x, f_n \rangle y_n \quad \text{for all} \quad x \in X . \tag{8.2}$$

We say that an $T \in L(X, Y)$ is pseudo $\lambda_\infty(P_0)$-nuclear if, in the definition of a $\lambda_\infty(P_0)$-nuclear map, we replace $\lambda_\infty(P_0)^\times$ by ℓ^∞. In view of the uniform boundedness theorem, this is equivalent to replacing the condition on $\{y_n\}$ by the requirement that $\{y_n\}$ be bounded in Y.

In view of (3.6.17), the nuclearity of $\lambda_\infty(P_0)$ ensures that an $T \in L(X, Y)$ is $\lambda_\infty(P_0)$-nuclear if and only if it is pseudo $\lambda_\infty(P_0)$-nuclear. In a more general sequence space λ, the notions of $\lambda_\infty(P_0)$-nuclearity and of pseudo $\lambda_\infty(P_0)$-nuclearity are quite different (see Dubinsky and Ramanujan [1]).

It is clear that the set of all $\lambda_\infty(P_0)$-nuclear maps from X into Y, denoted by $N_{\lambda_\infty(P_0)}(X, Y)$, is a proper subset of $N_{\ell^1}(X, Y)$, and that the composition of two continuous linear maps, in which one of them is $\lambda_\infty(P_0)$-nuclear, is $\lambda_\infty(P_0)$-nuclear; but the sum of two $\lambda_\infty(P_0)$-nuclear maps is, in general, not $\lambda_\infty(P_0)$-nuclear, this depends on the additivity of $\lambda_\infty(P_0)$ in the following sense: The space $\lambda_\infty(P_0)$ is said to be additive if for any two Banach spaces X and Y, $N_{\lambda_\infty(P_0)}(X, Y)$ is closed under the operation of addition.

The following characterization of $\lambda_\infty(P_0)$-nuclearity can be proved in the usual way.

(3.8.2) Theorem. <u>The following statements are equivalent.</u>

(a) E <u>is</u> $\lambda_\infty(P_o)$-<u>nuclear</u>.

(b) <u>For any continuous seminorm</u> p <u>on</u> E <u>there is a continuous seminorm</u> r <u>on</u> E <u>with</u> $p \leq r$ <u>such that</u> $Q_{p,r} \in N_{\lambda_\infty(P_o)}(\tilde{E}_r, \tilde{E}_p)$.

(c) <u>For any</u> $W \in \mathcal{U}_E$ <u>there is an</u> $V \in \mathcal{U}_E$ <u>which is absorbed by</u> W <u>such that</u> $[\alpha_n(V, W)] \in \lambda_\infty(P_o)$.

Subspaces, quotient spaces and completions of $\lambda_\infty(P_o)$-nuclear spaces are also $\lambda_\infty(P_o)$-nuclear; but stability of the class of $\lambda_\infty(P_o)$-nuclear spaces under the operations of taking cartesian product and direct sum depends on the additivity of $\lambda_\infty(P_o)$ (see Terzioǧlu [6]).

Kōmura's universal theorem for nuclear spaces has been extended to the $\lambda_\infty(P_o)$-nuclear spaces by Ramanujan [1] and Mori [1].

(3.8.3) Theorem. <u>Let</u> P_o <u>be a countable power set of infinite type such that</u> $\lambda_\infty(P_o)$ <u>is additive, and let</u> $\lambda_\infty(P_o)$ <u>be a nuclear space</u>. <u>Then a locally convex space</u> E <u>is</u> $\lambda_\infty(P_o)$-<u>nuclear if and only if it is topologically isomorphic to a subspace of the product space</u> $(\lambda_\infty(P_o)'_\beta)^\wedge$ <u>for some index set</u> $\wedge$, <u>where</u> $\lambda_\infty(P_o)'_\beta$ <u>is the strong dual of</u> $\lambda_\infty(P_o)$.

In extending a result of Dynin and Mitiagin [1], Mori [2] has shown that every $\lambda_\infty(P_o)$-nuclear Fréchet space with a basis is isomorphic to a $\lambda_\infty(P_o)$-nuclear Köthe space. By modifying the construction of Djakow and Mitiagin [1], Mori [3] also gives an example of a $\lambda_\infty(P_o)$-nuclear Fréchet space without basis.

We now turn our attention to the $\lambda_\infty(P_0)$-nuclearity of Köthe spaces. We first state the following Grothendieck-Pietsch-Köthe criterion (see Köthe [4]).

(3.8.4) Theorem. Let K be a Köthe set and let $\mathcal{T}_K$ be the natural topology determined by K. Then $(\lambda(K), \mathcal{T}_K)$ is $\lambda_\infty(P_0)$-nuclear if and only if for any $a = [a_n] \in K$ there exists an $b = [b_n] \in K$ and an injection $\sigma : \mathbb{N} \to \mathbb{N}$ such that

$$\sigma(\mathbb{N}) = \{n \in \mathbb{N} : a_n \neq 0\} \quad \text{and} \quad \left[\frac{a_{\sigma(n)}}{b_{\sigma(n)}} \right] \in \lambda_\infty(P_0) .$$

Following Köthe [4] a Köthe space $(\lambda(K), \mathcal{T}_K)$ is called a uniformly $\lambda_\infty(P_0)$-nuclear space if there is a bijection $\sigma : \mathbb{N} \to \mathbb{N}$ such that for any $[a_n] \in K$ there exist $[b_n] \in K$ and $[\zeta_n] \in \lambda_\infty(P_0)$ with

$$a_{\sigma(n)} \leq \zeta_n b_{\sigma(n)} \quad \text{for all} \quad n \in \mathbb{N} .$$

The following two results, due to Terzioğlu [5], should be compared with (3.6.12) and (3.6.18) respectively.

(3.8.5) Proposition. For an G_∞-space $\lambda_\infty(K)$, the following statements are equivalent.

(a) $\lambda_\infty(K)$ is $\lambda_\infty(P_0)$-nuclear.

(b) $\lambda_\infty(K)$ is uniformly $\lambda_\infty(P_0)$-nuclear.

(c) For any $[a_n] \in K$ there exists an $[b_n] \in K$ such that

$$a_n \leq b_n \quad \text{and} \quad \left[\frac{1}{b_n}\right] \in \lambda_\infty(P_0) .$$

(d) <u>There exists an</u> $[a_n] \in K$ <u>such that</u> $[\frac{1}{a_n}] \in \lambda_\infty(P_o)$.

From part (d) of (3.8.5), we see that $\lambda_\infty(P_o)$ is not a $\lambda_\infty(P_o)$-nuclear space (in particular, the space s is nuclear but not s-nuclear), and that if the G_∞-space $\lambda_\infty(K)$ is $\lambda_\infty(P_o)$-nuclear, then $\lambda_\infty(K) \subset \lambda_\infty(P_o)$. Furthermore, if $[\theta_n]$ is an increasing sequence of positive numbers, then the power series space $\lambda_\infty([\theta_n])$ of infinite type is $\lambda_\infty(P_o)$-nuclear if and only if there exists an $t \in (0, 1)$ such that

$$[t^{\theta_n}] \in \lambda_\infty(P_o).$$

This is a generalization of (3.6.14).

(3.8.6) Proposition. <u>For a</u> G_1-<u>space</u> $\lambda_1(Q)$ <u>the following statements are equivalent</u>.

(a) $\lambda_1(Q)$ <u>is</u> $\lambda_\infty(P_o)$-<u>nuclear</u>.

(b) $\lambda_1(Q)$ <u>is uniformly</u> $\lambda_\infty(P_o)$-<u>nuclear</u>.

(c) $Q \subset \lambda_\infty(P_o)$.

By combining (3.8.5) and (3.8.6), we see that the power series space $\lambda_\infty([n^m])$ of infinite type is $\lambda_\infty([\theta_n])$-nuclear, where $\theta_n = n^{m-\delta}$ and $0 < \delta \leq m$, and that the space $\Gamma = \lambda_\infty([n])$ of coefficients of power series expansions of entire functions is $\lambda_\infty([n^\delta])$-nuclear whenever $\delta \in [0, 1)$.

For the $\lambda_\infty(P_o)$-nuclearity of the strong duals of smooth sequence spaces, Terizoğlu [5] has obtained the following two results.

(3.8.7) Proposition. *Let $\lambda_\infty(K)$ be a nuclear G_∞-space which contains an element $[x_n]$ with $x_n \neq 0$ for all $n \in \mathbb{N}$. Then the strong dual of $\lambda_\infty(K)$ is isomorphic to a dense subspace of a uniformly $\lambda_\infty(K)$-nuclear G_1-space. Moreover, if in addition, $\lambda_\infty(K)$ is $\lambda_\infty(P_o)$-nuclear, then the strong dual of $\lambda_\infty(K)$ is also $\lambda_\infty(P_o)$-nuclear.*

(3.8.8) Proposition. *Let $\lambda_1(Q)$ be a nuclear G_1-space. Then the strong dual $\lambda_1(Q)'_\beta$ of $\lambda_1(Q)$ is isomorphic to a dense subspace of a nuclear G_∞-space. Moreover, if in addition, $\lambda_1(Q)'_\beta$ is $\lambda_\infty(P_o)$-nuclear, then so does $\lambda_1(Q)$.*

For the associated nuclearity of the strong duals of Köthe spaces, we have the following result which generalizes Terizoğlu [4, (3.1)] and Ramanujan [1, Prop.8].

(3.8.9) Proposition. *Let K be a Köthe set, let $\mathcal{T}_K$ be the natural topology determined by K, and let $\lambda_\infty(P_o)$ and $\lambda_\infty(P_1)$ be nuclear G_∞-spaces. If $(\lambda(K), \mathcal{T}_K)$ is uniformly $\lambda_\infty(P_o)$-nuclear and $\lambda_\infty(P_o)$ is $\lambda_\infty(P_1)$-nuclear, then the strong dual $\lambda(K)'_\beta$ of $(\lambda(K), \mathcal{T}_K)$ is $\lambda_\infty(P_1)$-nuclear.*

Proof. In view of (3.6.6)(b), the nuclearity of $(\lambda(K), \mathcal{T}_K)$ ensures that $\lambda(K)'_\beta$ is dense in the Köthe space $\lambda(L)$ equipped with the natural topology, where the Köthe set L is given by

$$L = \{\eta = [\eta_n] \in \lambda(K) : \eta_n \geq 0 \quad \text{for all} \quad n \geq 1\}.$$

Thus, for proving the $\lambda_\infty(P_1)$-nuclearity of $\lambda(K)'_\beta$, it suffices to verify the $\lambda_\infty(P_1)$-nuclearity of $\lambda(L)$. In view of (3.8.4), it is required to show that for any $x = [x_n] \in L$ there exist an $y = [y_n] \in L$ and an injection $\pi : \mathbb{N} \to \mathbb{N}$ such that

$$\pi(\mathbb{N}) = \{n \in \mathbb{N} : x_n \neq 0\} \quad \text{and} \quad [\frac{x_{\pi(n)}}{y_{\pi(n)}}] \in \lambda_\infty(P_1) .$$

In fact, the uniform $\lambda_\infty(P_0)$-nuclearity of $(\lambda(K), \mathcal{T}_K)$ ensures that there is a bijection $\sigma : \mathbb{N} \to \mathbb{N}$ such that for any $[a_n] \in K$ there exist an $[b_n] \in K$ and $[\zeta_n] \in \lambda_\infty(P_0)$ with

$$a_{\sigma(n)} \leq \zeta_n b_{\sigma(n)} \quad \text{for all} \quad n \in \mathbb{N} .$$

Since $\lambda_\infty(P_0)$ is $\lambda_\infty(P_1)$-nuclear, it follows from (3.8.5)(d) that there is an $[d_n] \in P_0$ such that $[\frac{1}{d_n}] \in \lambda_\infty(P_1)$, and hence that

$$\sum_{n=1}^{\infty} \zeta_n d_n < \infty .$$

Now for any $x = [x_n] \in L$ we have

$$\sum_{n=1}^{\infty} x_n b_n = \sum_{n=1}^{\infty} x_{\sigma(n)} b_{\sigma(n)} < \infty$$

because $[b_n] \in K$, thus

$$0 \leq \sum_n x_{\sigma(n)} a_{\sigma(n)} d_n \leq \sum_n x_{\sigma(n)} \zeta_n b_{\sigma(n)} d_n < \infty .$$

Since σ^{-1} is bijective, it follows that

$$\sum_n x_{\sigma(\sigma^{-1}(n))} a_{\sigma(\sigma^{-1}(n))} d_{\sigma^{-1}(n)} = \sum_n x_n a_n d_{\sigma^{-1}(n)} < \infty ,$$

and hence that $[x_n d_{\sigma^{-1}(n)}] \in L \subset \lambda(K)$ since $[a_n] \in K$ was arbitrary. Now the element $[y_n]$ in L, defined by

$$y_n = x_n d_{\sigma^{-1}(n)} \quad \text{for all} \quad n \geq 1 ,$$

has the required property.

CHAPTER 4. TENSOR PRODUCTS

4.1 The π-norm and the ε-norm on a tensor product

Let X, Y and Z be vector spaces over the same field, let $B^*(X, Y; Z)$ be the vector space of all bilinear maps from $X \times Y$ into Z, and $\omega \in B^*(X, Y; Z)$. We say that X and Y are ω-linearly disjoint if the subset $\{\omega(x_i, y_j) : 1 \leq i \leq m, 1 \leq j \leq n\}$ of Z is linearly independent whenever $\{x_i : 1 \leq i \leq m\}$ and $\{y_j : 1 \leq j \leq n\}$ and linearly independent subsets of X and Y respectively. It is not hard to show that the ω-linear disjointness of X and Y is equivalent to the following condition:

> Let $\Sigma_{i=1}^{n} \omega(u_i, v_i) = 0$. If $\{u_1, \ldots, u_n\}$ is linearly independent, then $v_i = 0$ $(i = 1, \ldots, n)$, and if $\{v_1, \ldots, v_n\}$ is linearly independent, then $u_i = 0$ $(i = 1, \ldots, n)$.

By a tensor product of X and Y we mean a pair (M, ω), where M is a vector space and $\omega \in B^*(X, Y, M)$, satisfying the following two conditions:

(i) M is the linear hull of $\omega(X \times Y)$, and

(ii) X and Y are ω-linearly disjoint.

For any two vector spaces X and Y there always exists a tensor product (for a proof, see, for instance, Greub [1] or Treves [1]). Let

(M, ω) be a tensor product of X and Y. From the condition (ii), it is easily seen that for any vector space Z, the map $T \mapsto T \circ \omega$ $(T \in L^*(M, Z))$ is an algebraic isomorphism from $L^*(M, Z)$ onto $B^*(X, Y; Z)$. In view of this result, it is easy to verify that the vector space M is determined by X and Y up to an algebraic isomorphism. The tensor product of X and Y is denoted by $X \otimes Y$; we write

$$\omega(x, y) = x \otimes y \quad \text{for any} \quad x \in X \quad \text{and} \quad y \in Y,$$

and ω is called the <u>canonical map</u>.

(4.1.1) <u>Examples</u>. (a) Let F be a locally convex space, let K be a compact Hausdorff space, let $C_f(K, F)$ be the vector space of all continuous maps from K into F whose images are finite-dimensional, and let

$$\omega(f, y) = f(\cdot)y \quad \text{for any} \quad f \in C(K) \quad \text{and} \quad y \in F.$$

Then $(C_f(K, F), \omega)$ is the tensor product of $C(K)$ and F.

<u>Proof</u>. It is easily seen that ω is bilinear, and that $C(K)$ and F are ω-linearly disjoint. Let $\psi \in C_f(K, F)$ and let $\{e_1, \ldots, e_m\}$ be a basis in the vector subspace of F generated by $\psi(K)$. For any $t \in K$ there exists uniquely $\{g_1(t), \ldots, g_m(t)\}$ such that

$$\psi(t) = \Sigma_{i=1}^{m} g_i(t)e_i \quad .$$

Let $v_i' \in F'$ be such that

$$<e_i, v_j'> = \delta_i^{(j)} \quad (i, j = 1, \ldots, m).$$

Then $\langle\psi(t), v'_j\rangle = g_j(t)$ $(j = 1, \ldots, m)$, hence $g_j \in C(K)$. As

$$\psi = \Sigma_{j=1}^{m} \omega(g_j, e_j) \quad ,$$

we conclude that $C_f(K, F)$ is the linear hull of $\omega(C(K) \times F)$, and hence that $C_f(K, F)$ and $C(K) \otimes F$ are algebraically isomorphic.

Let F be a locally convex space and let Λ be a non-empty set. A family $[y_\iota, \Lambda]$ in F is called a <u>null family</u> if for any o-neighbourhood U in F there exists an $\alpha \in \mathfrak{F}(\Lambda)$ such that

$$y_\iota \in U \quad \text{for all} \quad \iota \notin \alpha \quad .$$

In particular, a family $[\zeta_\iota, \Lambda]$ of numbers is a null family if for any $\delta > 0$ there is an $\alpha \in \mathfrak{F}(\Lambda)$ such that

$$|\zeta_\iota| < \delta \quad \text{for all} \quad \iota \notin \alpha \ .$$

The set consisting of all null families in F (respectively $\mathbb{C}$) with the same index set Λ is a vector space, and denoted by $c_o(\Lambda, F)$ (respectively c_Λ) . Denote by $c_{of}(\Lambda, F)$ the vector subspace of $c(\Lambda, F)$ consisting of elements each of which is contained in a finite-dimensional subspace of F , and let us define

$$\omega([\zeta_\iota, \Lambda], y) = [\zeta_\iota y, \Lambda] \quad \text{for any} \quad [\zeta_\iota, \Lambda] \in c_\Lambda \quad \text{and} \quad y \in F. \qquad (1.1)$$

Then ω is clearly a bilinear map from $c_\Lambda \times F$ into $c_{of}(\Lambda, F)$. Furthermore, we have:

(b) $(c_{of}(\Lambda, F), \omega)$ is the tensor product of c_Λ and F .

<u>Proof</u>. Let $\{[\zeta_{\iota}^{(k)}, \wedge] : k = 1, \ldots, m\}$ and $\{y^{(1)}, \ldots, y^{(n)}\}$ be linearly independent subsets of $c_\wedge$ and F respectively, and let

$$\sum_{j=1}^{n} \sum_{k=1}^{m} \eta_j^{(k)} \, \omega([\zeta_\iota^{(k)}, \wedge], y^{(j)}) = 0 .$$

Then

$$\sum_{j=1}^{n} \sum_{k=1}^{m} \eta_j^{(k)} \zeta_\iota^{(k)} y^{(j)} = 0 \text{ for all } \iota \in \wedge ,$$

hence

$$\sum_{k=1}^{m} \eta_j^{(k)} \zeta_\iota^{(k)} = 0 \quad \text{for all } \iota \in \wedge \text{ and } j = 1, \ldots, n$$

since $y^{(1)}, \ldots, y^{(n)}$ are linearly independent, and thus

$$\sum_{k=1}^{m} \eta_j^{(k)} [\zeta_\iota^{(k)}, \wedge] = 0 \text{ for all } j = 1, \ldots, n ,$$

which implies that

$$\eta_j^{(k)} = 0 \quad \text{for all } \quad j = 1, \ldots, n \text{ and } k = 1, \ldots, m$$

because $\{[\zeta_\iota^{(k)}, \wedge] : k = 1, \ldots, m\}$ is linearly independent. Therefore $c_\wedge$ and F are ω-linearly disjoint.

Finally, let $[y_\iota, \wedge] \in c_{of}(\wedge, F)$ and let $\{e_1, \ldots, e_m\}$ be a basis in the vector subspace of F generated by the set $\{y_\iota : \iota \in \wedge\}$. Then each y_ι has a unique representation:

$$y_\iota = \sum_{r=1}^{m} \zeta_\iota^{(r)} e_r . \qquad (1.2)$$

Let $v_1', \ldots, v_m'$, in F', be such that $\langle e_r, v_k' \rangle = \delta_r^{(k)}$. Then

$$\langle y_\iota, v_k' \rangle = \zeta_\iota^{(k)} \quad \text{for all } k = 1, \ldots, m \text{ and } \iota \in \wedge . \qquad (1.3)$$

For any $\delta > 0$, the set

$$U = \{y \in F : |\langle y, v'_k\rangle| \leq \delta \quad \text{for all} \quad k = 1, \ldots, m\}$$

is a o-neighbourhood in F , hence there is an $\alpha \in \mathfrak{F}(\Lambda)$ such that

$$y_i \in U \quad \text{for all} \quad i \notin \alpha .$$

It then follows from (1.3) that

$$|\zeta_i^{(k)}| \leq \delta \quad \text{for all} \quad i \notin \alpha \quad \text{and} \quad k = 1, \ldots, m$$

or, equivalently, $[\zeta_i^{(k)}, \Lambda] \in c_\Lambda$ for all $k = 1, 2, \ldots, m$. In view of (1.2) and (1.1),

$$[y_i , \Lambda] = \Sigma_{r=1}^{m} \omega([\zeta_i^{(r)}, \Lambda], e_r) ,$$

thus $c_{of}(\Lambda, F)$ is the linear hull of $\omega(c_\Lambda \times F)$. Consequently $c_{of}(\Lambda, F)$ and $c_\Lambda \otimes F$ are algebraically isomorphic.

Let $\ell_f^1[\Lambda, F]$ be the vector subspace of $\ell^1[\Lambda, F]$ consisting of all summable families in F with the index set Λ each of which is contained in a finite-dimensional subspace of F , and let ℓ_Λ^1 be the vector space of all summable families of numbers with the index set Λ . A similar argument given in the proof of (b) yields the following:

(c) $\ell_f^1[\Lambda, F]$ and $\ell_\Lambda^1 \otimes F$ are algebraically isomorphic.

Let λ be a perfect sequence space, let $\lambda_f[F]$ be the vector space consisting of all absolutely λ-summable sequences in F each of which is contained in a finite-dimensional subspace of F , and let

$$\omega([\zeta_n], y) = [\zeta_n y] \quad \text{for all} \quad [\zeta_n] \in \lambda \quad \text{and} \quad y \in F . \tag{1.4}$$

Then ω is clearly a bilinear map from $\lambda \times F$ into $\lambda_f[F]$. By a similar argument given in the proof of (b), it can be shown easily that λ and F are ω-linearly disjoint. Furthermore, we have the following

(d) $\lambda_f[F]$ and $\lambda \otimes F$ are algebraically isomorphic.

<u>Proof</u>. It suffices to show that $\lambda_f[F]$ is the linear hull of $\omega(\lambda, F)$. To this end, let $[y_n] \in \lambda_f[F]$ and let $\{e_1, \ldots, e_m\}$ be a basis in the vector subspace of F generated by the set $\{y_n : n \in \mathbb{N}\}$. Then each y_n has a unique representation

$$y_n = \Sigma_{r=1}^{m} \zeta_n^{(r)} e_r . \tag{1.5}$$

Let $v'_1, \ldots, v'_m$, in F' , be such that $\langle e_r, v'_k \rangle = \delta_r^{(k)}$. Then

$$\langle y_n, v'_k \rangle = \zeta_n^{(k)} \quad \text{for all} \quad k = 1, \ldots, m \quad \text{and} \quad n \in \mathbb{N} . \tag{1.6}$$

The set

$$U = \{y \in F : |\langle y, v'_k \rangle| \leq 1 \quad \text{for all} \quad k = 1, \ldots, m\}$$

is an absolutely convex o-neighbourhood in F , hence the gauge of U , denoted by q_U , is a continuous seminorm on F . As $[y_n] \in \lambda_f[F]$, it follows that $[q_U(y_n)] \in \lambda$, and hence from (1.6) that

$$|\zeta_n^{(k)}| \leq q_U(y_n) \quad \text{for all} \quad k = 1, \ldots, m \quad \text{and} \quad n \in \mathbb{N} .$$

Therefore $[\zeta_n^{(k)}] \in \lambda$ for all $k = 1, \ldots, m$ since λ is perfect, and thus $\lambda_f[F]$ is the linear hull of $\omega(\lambda, F)$ by (1.4) and (1.5).

(e) Let X and Y be vector spaces and let X' (resp. Y') be a vector subspace of X^* (resp. Y^*) which separates points of X (resp. Y). The map $\psi : X \times Y \rightarrow L^f(X, Y)$, defined by

$$\omega(x, y)(x') = \langle x, x' \rangle y \quad \text{for all} \quad x' \in X' ,$$

is clearly a bilinear map; X and Y are ω-linearly disjoint, and $L^f(X, Y)$ is the linear hull of $\omega(X \times Y)$; thus $L^f(X, Y)$ and $X \otimes Y$ are algebraically isomorphic. Also the map $\omega : X \times Y \rightarrow (B^*(X, Y))^*$, defined by

$$\omega(x, y)(f) = f(x, y) \quad \text{for all} \quad f \in B^*(X, Y) , \tag{1.7}$$

is bilinear such that X and Y are ω-linearly disjoint, hence the vector subspace of $(B^*(X, Y))^*$ generated by $\omega(X \times Y)$ is algebraically isomorphic with $X \otimes Y$. Consequently $\Sigma_{i=1}^{n} x_i \otimes y_i = \Sigma_{j=1}^{m} u_j \otimes v_j$ if and only if

$$\Sigma_{i=1}^{n} f(x_i, y_i) = \Sigma_{j=1}^{m} f(u_j, v_j) \quad \text{for all} \quad f \in B^*(X, Y)$$

or, equivalently,

$$\Sigma_{i=1}^{n} \langle y_i , Tx_i \rangle = \Sigma_{j=1}^{m} \langle v_j , Tu_j \rangle \quad \text{for all} \quad T \in L^*(X, Y^*)$$

since the map $T \mapsto b_T$ is an algebraic isomorphism from $L^*(X, Y^*)$ onto $B^*(X, Y)$. Finally, the map $\omega : X \times Y \rightarrow B^*(X', Y')$, defined by

$$\omega(x, y)(x', y') = \langle x, x' \rangle \langle y, y' \rangle \quad \text{for all} \quad x' \in X' \quad \text{and} \quad y' \in Y' , \tag{1.8}$$

is bilinear such that X and Y are ω-linearly disjoint, therefore the vector subspace of $B^*(X', Y')$ generated by $\omega(X \times Y)$ is algebraically isomorphic with $X \otimes Y$. Consequently, $\Sigma_{i=1}^{n} x_i \otimes y_i = \Sigma_{j=1}^{m} u_j \otimes v_j$ if and only if

$$\Sigma_{i=1}^{n} \langle x_i , x'\rangle y_i = \Sigma_{j=1}^{m} \langle u_j , x'\rangle v_j \quad \text{for all} \quad x' \in X' ,$$

and this is the case if and only if

$$\Sigma_{i=1}^{n} \langle y_i , y'\rangle x_i = \Sigma_{j=1}^{m} \langle v_j , y'\rangle u_j \quad \text{for all} \quad y' \in Y' .$$

From now on, (X, p) and (Y, q) will denote normed spaces whose Banach duals are denoted by (X', p^*) and (Y', q^*) respectively, where p^* (resp. q^*) is the dual norm of p (resp. q), while the operator norm will be denoted by $|||\cdot|||$. That, E and F are algebraically isomorphic, will be denoted by $E = F$, that, E and F are topologically isomorphic, will be denoted by $E \tilde{=} F$, and that, X and Y are isometrically isomorphic, will be denoted by $X \equiv Y$.

Clearly, an $\psi \in B^*(X, Y)$ is continuous if and only if it is <u>bounded</u> in the sense

$$|\psi(x, y)| \leq \eta p(x) q(y) \quad \text{for all} \quad (x, y) \in X \times Y$$

for some $\eta \geq 0$. Therefore we define for any $\psi \in B(X, Y)$ that

$$||\psi|| = \sup\{|\psi(x, y)| : p(x) \leq 1 \text{ and } q(y) \leq 1\} .$$

$||\cdot||$ is actually a norm which is called the <u>bilinear norm</u>. It is easily seen that the algebraic isomorphism $T \mapsto b_T$ from $L^*(X, Y^*)$ onto $B^*(X, Y)$ induces an isometric isomorphism from the Banach space $(L(X, Y'), |||\cdot|||)$ onto $(B(X, Y), ||\cdot||)$ because of

$$\begin{aligned} |||T||| &= \sup\{q^*(Tx) : p(x) \leq 1\} \\ &= \sup\{|\langle y, Tx\rangle| : p(x) \leq 1 \text{ and } q(y) \leq 1\} = ||b_T|| . \end{aligned}$$

As $B(X, Y)$ is a vector subspace of $B^*(X, Y)$, it follows from (1.7) that each element $x \otimes y$ in $X \otimes Y$ can be regarded as a linear functional on $B(X, Y)$. Moreover,

$$\sup\{|<\psi, x \otimes y>| : \|\psi\| \leq 1\} = \sup\{|\psi(x, y)| : \|\psi\| \leq 1\} \leq p(x)q(y)$$

and

$$\begin{aligned}\sup\{|<\psi, x \otimes y>| &: \|\psi\| \leq 1\} \\ &\geq \sup\{|<x, x'><y, y'>| : p^*(x') \leq 1 \text{ and } q^*(y') \leq 1\} \\ &= p(x)q(y) .\end{aligned}$$

Therefore, $x \otimes y$ can be identified with a continuous linear functional on $(B(X, Y), \|\cdot\|)$ such that

$$\|x \otimes y\|^* = \sup\{|<\psi, x \otimes y>| : \|\psi\| \leq 1\} = p(x)q(y) , \tag{1.9}$$

where $\|\cdot\|^*$ is the dual norm of the bilinear norm. On the other hand, (1.8) shows that each $x \otimes y$ in $X \otimes Y$ can be regarded as a bilinear form on $X' \times Y'$. Moreover,

$$\|x \otimes y\| = \sup\{|<x, x'><y, y'>| : p^*(x') \leq 1, q^*(y') \leq 1\} = p(x)q(y), \tag{1.10}$$

hence we identify $x \otimes y$ with a continuous bilinear form on $X' \times Y'$.

These observations lead to the following definition: For any $u = \Sigma_{i=1}^{n} x_i \otimes y_i \in X \otimes Y$, we define

$$p \otimes_\pi q(u) = \sup\{|<\psi, u>| : \|\psi\| \leq 1, \psi \in B(X, Y)\}$$

and

$$p \otimes_\varepsilon q(u) = \sup\{|\Sigma_{i=1}^{n} <x_i, x'><y_i, y'>| : p^*(x') \leq 1 \text{ and } q^*(y') \leq 1\} .$$

Namely, $p \otimes_\pi q$ is the restriction to $X \otimes Y$ of the dual norm of the bilinear norm, and $p \otimes_\varepsilon q$ is the restriction to $X \otimes Y$ of the bilinear norm on $B(X', Y')$. Therefore, they are all norms on $X \otimes Y$.

$p \otimes_\pi q$ is called the π-norm of p and q (or shortly π-norm), while $p \otimes_\varepsilon q$ is referred to as the ε-norm of p and q (or shortly ε-norm). Denote by $X \otimes_\pi Y$ (resp. $X \otimes_\varepsilon Y$) the normed space $(X \otimes Y, p \otimes_\pi q)$ (resp. $(X \otimes Y, p \otimes_\varepsilon q)$). $X \tilde{\otimes}_\pi Y$ and $X \tilde{\otimes}_\varepsilon Y$ will denote the completion of $X \otimes_\pi Y$ and $X \otimes_\varepsilon Y$ respectively, and the complete norm on $X \tilde{\otimes}_\pi Y$ (resp. $X \tilde{\otimes}_\varepsilon Y$) will be denoted by $p \tilde{\otimes}_\pi q$ (resp. $p \tilde{\otimes}_\varepsilon q$).

From the definition of the ε-norm, it is easily seen that

$$p \otimes_\varepsilon q(\Sigma_{i=1}^n x_i \otimes y_i) = \sup\{q(\Sigma_{i=1}^n \langle x_i, x'\rangle y_i) ; \ p^*(x') \leq 1\}$$
$$= \sup\{p(\Sigma_{i=1}^n \langle y_i, y'\rangle x_i) : q^*(y') \leq 1\} = |||\Sigma_{i=1}^n x_i \otimes y_i||| , \qquad (1.11)$$

where $\Sigma_{i=1}^n x_i \otimes y_i$ is regarded as the continuous linear map from X' into Y with finite rank, obtained by setting

$$\Sigma_{i=1}^n x_i \otimes y_i : x' \longmapsto \Sigma_{i=1}^n \langle x_i, x'\rangle y_i \quad \text{for all } x' \in X' .$$

Therefore, $X \otimes_\varepsilon Y \equiv (L^f(X', Y), |||\cdot|||)$. On the other hand, since a normed space (X, p) is always isometrically isomorphic to a subspace of (X'', p^{**}), we obtain a simpler description of $p^* \otimes_\varepsilon q^*$ than the definition of ε-norms as follows

$$p^* \otimes_\varepsilon q^*(\Sigma_{i=1}^n x_i' \otimes y_i') = \sup\{|\Sigma_{i=1}^n \langle x, x_i'\rangle\langle y, y_i'\rangle| : p(x) \leq 1 \text{ and } q(y) \leq 1\}$$
$$= |||\Sigma_{i=1}^n x_i' \otimes y_i'||| \quad \text{for all } \Sigma_{i=1}^n x_i' \otimes y_i' \in X' \otimes Y' . \qquad (1.12)$$

It is also clear that $X' \otimes_\varepsilon Y \equiv (L^f(X, Y), |||\cdot|||)$ under the map $\Sigma_{i=1}^n x_i' \otimes y_i \longmapsto T$, defined by

$$T : x \mapsto \Sigma_{i=1}^n \langle x, x_i'\rangle y_i \quad \text{for all} \quad x \in X .$$

Therefore, we identify each $\Sigma_{i=1}^n x_i' \otimes y_i \in X' \otimes Y$ with the linear map T defined above. If Y is complete, then $X' \tilde{\otimes}_\varepsilon Y$ can be identified with the closure of $L^f(X, Y)$ in $(L^c(X, Y), |||\cdot|||)$. Furthermore, if X and Y are complete, and if either X' or Y has the approximation property, then

$$X' \tilde{\otimes}_\varepsilon Y \equiv (L^c(X, Y), |||\cdot|||) \tag{1.13}$$

as shown by Grothendieck (see Schaefer [1, p.113]).

A similar argument given in the proof of (1.10) shows that $X' \otimes Y'$ can be identified with a subspace of $(B(X, Y), ||\cdot||)$ such that

$$||x' \otimes y'|| = \sup\{|\langle x, x'\rangle\langle y, y'\rangle| : p(x) \leq 1, q(y) \leq 1\} = p^*(x')q^*(y').$$

Under this identification, the ε-norm $p \otimes_\varepsilon q$ is the gauge of the polar, taken in $X \otimes Y$, of the set

$$\{x' \otimes y' \in X' \otimes Y' : ||x' \otimes y'|| \leq 1\} = \Sigma_X^\circ \otimes \Sigma_Y^\circ ,$$

where Σ_X° is the dual ball in X' of the closed unit ball Σ_X in X, and similarly for Σ_Y°. From this we conclude from the definition of the π-norm that

$$p \otimes_\varepsilon q \leq p \otimes_\pi q \quad \text{on} \quad X \otimes Y .$$

On the other hand, (1.9) and (1.10) show that the ε-norm and the π-norm have the cross-property:

$$p \otimes_{\varepsilon} q(x \otimes y) = p \otimes_{\pi} q(x \otimes y) = p(x)q(y) .$$

Before giving another expression of the π-norm, we first prove the following interesting result.

(4.1.2) Lemma. The following statements hold:

(a) $(X \tilde{\otimes}_{\pi} Y)' = (B(X, Y), \|\cdot\|) = (L(X, Y'), |||\cdot|||)$; consequently,

$$(p \tilde{\otimes}_{\pi} q)^*(f) = \|f \circ \omega\| = |||T_f||| ,$$

where $\omega : X \times Y \to X \otimes Y$ is the canonical map and T_f is defined by

$$\langle y, T_f x\rangle = f(x \otimes y) .$$

(b) $X' \otimes Y' \subset (X \tilde{\otimes}_{\pi} Y)'$.

(c) For any $f \in (X \tilde{\otimes}_{\pi} Y)'$,

$$(p \otimes_{\pi} q)^*(f) = \sup\{|f(x \otimes y)| : p(x) \leq 1, q(y) \leq 1\} ; \quad (1.14)$$

in particular,

$$(p \otimes_{\pi} q)^*(\Sigma_{i=1}^{n} x_i' \otimes y_i') = \sup\{|\Sigma_{i=1}^{n} \langle x, x_i'\rangle\langle y, y_i'\rangle| : p(x) \leq 1, q(y) \leq 1\}$$

$$= |||\Sigma_{i=1}^{n} x_i' \otimes y_i'||| \quad \text{for all} \quad \Sigma_{i=1}^{n} x_i' \otimes y_i' \in X' \otimes Y' . \quad (1.15)$$

Proof. For any $f \in (X \tilde{\otimes}_{\pi} Y)'$,

$$|f \circ \omega(x, y)| = |f(x \otimes y)| \leq (p \otimes_{\pi} q)^*(f)(p(x)q(y)) ,$$

hence $f \circ \omega \in B(X, Y)$ and

$$\|f \circ \omega\| \leq (p \otimes_\pi q)^*(f) \ . \tag{1.16}$$

On the other hand, since $p \otimes_\pi q$ is the dual norm $\|\cdot\|^*$ of $\|\cdot\|$ and since $X \otimes Y$ is a vector subspace of $(B(X, Y), \|\cdot\|)'$, it follows that

$$\|f \circ \omega\| = \sup\{|\langle f \circ \omega, \psi'\rangle| : \psi' \in (B(X, Y))^*, \|\psi'\|^* \leq 1\}$$

$$\geq \sup\{|\langle u, f\rangle| : u \in X \otimes Y, p \otimes_\pi q(u) \leq 1\} = (p \tilde{\otimes}_\pi q)^*(f) \ , \tag{1.17}$$

and hence from (1.16) that

$$(p \otimes_\pi q)^*(f) = \|f \circ \omega\| = \sup\{|f(x \otimes y)| : p(x) \leq 1 \text{ and } q(y) \leq 1\} \ .$$

Conversely, if $\psi \in B(X, Y)$, then there is a unique $g \in (X \otimes Y)^*$ such that $\psi = g \circ \omega$. According to (1.17), $g \in (X \tilde{\otimes}_\pi Y)'$. Therefore, the map $f \mapsto f \circ \omega$ is also an isometric isomorphism from $(X \tilde{\otimes}_\pi Y)'$ onto $(B(X, Y), \|\cdot\|)$.

Finally, (1.15) follows from (1.14) by making use of part (b). Therefore the proof is complete.

In the sequel we denote by Σ_X and Σ_Y the closed unit balls in X and Y respectively. Let us define

$$\Sigma_X \otimes \Sigma_Y = \{x \otimes y \in X \otimes Y : x \in \Sigma_X \text{ and } y \in \Sigma_Y\} \ .$$

As the π-norm is a norm, it follows from (1.9) that if $\Sigma_{i=1}^n x_i \otimes y_i$ is any representation of u, then

$$p \otimes_\pi q(u) = p \otimes_\pi q(\Sigma_{i=1}^n x_i \otimes y_i) \leq \Sigma_{i=1}^n p \otimes_\pi q(x_i \otimes y_i) = \Sigma_{i=1}^n p(x_i)q(y_i) \ ,$$

hence

$$p \otimes_\pi q(u) \leq \inf\{\Sigma_{i=1}^n p(x_i)q(y_i) : u = \Sigma_{i=1}^n x_i \otimes y_i\} .$$

In fact, they are equal as a part of the following result shows.

(4.1.3) Theorem. (a) $p \otimes_\pi q(u) = \inf\{\Sigma_{i=1}^n p(x_i)q(y_i) : u = \Sigma_{i=1}^n x_i \otimes y_i\}$ <u>and</u> $p \otimes_\pi q$ <u>is the gauge of the absolutely convex hull</u> $\Gamma(\Sigma_X \otimes \Sigma_Y)$ <u>of</u> $\Sigma_X \otimes \Sigma_Y$.

(b) <u>If</u> Σ_ε <u>is the closed unit ball in</u> $X \otimes_\varepsilon Y$ <u>and</u> Σ_π' <u>is the closed unit ball in</u> $X' \otimes_\pi Y'$, <u>then</u> $p \otimes_\varepsilon q$ <u>is the gauge of</u> $(\Gamma(\Sigma_X^o \otimes \Sigma_Y^o))^o$ <u>and</u>

$$\Sigma_\varepsilon = (\Sigma_\pi')^o = (\Sigma_X^o \otimes \Sigma_Y^o)^o .$$

<u>Proof</u>. Let us define

$$r(u) = \inf\{\Sigma_{i=1}^n p(x_i)q(y_i) : u = \Sigma_{i=1}^n x_i \otimes y_i\}$$

$$V_0 = \{u \in X \otimes Y : r(u) < 1\} \quad \text{and} \quad V_1 = \{u \in X \otimes Y : r(u) \leq 1\} .$$

We first show that r is the gauge of $\Gamma(\Sigma_X \otimes \Sigma_Y)$ or, equivalently,

$$V_0 \subset \Gamma(\Sigma_X \otimes \Sigma_Y) \subset V_1 .$$

In fact, the inclusion $\Gamma(\Sigma_X \otimes \Sigma_Y) \subset V_1$ is obvious by making use of the definition of r . To prove the inclusion $V_0 \subset \Gamma(\Sigma_X \otimes \Sigma_Y)$, let $u \in V_0$. Then there exist $x_i \in X$ and $y_i \in Y$ $(i = 1, 2, \ldots, n)$ such that

$$u = \Sigma_{i=1}^n x_i \otimes y_i \quad \text{and} \quad \Sigma_{i=1}^n p(x_i)q(y_i) < 1 .$$

Choose $\delta_i > 0$ satisfying $\Sigma_{i=1}^{n} (p(x_i) + \delta_i)(q(y_i) + \delta_i) < 1$, and set

$$\zeta_i = (p(x_i) + \delta_i)(p(y_i) + \delta_i) \quad (i = 1, 2, \ldots, n) .$$

Then $\Sigma_{i=1}^{n} \zeta_i < 1$,

$$\bar{x}_i = (p(x_i) + \delta_i)^{-1} x_i \in \Sigma_X \text{ and } \bar{y}_i = (q(y_i) + \delta_i)^{-1} y_i \in \Sigma_Y ,$$

and $u = \Sigma_{i=1}^{n} \zeta_i (\bar{x}_i \otimes \bar{y}_i) \in \Gamma(\Sigma_X \otimes \Sigma_Y)$, which obtains our assertion.

Therefore r is a seminorm on $X \otimes Y$. It is easily seen that

$$r(x \otimes y) = p(x)q(y) \quad \text{for all } x \in X \text{ and } y \in Y , \tag{1.18}$$

and that $p \otimes_\pi q(u) \leq r(u)$. Therefore r is a norm on $X \otimes Y$ and $(X \tilde{\otimes}_\pi Y)' \subset (X \otimes Y, r)'$. Furthermore, we claim that the map $f \longmapsto f \circ \omega$ is an isometric isomorphism from $(X \otimes Y, r)'$ onto $(B(X, Y), \|\cdot\|)$; consequently, we obtain from (4.1.2) that

$$(X \tilde{\otimes}_\pi Y)' \equiv (X \otimes Y, r)' \quad \text{and} \quad r^* = (p \otimes_\pi q)^* . \tag{1.19}$$

To do this, let $f \in (X \otimes Y, r)'$. Then the formula (1.18) shows that

$$\begin{aligned} r^*(f) &= \sup\{|f(u)| : u \in X \otimes Y, r(u) \leq 1\} \\ &\geq \sup\{|f(x \otimes y)| : r(x \otimes y) \leq 1\} = \|f \circ \omega\| , \end{aligned} \tag{1.20}$$

so that $f \circ \omega \in B(X, Y)$. Consequently, we obtain from (1.20) and (4.1.2) that $f \in (X \tilde{\otimes}_\pi Y)'$ and $(p \tilde{\otimes}_\pi q)^*(f) = \|f \circ \omega\| \leq r^*(f)$. On the other hand, since $p \otimes_\pi q \leq r$, it follows that

$$\|f \circ \omega\| = (p \otimes_{\pi} q)^*(f) = \sup\{|f(u)| : u \in X \otimes Y,\ p \otimes_{\pi} q(u) \leq 1\}$$

$$\geq \sup\{|f(u)| : u \in X \otimes Y,\ r(u) \leq 1\} = r^*(f) .$$

Therefore our assertion is proved.

To complete the proof of part (a), it is sufficient to show that $p \otimes_{\pi} q = r$ because r is the gauge of $\Gamma(\Sigma_X \otimes \Sigma_Y)$. But this is obvious by making use of (1.19).

According to part (a), $p^* \otimes_{\pi} q^*$ is the gauge of $\Gamma(\Sigma_X^o \otimes \Sigma_Y^o)$, hence

$$\text{Int}\, \Sigma'_{\pi} \subset \Gamma(\Sigma_X^o \otimes \Sigma_Y^o) \subset \Sigma'_{\pi} .$$

From these inclusions, it follows that

$$(\Sigma'_{\pi})^o = (\Gamma(\Sigma_X^o \otimes \Sigma_Y^o))^o = (\Sigma_X^o \otimes \Sigma_Y^o)^o ,$$

and hence from (1.11) that $p \otimes_{\varepsilon} q$ is the gauge of $(\Gamma(\Sigma_X^o \otimes \Sigma_Y^o))^o$. Therefore

$$\Sigma_{\varepsilon} = (\Sigma'_{\pi})^o = (\Sigma_X^o \otimes \Sigma_Y^o)^o .$$

<u>Remark</u>. It is known that $X' \otimes Y$ can be identified with $L^f(X, Y)$ by identifying $\Sigma_{i=1}^m x' \otimes y_i$ with $T \in L^f(X, Y)$ obtained from the equation

$$T(x) = \Sigma_{i=1}^m \langle x, x'_i \rangle y_i \qquad \text{for all } x \in X .$$

Under this identification, each $\Sigma_{i=1}^m x'_i \otimes y_i$ is a continuous linear map from X into Y with finite rank. As elements in $L^f(X, Y)$ are nuclear, each $\Sigma_{i=1}^m x'_i \otimes y_i$ is of the form

$$\Sigma_{i=1}^{m} \langle x, x_i' \rangle y_i = \Sigma_{k=1}^{\infty} f_k(x) v_k \quad \text{for all} \quad x \in X ,$$

for some sequences $\{f_k\}$ and $\{v_k\}$ in X' and Y respectively such that $\Sigma_{k=1}^{\infty} p^*(f_k) q(y_k) < +\infty$. Therefore (4.1.3)(a) and the definition of the nuclear norm show that

$$\|\Sigma_{i=1}^{m} x_i' \otimes y_i\|_{(n)} \leq p^* \otimes_\pi q(\Sigma_{i=1}^{m} x_i' \otimes y_i) .$$

We shall show in §4.3 that they are equal (on $L^f(X, Y)$) provided that either X' or Y has the metric approximation property.

In view of (4.1.3)(b), it is natural to ask whether the relation $\Sigma_\pi = (\Sigma_\varepsilon')^0$ holds, where Σ_π is the closed unit ball in $X \otimes_\pi Y$ and Σ_ε' is the closed unit ball in $X' \otimes_\varepsilon Y'$. This relation does not hold in general, but is related to the metric approximation property of X in the following sense.

A Banach space X is said to have the <u>metric approximation property</u> if the identify map on X belongs to the closure of $\{S \in L^f(X) : |||S||| \leq 1\}$ in $L_p(X)$, where $L_p(X)$ is the locally convex space of $L(X)$ equipped with the topology of precompact convergence.

Clearly, any Banach space, that has the metric approximation property, must have the approximation property. The Banach space $C(K)$ and its Banach dual, where K is a compact Hausdorff space, have the metric approximation property. If (Ω, μ) is a finite measure space, then $L_\mu^p(\Omega)$ $(1 \leq p \leq \infty)$ have the metric approximation property.

(4.1.4) Proposition. <u>For a Banach space</u> (X, p) , <u>the following statements are equivalent</u>.

(a) X <u>has the metric approximation property</u>.

(b) <u>The closed unit ball in</u> $X' \otimes_{\varepsilon} X$ <u>is dense in the closed unit ball in</u> $(L(X), |||\cdot|||)$ <u>for the topology of pointwise convergence</u>.

(c) <u>For any Banach space</u> (Y, q) , <u>the closed unit ball</u> Σ'_{ε} <u>in</u> $X' \otimes_{\varepsilon} Y'$ <u>is</u> $\sigma(B(X, Y), X \otimes_{\pi} Y)$<u>-dense in the closed unit ball in</u> $(B(X, Y), \|\cdot\|)$.

(d) <u>For any Banach space</u> (Y, q) <u>there is</u> $\Sigma_{\pi} = (\Sigma'_{\varepsilon})^{\circ}$.

<u>Proof</u>. Since two topologies of pointwise convergence and of precompact convergence coincide on the closed unit ball in $(L(X), |||\cdot|||)$, it follows from $X' \otimes_{\varepsilon} X = (L^f(X), |||\cdot|||)$ that (a) and (b) are equivalent. As $(X \otimes_{\pi} Y)' = (B(X, Y), \|\cdot\|)$ and Σ°_{π} being the closed unit ball in $(B(X, Y), \|\cdot\|)$, it follows from the bipolar theorem that (c) and (d) are equivalent. Therefore we complete the proof by showing that (b) and (c) are equivalent.

(b) $\Rightarrow$ (c) : We first identify $(L(X, Y'), |||\cdot|||)$ with $(B(X, Y), \|\cdot\|)$. For any finite subset A of X and $\delta > 0$, there exists $\Sigma_{i=1}^{n} u'_i \otimes u_i \in X' \otimes X$ with $p^* \otimes_{\varepsilon} p(\Sigma_{i=1}^{n} u'_i \otimes u_i) \leq 1$ such that

$$p(\Sigma_{i=1}^{n} \langle x, u'_i\rangle u_i - x) \leq \delta \quad \text{for all} \quad x \in A \tag{1.21}$$

because of $|||I||| \leq 1$. For any $T \in L(X, Y')$ with $|||T||| \leq 1$, the composition $T \circ (\Sigma_{i=1}^{n} u'_i \otimes u_i)$ belongs to $L^f(X, Y')$ and

$$T \circ (\Sigma_{i=1}^{n} u_i' \otimes u_i) = \Sigma_{i=1}^{n} u_i' \otimes (Tu_i) \in X' \otimes Y' , \tag{1.22}$$

hence

$$p^* \otimes_\varepsilon q^*(\Sigma_{i=1}^{n} u_i' \otimes (Tu_i)) = |||T \circ (\Sigma_{i=1}^{n} u_i' \otimes u_i)||| \leq |||T||| \; |||\Sigma_{i=1}^{n} u_i' \otimes u_i|||$$

$$= |||T|||p^* \otimes_\varepsilon q(\Sigma_{i=1}^{n} u_i' \otimes u_i) \leq 1 ,$$

which shows that $\Sigma_{i=1}^{n} u_i' \otimes (Tu_i) \in \Sigma_\varepsilon'$. As

$$(T \circ (\Sigma_{i=1}^{n} u_i' \otimes u_i))x - Tx = T(\Sigma_{i=1}^{n} \langle x, u_i' \rangle u_i - x) \quad (x \in X) ,$$

and $|||T||| \leq 1$, it follows from (1.21) and (1.22) that

$$q^*((\Sigma_{i=1}^{n} u_i' \otimes Tu_i)x - Tx) \leq |||T|||p(\Sigma_{i=1}^{n} \langle x, u_i' \rangle u_i - x) \leq \delta \quad \text{for all} \quad x \in A ,$$

and hence that Σ_ε' is $\sigma(L(X, Y'), X \otimes_\pi Y)$-dense in the closed unit ball in $(L(X, Y'), |||\cdot|||)$. Translating back to $B(X, Y)$, we see that (b) implies (c).

(c) ⇒ (b) : Denote by U_ε the closed unit ball in $X' \otimes_\varepsilon X$ and by U_ε'' the closed unit ball in $X' \otimes_\varepsilon X''$. In view of Schaefer [2, p.257], U_ε is $\sigma(X' \otimes X'', X \otimes X')$-dense in the unit ball in $L(X, X'')$.

Let $T \in L(X)$ with $|||T||| \leq 1$ and let $x \in X$ be fixed. Then, for any $\delta > 0$ and $x' \in X'$, there exist $u'' \in U_\varepsilon''$ and $u = \Sigma_{i=1}^{n} x_i' \otimes x_i \in U_\varepsilon$ such that

$$|\langle x \otimes x', T \rangle - \langle x \otimes x', u'' \rangle| \leq \delta/_2 \quad \text{and} \quad |\langle x \otimes x', u \rangle - \langle x \otimes x', u'' \rangle| \leq \delta/_2 .$$

Consequently, we obtain

$$|\langle x \otimes x', T \rangle - \langle x \otimes x', u \rangle| \leq \delta . \tag{1.23}$$

As

$$\langle x \otimes x', T\rangle - \langle x \otimes x', u\rangle = \langle Tx, x'\rangle - \Sigma_{i=1}^{n} \langle x, x_i'\rangle\langle x_i , x'\rangle$$

$$= \langle Tx, \Sigma_{i=1}^{n} \langle x, x_i'\rangle x_i - x'\rangle ,$$

it follows from (1.23) that

$$p(Tx - \Sigma_{i=1}^{n} \langle x, x_i'\rangle x_i) \leq \delta$$

because x' is arbitrary. Therefore the above inequality shows that U_ε is dense in the closed unit ball in $L(X)$ for the topology of pointwise convergence.

The following example gives the relationship between the π-norm and the π-topology as well as the relationship between the ε-norm and the ε-topology.

(4.1.5) Examples. (a) <u>For any Banach space</u> (Y, q) , <u>there are</u>

$$\ell^1 \tilde{\otimes}_\pi Y = \ell^1_\pi[Y] \quad \underline{\text{and}} \quad \ell^1 \tilde{\otimes}_\varepsilon Y = \ell^1_\varepsilon(Y) .$$

<u>Proof.</u> Denote by p the usual norm on ℓ^1 and by $\ell^1_f[Y]$ the vector space consisting of all summable sequences in Y each of which is contained in a finite-dimensional subspace of Y . (4.1.1)(c) shows that the map

$$\Sigma_{r=1}^{k} (\beta_n^{(r)}) \otimes y^{(r)} \longmapsto [\Sigma_{r=1}^{k} \beta_n^{(r)} y^{(r)}, \mathbb{N}] = [x_n, \mathbb{N}]$$

is an algebraic isomorphism from $\ell^1 \otimes Y$ onto $\ell^1_f[Y]$. By using (2.1.1) and (4.1.3), it can be shown that

$$q_\pi([x_n]) = \inf\{\Sigma_{r=1}^k p((\beta_n^{(r)}))q(y^{(r)}) : x_n = \Sigma_{r=1}^k \beta_n^{(r)} y^{(r)}\} .$$

Therefore, $\ell^1 \otimes_\pi Y$ is isometrically isomorphic to the subspace $\ell_f^1[Y]$ of $\ell_\pi^1[Y]$. On the other hand, for any $[y_n] \in \ell_\pi^1[Y]$, (2.1.4)(b) shows that the net $\{[y_n^{(\alpha)}], \alpha \in \mathcal{F}(\mathbb{N})\}$ in $\ell_f^1[Y]$ associated with $[y_n]$ converges to $[y_n]$ in $\ell_\pi^1[Y]$. Therefore the completeness of $\ell_\pi^1[Y]$ ensures that $\ell^1 \tilde{\otimes}_\pi Y = \ell_\pi^1[Y]$.

To prove $\ell^1 \tilde{\otimes}_\varepsilon Y = \ell_\varepsilon^1(Y)$, let Σ_Y be the closed unit ball in Y and Σ the closed unit ball in ℓ^1 . Any $[x_n] \in \ell_f^1[Y]$ is of the form

$$x_n = \Sigma_{r=1}^k \beta_n^{(r)} y^{(r)} \quad \text{with} \quad (\beta_n^{(r)}) \in \ell^1 \quad \text{and} \quad y^{(r)} \in Y \ (r=1, 2, \dots, k),$$

hence

$$\begin{aligned} q_\varepsilon([x_n]) &= \sup\{\Sigma_n |\langle x_n, x'\rangle| : x' \in \Sigma_Y^0\} = \sup\{|\Sigma_n \zeta_n \langle x_n, f\rangle| : |\zeta_n| \leq 1, f \in \Sigma_Y^0\} \\ &= \sup\{|\Sigma_{r=1}^k \langle (\beta_n^{(r)}), (\zeta_n)\rangle\langle y^{(r)}, f\rangle| : (\zeta_n) \in \Sigma^0, f \in \Sigma_Y^0\} \\ &= p \otimes_\varepsilon q(\Sigma_{r=1}^k (\beta_n^{(r)}) \otimes y^{(r)}) . \end{aligned}$$

Therefore, $\ell^1 \otimes_\varepsilon Y$ is isometrically isomorphic to a subspace of $\ell_\varepsilon^1(Y)$. Since $\ell_\varepsilon^1(Y)$ is complete, we conclude from the same consideration as $\ell_\pi^1[Y]$ that $\ell^1 \tilde{\otimes}_\varepsilon Y = \ell_\varepsilon^1(Y)$.

(b) <u>Let</u> (Y, q) <u>be a Banach space and let</u> K_ι $(\iota = 1, 2)$ <u>be compact Hausdorff spaces. Denote by</u> $C(K_1, Y)$ <u>the Banach space of all continuous mappings from</u> K_1 <u>into</u> Y <u>equipped with the sup-norm. Then</u>

$$C(K_1) \tilde{\otimes}_\varepsilon Y = C(K_1, Y) \quad \underline{\text{and}} \quad C(K_1) \tilde{\otimes}_\varepsilon C(K_2) = C(K_1 \times K_2) .$$

Proof. By using the Riesz representation theorem and the Krein-Milman theorem, the map $\Sigma_{r=1}^{k} f_r \otimes y_r \longmapsto \Sigma_{r=1}^{k} f_r(\cdot) y_r$ is an isometric isomorphism from $C(K_1) \otimes_\varepsilon Y$ into $C(K_1, Y)$. On the other hand, for any $f \in C(K_1, Y)$ and $\delta > 0$, the uniform continuity of f ensures that there is a finite open covering $\{G_j : j = 1, \ldots, n\}$ of K_1 such that

$$q(f(s) - f(t)) < \delta \quad \text{for all} \quad (s, t) \in G_j \times G_j \quad (j = 1, 2, \ldots, n) .$$

Now let f_j $(j = 1, 2, \ldots, n)$ be positive elements in $C(K_1)$ such that

$$\Sigma_{j=1}^{n} f_j(t) = 1 \quad (t \in K_1) \quad \text{and} \quad f_j(t) = 0 \quad \text{for all} \quad t \notin G_j \quad (j = 1, \ldots, n).$$

If $y_j \in f(G_j)$ $(j = 1, 2, \ldots, n)$, then

$$\|\Sigma_{j=1}^{n} f_j \otimes y_j - f\| < \delta ,$$

hence $C(K_1) \otimes Y$ is dense in $C(K_1, Y)$. Therefore $C(K_1) \tilde{\otimes}_\varepsilon Y \equiv C(K_1, Y)$.

To prove $C(K_1) \tilde{\otimes}_\varepsilon C(K_2) \equiv C(K_1 \times K_1)$, let $\|\cdot\|$ be the sup-norm on $C(K_1 \times K_2)$, let p and q be the sup-norms on $C(K_1)$ and $C(K_2)$ respectively. Any $(f, g) \in C(K_1) \times C(K_2)$ associates a unique element in $C(K_1 \times K_2)$ by the following equation

$$(f, g)(t, s) = f(t)g(s) \quad \text{for all} \quad (t, s) \in K_1 \times K_2 .$$

Hence the universal property of algebraic tensor product ensures that there is a unique linear map $T : C(K_1) \otimes C(K_2) \to C(K_1 \times K_2)$ such that

$$T(f \otimes g)(t, s) = f(t)g(s) \quad \text{for all} \quad (t, s) \in K_1 \times K_2 .$$

It is straightforward to check that

$$\|T(\Sigma_{j=1}^{n} f_j \otimes g_j)\| = \sup\{|\Sigma_{j=1}^{n} f_j(t)g_j(s)| : (t, s) \in K_1 \times K_2\}$$

$$= \sup\{|\Sigma_{j=1}^{n} <f_j, \psi><g_j, \phi>| : p^*(\psi) \leq 1, q^*(\phi) \leq 1\}$$

$$= p \otimes_\varepsilon q(\Sigma_{j=1}^{n} f_j \otimes g_j) .$$

Therefore T can be extended to an isometry from $C(K_1) \tilde{\otimes}_\varepsilon C(K_2)$ into $C(K_1 \times K_2)$, which is denoted by $\tilde{T}$. By the Stone-Weierstrass theorem, the range of $\tilde{T}$ is a dense vector subspace of $C(K_1 \times K_2)$, thus $\tilde{T}$ is surjective by the Banach open mapping theorem. Therefore $C(K_1) \tilde{\otimes}_\varepsilon C(K_2) = C(K_1 \times K_2)$ under the map $\tilde{T}$.

(c) Let Y be a Banach space. It is well-known that ℓ^1 is the Banach dual of c_0, and that ℓ^1 has the approximation property. It follows from (a) and (1.13) that

$$\ell^1_\varepsilon(Y) = \ell^1 \tilde{\otimes}_\varepsilon Y = (L^c(c_0, Y), |||\cdot|||) ;$$

namely, <u>summable sequences in</u> Y <u>are precisely the compact operators from</u> c_0 <u>into</u> Y.

On the other hand, if Ω is a compact Hausdorff space and μ is a strictly positive Radon measure on Ω, then Kakutani's representation theorem ensures that there exists a compact Hausdorff Stonien space K_1 such that $C(K_1)$ is the Banach dual of $L^1_\mu(\Omega)$. As $C(K_1)$ has the approximation property, it follows from (b) and (1.13) that

$$C(K_1, Y) = C(K_1) \tilde{\otimes}_\varepsilon Y = (L^c(L^1_\mu(\Omega), Y), |||\cdot|||) ;$$

namely, <u>continuous mappings from</u> K_1 <u>into</u> Y <u>are precisely the compact operators from</u> $L^1_\mu(\Omega)$ <u>into</u> Y. In particular, if K_2 is a compact Hausdorff space, then

$$C(K_1 \times K_2) \cong C(K_1) \tilde{\otimes}_\varepsilon C(K_2) \cong (L^c(L^1_\mu(\Omega), C(K_2)), |||\cdot|||) ;$$

namely, <u>continuous mappings on</u> $K_1 \times K_2$ <u>are precisely the compact operators from</u> $L^1_\mu(\Omega)$ <u>into</u> $C(K_2)$.

(d) Let Y be a Banach space which has the approximation property. Then part(a) of this example, together with a result of Grothendieck (see (4.3.5)) show that

$$\ell^1_\pi[Y] \cong \ell^1 \tilde{\otimes}_\pi Y \cong (N_{\ell^1}(c_o, Y), \|\cdot\|_{(n)}) ; \tag{1.24}$$

namely, <u>absolutely summable sequences in</u> Y <u>are precisely the nuclear operators from</u> c_o <u>into</u> Y.

On the other hand, we have by (1.24) that

$$(\ell^1_\pi[Y])' \cong (N_{\ell^1}(c_o, Y), \|\cdot\|_{(n)})' \cong (L(\ell^1, Y'), |||\cdot|||) .$$

It then follows from (2.1.5) that <u>equicontinuous sequences in</u> Y' <u>are precisely continuous linear maps from</u> ℓ^1 <u>into</u> Y'.

As $C(K_2)$ has the approximation property, where K_2 is a compact Hausdorff space, it follows that absolutely summable sequences in $C(K_2)$ are precisely nuclear linear maps from c_o in $C(K_2)$, and equicontinuous sequences in $R(K_2)$ are precisely continuous linear mappings from ℓ^1 into $R(K_2)$, where $R(K_2)$ is the Banach space of all Radon measures on K_2.

We return now to consider the Banach dual of $X \tilde{\otimes}_\varepsilon Y$. Observe that

$$X' \otimes Y' \subset (X \tilde{\otimes}_\varepsilon Y)' \subset (X \tilde{\otimes}_\pi Y)' = (B(X, Y), \|\cdot\|) = (L(X, Y'), |||\cdot|||) \quad (1.25)$$

and that

$$\|f \circ \omega\| = (p \tilde{\otimes}_\pi q)^*(f) \leq (p \tilde{\otimes}_\varepsilon q)^*(f) \quad \text{for all} \quad f \in (X \tilde{\otimes}_\varepsilon Y)' ,$$

where $\omega : X \times Y \to X \otimes Y$ is the canonical map. Therefore, we define naturally

$$\mathcal{J}(X, Y) = \{f \circ \omega \in B(X, Y) : f \in (X \tilde{\otimes}_\varepsilon Y)'\} \quad \text{and}$$

$$\|f \circ \omega\|_{(i)} = (p \tilde{\otimes}_\varepsilon q)^*(f) \quad \text{for all} \quad f \in (X \tilde{\otimes}_\varepsilon Y)' .$$

Elements in $\mathcal{J}(X, Y)$ are called <u>integral bilinear forms</u> on $X \times Y$, and $\|\cdot\|_{(i)}$ is referred to as the <u>integral norm</u>.

Formula (1.25) shows that elements in $X' \otimes Y'$ are integral bilinear forms on $X \times Y$. Clearly an $\psi \in B(X, Y)$ is integral if and only if

$$\|\psi\|_{(i)} = \sup\{|\Sigma_{i=1}^{n} \psi(x_i, y_i)| : p \otimes_\varepsilon q(\Sigma_{i=1}^{n} x_i \otimes y_i) \leq 1\} < +\infty . \quad (1.26)$$

Therefore, the map $f \mapsto f \circ \omega$ is an isometric isomorphism from $(X \tilde{\otimes}_\varepsilon Y)'$ onto $(\mathcal{J}(X, Y), \|\cdot\|_{(i)})$, thus we identity $(\mathcal{J}(X, Y), \|\cdot\|_{(i)})$ with the Banach dual $(X \tilde{\otimes}_\varepsilon Y)'$ of $X \tilde{\otimes}_\varepsilon Y$, and integral bilinear forms on $X \times Y$ are regarded as continuous linear functionals on $X \tilde{\otimes}_\varepsilon Y$.

As $(L(X, Y), |||\cdot|||)$ being canonically isomorphic to a subspace of

$(B(X, Y'), \|\cdot\|)$, the corresponding definition of integral bilinear forms can be possibly transferred to the elements in $L(X, Y)$, namely, we define an $T \in L(X, Y)$ to be an <u>integral operator</u> if the bilinear form u_T , defined by

$$u_T(x, y') = \langle Tx, y'\rangle \quad \text{for all} \quad (x, y') \in X \times Y' ,$$

is an integral bilinear form on $X \times Y'$, and the integral norm $\|T\|_{(i)}$ is defined to be the integral norm of u_T . Therefore, an $T \in L(X, Y)$ is integral if and only if

$$\|T\|_{(i)} = \sup\{|\Sigma_{i=1}^{n} \langle Tx_i , y_i'\rangle| : p \otimes_\varepsilon q^*(\Sigma_{i=1}^{n} x_i \otimes y_i') \leq 1\} < +\infty . \quad (1.27)$$

Denote by $L^i(X, Y)$ the set consisting of all integral operators from X into Y . Then the map $T \longmapsto u_T$ is clearly an isometric isomorphism from the normed space $(L^i(X, Y), \|\cdot\|_{(i)})$ onto a subspace of $(\mathcal{J}(X, Y'), \|\cdot\|_{(i)})$. Moreover, if Y is complete, then $(L^i(X, Y), \|\cdot\|_{(i)})$ is isometrically isomorphic to a closed subspace of $(\mathcal{J}(X, Y'), \|\cdot\|_{(i)})$.

(4.1.6) Lemma. <u>Let</u> X, Y <u>and</u> Z <u>be normed spaces, let</u> $T \in L(X, Y)$ <u>and</u> $S \in L(Y, Z)$. <u>If one of them is integral, then</u> $S \circ T \in L^i(X, Z)$ <u>and</u>

$$\|S \circ T\|_{(i)} \leq |||S||| \, \|T\|_{(i)} \quad \text{whenever} \quad T \in L^i(X, Y) ,$$

$$\|S \circ T\|_{(i)} \leq \|S\|_{(i)} |||T||| \quad \text{whenever} \quad S \in L^i(Y, Z) .$$

<u>Proof</u>. Let us define

$$f_T(x \otimes y') = \langle Tx, y'\rangle \text{ and } g_S(y \otimes z') = \langle Sy, z'\rangle \text{ and } h_{ST}(x \otimes z') = \langle S(Tx), z'\rangle.$$

Denote by I the identify map. Then the tensor product map $I \otimes S'$ of I and S', defined by

$$I \otimes S'(\Sigma_{i=1}^{n} x_i \otimes z_i') = \Sigma_{i=1}^{n} x_i \otimes (Sz_i') \quad \text{for all} \quad \Sigma_{i=1}^{n} x_i \otimes z_i' \in X \otimes Z' ,$$

is a continuous linear map from $X \otimes_\varepsilon Z'$ into $X \otimes_\varepsilon Y'$ with

$$|||I \otimes S'||| \leq |||I||| \, |||S'||| = |||S'||| ,$$

and $T \otimes I : X \otimes_\varepsilon Z' \to Y \otimes_\varepsilon Z'$ is continuous with

$$|||T \otimes I||| \leq |||T||| .$$

Furthermore, it is clear that

$$h_{ST} = f_T \circ (I \otimes S') = g_S \circ (T \otimes I) . \tag{1.28}$$

Suppose $T \in L^i(X, Y)$. Then $f_T \in (X \otimes_\varepsilon Y')'$, thus $h_{SoT} \in (X \otimes_\varepsilon Z')'$ and

$$||S\,T||_{(i)} \leq ||T||_{(i)} |||I \otimes S'||| \leq |||S||| \, ||T||_{(i)}$$

by virtue of (1.28). A similar argument yields another inequality whenever $S \in L^i(Y, Z)$.

Suppose that K is a compact Hausdorff space and that μ is a positive Radon measure on K. Then the bilinear form on $C(K) \times C(K)$, defined by

$$\psi : (f, g) \mapsto \int_K f(t)g(t)d\mu(t) \quad \text{for all} \quad (f, g) \in C(K) \times C(K) ,$$

associates a linear functional on $C(K) \otimes C(K)$, which is the restriction to $C(K) \otimes C(K)$ of the Radon measure $\nu : h \mapsto \int h(s, s)d\mu(s)$ on $K \times K$. It is known from (4.1.5)(b) that $C(K) \otimes_{\varepsilon} C(K)$ is isometrically isomorphic to a subspace of $C(K \times K)$, it then follows that ψ is integral. On the other hand, it is clear that ψ is the associated bilinear form of the embedding $J : C(K) \to C(K)'$, defined by

$$\langle f, Jg \rangle = \int_K f(t)g(t)\,dt \quad \text{for all} \quad f \in C(K) .$$

Therefore, J is an integral linear map. This remark together with Kakutani's representation theorem yield the following typical example of integral linear maps.

(4.1.7) Example. Let K be a compact Hausdorff space and μ a positive Radon measure on K. Then the canonical embedding $J : C(K) \to L^1_\mu(K)$ is integral and

$$\|J\|_{(i)} = |||J||| = \mu(1) .$$

We now state a criterion for a bilinear form to be integral. This will be useful in verifying that various examples of continuous linear maps are integral.

(4.1.8) Theorem. An $\psi \in B(X, Y)$ is integral with $\|\psi\|_{(i)} \leq 1$ if and only if there exists a unique Radon measure μ on the $\sigma(X', X) \times \sigma(Y', Y)$-compact set $\Sigma^o_X \times \Sigma^o_Y$ whose total mass ≤ 1 such that

$$\psi(x, y) = \int_{\Sigma^o_X \times \Sigma^o_Y} \langle x, x' \rangle \langle y, y' \rangle \, d\mu(x', y') \quad \text{for all} \quad (x, y) \in X \times Y . \tag{1.29}$$

Proof. The sufficiency follows from (1.26) and

$$|\Sigma_{i=1}^{n}\psi(x_i, y_i)| \leq \int_{\Sigma_X^o \times \Sigma_Y^o} |\Sigma_{i=1}^{n}\langle x_i, x'\rangle\langle y_i, y'\rangle| d|\mu|(x', y')$$

$$\leq \|\mu\| p \otimes_\varepsilon q(\Sigma_{i=1}^{n} x_i \otimes y_i) < \infty .$$

To prove the necessity, we first consider, by means of the evaluation, (X, p) and (Y, q) as normed subspaces of $C(\Sigma_X^o)$ and $C(\Sigma_Y^o)$ respectively, then it is easy to verify that the embedding map $J : X \otimes_\varepsilon Y \to C(\Sigma_X^o) \tilde{\otimes}_\varepsilon C(\Sigma_Y^o)$ is an isometry (into). In view of (4.1.5)(b), $C(\Sigma_X^o) \tilde{\otimes}_\varepsilon C(\Sigma_Y^o) \equiv C(\Sigma_X^o \times \Sigma_Y^o)$, the adjoint map J' of J is a metric homomorphism from $(C(\Sigma_X^o \times \Sigma_Y^o))'$ onto $(X \tilde{\otimes}_\varepsilon Y)'$ by the Banach open mapping theorem because the image of J' is weakly dense in $(X \tilde{\otimes}_\varepsilon Y)'$, thus ψ can be regarded as a continuous linear functional on $C(\Sigma_X^o \times \Sigma_Y^o)$. We now apply the Riesz representation theorem to get a Radon measure μ on $\Sigma_X^o \times \Sigma_Y^o$ with $\|\mu\| \leq \|\psi\|_{(i)} \leq 1$ such that (1.29) holds.

Formula (1.29) can be written symbolically by

$$\psi = \int x' \otimes y' \ d\mu(x', y')$$

which is called the <u>weak integral</u>.

(4.1.9) Corollary. <u>Let</u> $T \in L(X, Y)$ <u>and let</u> $e_Y : Y \to Y''$ <u>be the evaluation map. Then the following statements are equivalent.</u>

(a) $T \in L^i(X, Y)$ <u>with</u> $\|T\|_{(i)} \leq 1$.

(b) <u>There exists a Radon measure</u> μ <u>on the</u> $\sigma(X', X) \times \sigma(Y'', Y')$-<u>compact set</u> $\Sigma_X^o \times \Sigma_Y^{oo}$ <u>whose total mass</u> ≤ 1 <u>such that</u>

$$\langle Tx, y'\rangle = \int_{\Sigma_X^o \times \Sigma_Y^{oo}} \langle x, x'\rangle\langle y', y''\rangle d\mu(x', y'') \text{ for all } (x, y') \in X \times Y' ;$$

<u>in this case, the above equation is written symbolically by</u>

$$Tx = \int_{\Sigma_X^o \times \Sigma_Y^{oo}} \langle x, x'\rangle y'' d\mu(x', y'') .$$

(c) <u>There exist a compact Hausdorff space</u> K <u>and a Radon measure</u> μ on K <u>such that</u> $e_Y \circ T : X \to Y''$ <u>permits a factoring</u>

$$X \xrightarrow{S} C(K) \xrightarrow{J} L^1_\mu(K) \xrightarrow{R} Y'' ,$$

<u>where</u> S <u>and</u> R <u>are continuous linear maps with norms</u> ≤ 1 <u>respectively</u>, <u>and</u> J <u>is the canonical embedding</u> .

<u>Proof</u>. The equivalence of (a) and (b) follows from (4.18) and the definition of integral linear maps. To prove the implication (b) $\Rightarrow$ (c), let $K = \Sigma_X^o \times \Sigma_Y^{oo}$ and let us define

$$(Sx)(x', y'') = \langle x, x'\rangle \quad \text{for all} \quad (x', y'') \in K ,$$

$$R(f) = \int_{\Sigma_X^o \times \Sigma_Y^{oo}} f(x', y'')y'' d\mu(x', y'') \quad \text{for all} \quad f \in L^1_\mu(K) .$$

Then S and R are continuous linear maps with norms ≤ 1 and

$$R(T(Sx)) = \int_{\Sigma_X^o \times \Sigma_Y^{oo}} (Sx)(x', y'')y'' d\mu(x', y'') = \int_{\Sigma_X^o \times \Sigma_Y^{oo}} \langle x, x'\rangle y'' d\mu(x', y'')$$

$$= Tx \quad \text{for all} \quad x \in X .$$

Therefore, the statement (c) holds.

Finally, we show that (c) implies (a). In view of (4.1.6) and (4.1.7), $e_Y T \in L^i(X, Y'')$ with $\|e_Y T\|_{(i)} \leq 1$, hence the linear functional, defined by

$$f(x \otimes y''') = \langle (e_Y T)x, y'''\rangle \quad \text{for all} \quad x \otimes y''' \in X \otimes_\varepsilon Y''' ,$$

is continuous on $X \otimes_\varepsilon Y'''$ with norm ≤ 1; consequently, the restriction of f to $X \otimes_\varepsilon Y'$, denoted by g, is a continuous linear functional of norm ≤ 1 since $X \otimes_\varepsilon Y'$ can be identified with a subspace of $X \otimes_\varepsilon Y'''$. As

$$\langle Tx, y'\rangle = \langle (e_Y T)x, y'\rangle = g(x \otimes y') \quad \text{for all} \quad x \in X \quad \text{and} \quad y' \in Y' ,$$

we conclude that $T \in L^i(X, Y)$ with $\|T\|_{(i)} \leq 1$.

In view of the preceding result, it can be shown that each integral linear map from X into Y is weakly compact.

(4.1.10) Corollary. <u>Let (Ω, ν) be a finite measure space. Then the canonical embedding $J_\infty : L^\infty_\nu(\Omega) \to L^1_\nu(\Omega)$ is integral and</u> $\|J_\infty\|_{(i)} = |||J_\infty||| = \mu(\Omega)$.

<u>Proof.</u> There exists a compact Hausdorff space K such that $L^\infty_\nu(\Omega)$ and $C(K)$ are isometrically isomorphic. Let S be the isometric isomorphism from $L^\infty_\nu(\Omega)$ onto $C(K)$, and suppose that

$$\mu(f) = \int_\Omega (S^{-1}f)(t)\,d\nu(t) \quad \text{for all} \quad f \in C(K) .$$

Then μ is a positive Radon measure on K by the Riesz representation theorem, and

$$\|Jf\|_1 = \int_K |f|\, d\mu = \int_\Omega |(J_\infty S^{-1})f|\, d\nu = \|(J_\infty S^{-1})f\|_1 \quad \text{for all} \quad f \in C(K) ,$$

where $J : C(K) \to L^1_\mu(K)$ is the canonical embedding. Since J_∞ is the canonical embedding, it follows that $\operatorname{Ker}(J_\infty S^{-1}) = \operatorname{Ker} J$, and hence by the density of $J(C(K))$ in $L^1_\mu(K)$ and the completeness of $L^1_\mu(K)$ that there is a continuous linear map $R : L^1_\mu(K) \to L^1_\nu(\Omega)$ with $|||R||| \leq 1$ such that

$$RJ = J_\infty S^{-1} .$$

The result then follows from (4.1.9)(c).

(4.1.11) Corollary. <u>Let</u> $T \in L(X, Y)$ <u>and let</u> $e_Y : Y \to Y''$ <u>be the evaluation map</u>. <u>Then the following statements are equivalent.</u>

(a) $T \in L^i(X, Y)$.

(b) <u>There exist a compact Hausdorff space</u> K <u>and a Radon measure</u> μ <u>on</u> K <u>such that</u> $e_Y T : X \to Y''$ <u>permits a factoring</u>

$$X \xrightarrow{S} L^\infty_\mu(K) \xrightarrow{J_\infty} L^1_\mu(K) \xrightarrow{R} Y'' ,$$

<u>where</u> S <u>and</u> R <u>are continuous with norms</u> ≤ 1 , <u>and</u> J_∞ <u>is the canonical embedding</u>.

(c) <u>There exists a finite measure space</u> (Ω, ν) <u>such that the linear map</u> $e_Y T : X \to Y''$ <u>permits a factoring</u>

$$X \xrightarrow{S} L^\infty_\nu(\Omega) \xrightarrow{J_\infty} L^1_\nu(\Omega) \xrightarrow{R} Y'' ,$$

<u>where</u> S <u>and</u> R <u>are continuous with norms</u> ≤ 1 , <u>and</u> J_∞ <u>is the canonical embedding</u>.

Proof. The implication (a) $\Rightarrow$ (b) follows from (4.1.9), and the implication (b) $\Rightarrow$ (c) is obvious. In view of (4.1.10) and (4.1.6), a similar argument given in the proof of (c) $\Rightarrow$ (a) of (4.1.9) shows that (c) implies (a).

It is known that $(L(X, Y'), |||\cdot|||)$ and $(B(X, Y), ||\cdot||)$ are isometrically isomorphic under the map $T \mapsto b_T$, where b_T is the associated bilinear form of T. On the other hand, the definition of integral linear maps shows that an $T \in L(X, Y')$ is integral if and only if the bilinear form u_T on $X \times Y''$, defined by

$$u_T(x, y'') = \langle Tx, y''\rangle \quad \text{for all} \quad (x, y'') \in X \times Y'' ,$$

is integral on $X \times Y''$. The following result shows that $T \in L^i(X, Y')$ if and only if $b_T \in \mathcal{J}(X, Y)$.

(4.1.12) Theorem. Let $T \in L(X, Y')$. Then $T \in L^i(X, Y')$ if and only if the associated bilinear form b_T of T is integral on $X \times Y$; in this case,

$$\|b_T\|_{(i)} = \|u_T\|_{(i)} . \tag{1.30}$$

Proof. Necessity. We identify u_T with a continuous linear functional on $X \tilde{\otimes}_\varepsilon Y''$. As $X \tilde{\otimes}_\varepsilon Y$ being a closed subspace of $X \tilde{\otimes}_\varepsilon Y''$, and b_T, viewing as a linear functional on $X \otimes_\varepsilon Y$, is the restriction of u_T to $X \otimes_\varepsilon Y$, it follows that b_T is continuous on $X \tilde{\otimes}_\varepsilon Y$ and

$$\|b_T\|_{(i)} \leq \|u_T\|_{(i)} . \tag{1.31}$$

Sufficiency. Suppose $b_T \in \mathcal{J}(X, Y)$ and $\|b_T\|_{(i)} = 1$.

Then (4.1.8) shows that there exists a unique Radon measure μ on $\Sigma_X^o \times \Sigma_Y^o$ whose total mass ≤ 1 such that

$$\langle y, Tx\rangle = \int_{\Sigma_X^o \times \Sigma_Y^o} \langle x, x'\rangle \langle y, y'\rangle d\mu(x', y') \quad \text{for all} \quad (x, y) \in X \times Y .$$

Proceeding as in the proof of (a) $\Rightarrow$ (b) of (4.1.9), T permits a factoring

$$X \xrightarrow{S} C(K) \xrightarrow{J} L^1_\mu(K) \xrightarrow{R} Y' ,$$

where J is the canonical embedding, hence $T \in L^i(X, Y')$ by (4.1.9), and

$$\|T\|_{(i)} = \|u_T\|_{(i)} \leq |||S||| \, \|J\|_{(i)} \, |||R||| \leq 1$$

by (4.1.6), thus $\|u_T\|_{(i)} \leq \|b_T\|_{(i)}$. Consequently, (1.30) holds by (1.31).

(4.1.13) Corollary. <u>For normed spaces</u> X <u>and</u> Y , <u>we have</u>

$$(X \tilde{\otimes}_\varepsilon Y)' \equiv (\mathcal{J}(X, Y), \|\cdot\|_{(i)}) \equiv (L^i(X, Y'), \|\cdot\|_{(i)}) . \tag{1.32}$$

<u>Moreover, if</u> X <u>and</u> Y <u>are complete and if either</u> X' <u>or</u> Y <u>has the approximation property, then</u>

$$(L^c(X, Y), |||\cdot|||)' \equiv (L^i(X', Y'), \|\cdot\|_{(i)}) . \tag{1.33}$$

<u>Proof</u>. As $L(X, Y')$ being algebraically isomorphic with $B(X, Y)$ under the map $T \mapsto b_T$, where b_T is the associated bilinear form of T , the formula (1.32) follows from (4.1.12).

Finally, if either X' or Y has the approximation property, then

$X' \tilde{\otimes}_\varepsilon Y \equiv (L^c(X, Y), |||\cdot|||)$ by formula (1.13), thus (1.33) follows from (1.32).

(4.1.14) Corollary. Let $T \in L(X, Y)$. Then $T \in L^i(X, Y)$ if and only if $T' \in L^i(Y', X')$; in this case, $\|T\|_{(i)} = \|T'\|_{(i)}$.

Proof. In view of (4.1.13), $(L^i(Y', X'), \|\cdot\|_{(i)}) \equiv (Y' \tilde{\otimes}_\varepsilon X)'$. On the other hand, it is known from the definition of integral linear maps that $T \in L^i(X, Y)$ if and only if the bilinear form u_T, defined by

$$u_T(x, y') = \langle Tx, y' \rangle \quad \text{for all} \quad (x, y') \in X \times Y',$$

is integral on $X \times Y'$, hence u_T is identified with a continuous linear functional on $X \tilde{\otimes}_\varepsilon Y'$. It is easily seen that $X \otimes_\varepsilon Y'$ and $Y' \otimes_\varepsilon X$ are isometrically isomorphic, the result then follows.

(4.1.15) Corollary. Let $T \in L(X, Y)$ and let $e_Y : Y \to Y''$ be the evaluation map. If $e_Y T \in L^i(X, Y'')$, then $T' \in L^i(Y', X')$ and $T \in L^i(X, Y)$ and

$$\|T\|_{(i)} = \|T'\|_{(i)} = \|e_Y T\|_{(i)} .$$

Proof. Denote by $e_{Y'} : Y' \to Y'''$ the evaluation. For any $x \in X$ and $y' \in Y'$,

$$\langle x, T'y' \rangle = \langle Tx, y' \rangle = \langle y', (e_Y T)x \rangle = \langle (e_Y T)x, e_{Y'}(y') \rangle$$

$$= \langle x, ((e_Y T)' \circ e_{Y'})y' \rangle ,$$

hence $T' = (e_Y T)' \circ e_{Y'}$. In view of (4.1.4) and (4.1.6), $T' \in L^i(Y', X')$ and

$$\|T'\|_{(i)} \leq \|(e_Y Y)'\|_{(i)} |||e_{Y'}||| = \|e_Y T\|_{(i)} .$$

As $T' \in L^i(Y', X')$, it follows from (4.1.14), (4.1.6) and the above inequalities that $T \in L^i(X, Y)$ and

$$\|T\|_{(i)} = \|T'\|_{(i)} \leq \|e_Y T\|_{(i)} \leq |||e_Y||| \, \|T\|_{(i)} = \|T\|_{(i)} .$$

Therefore we obtain the required equalities.

(4.1.16) Example. Let K_2 be a compact Hausdorff space. By Riesz's representation theorem, the Banach dual of $C(K_2)$ can be identified with the Banach space $R(K_2)$ of all Radon measures on K_2 . It then follows from (4.1.5)(c) and (4.1.13) that

$$(\ell^1_\varepsilon(C(K_2)))' \equiv (L^i(\ell^1, R(K_2)), \|\cdot\|_{(i)}) ,$$

and hence from (2.1.8) that <u>prenuclear sequences in</u> $R(K_2)$ <u>are precisely the integral linear maps from</u> ℓ^1 <u>into</u> $R(K_2)$.

On the other hand, Example (4.1.5)(c), together with (1.33) of (4.1.13) also show that

$$(C(K_1 \times K_2))' \equiv (L^i(C(K_1), R(K_2)), \|\cdot\|_{(i)}) .$$

As $(C(K_1 \times K_2))'$ is identified with the Banach space of all Radon measures on $K_1 \times K_1$, it follows that <u>Radon measures on</u> $K_1 \times K_2$ <u>are precisely the integral operators from</u> $C(K_1)$ <u>into</u> $R(K_2)$.

(4.1.17) Proposition. <u>For two normed spaces</u> (X, p) <u>and</u> (Y, q) , we have

$$N_{\ell^1}(X, Y) \subset L^i(X, Y) \quad \underline{\text{and}} \quad \|T\|_{(i)} \leq \|T\|_{(n)} \quad \underline{\text{for all}} \quad T \in N_{\ell^1}(X, Y) .$$

Proof. Let $T \in N_{\ell^1}(X, Y)$ and $\delta > 0$. There exist two sequences $\{f_n\}$ and $\{y_n\}$ in X' and Y respectively such that

$$\Sigma_{n=1}^{\infty} p^*(f_n) q(y_n) < \|T\|_{(n)} + \delta \quad \text{and} \quad Tx = \Sigma_{n=1}^{\infty} f_n(x) y_n \quad (x \in X) .$$

Without loss of generality one can assume that $p^*(f_n) \neq 0$ and $q(y_n) \neq 0$ for all n. For any $\Sigma_{j=1}^{m} u_j \otimes v_j' \in X \otimes_\varepsilon Y'$,

$$\begin{aligned} \Sigma_{j=1}^{m} \langle Tu_j , v_j'\rangle &= \Sigma_{j=1}^{m} \Sigma_{n=1}^{\infty} f_n(u_j) \langle y_n , v_j'\rangle \\ &= \Sigma_{n=1}^{\infty} p^*(f_n) g(y_n) (\Sigma_{j=1}^{m} \langle u_j , x_n'\rangle \langle v_n , v_j'\rangle) , \end{aligned}$$

where $x_n' = f_n/p^*(f_n)$ and $v_n = y_n/q(y_n)$ belong to the closed unit balls respectively. It follows that

$$\begin{aligned} |\Sigma_{j=1}^{m} \langle Tu_j , v_j'\rangle| &\leq (\Sigma_{n=1}^{\infty} p^*(f_n) q(y_n)) \; p \otimes_\varepsilon q^* (\Sigma_{j=1}^{m} u_j \otimes v_j') \\ &< (\|T\|_{(n)} + \delta) \; p \otimes_\varepsilon q^* (\Sigma_{j=1}^{m} u_j \otimes v_j') , \end{aligned}$$

and hence that $T \in L^i(X, Y)$ and $\|T\|_{(i)} \leq \|T\|_{(n)}$.

The remark after (3.1.1) together with Examples (2.2.7) and (4.1.7) show that integral linear maps need not be nuclear. But, if X and Y are Hilbert spaces, then $(N_{\ell^1}(X, Y), \|\cdot\|_{(n)}) = (L^i(X, Y), \|\cdot\|_{(i)})$ (see Treves [1, p.506]).

We shall show in the next section that integral linear maps are absolutely summing, hence the composite of two integral linear maps is nuclear by a result of Grothendieck (see Pietsch [1, p.66]). More general, Grothendieck has verified, using a strong version of the Dunford-Pettis theorem the following somewhat surprising result: The composite of two continuous linear maps, in which one of them is integral and the another is weakly compact, is integral. By using this result, it is easily seen that if X and Y are Banach spaces for which Y is reflexive, then $L^i(X, Y) = N_{\ell^1}(X, Y)$ because the evaluation map $Y \to Y''$ is weakly compact.

It is known that $N_{\ell^1}(X, Y)$ is a proper vector subspace of $L^i(X, Y)$. However, every integral linear map is the limit in $L_\sigma(X, Y_\sigma)$ of a bounded net in $(N_{\ell^1}(X, Y), \|\cdot\|_{(n)})$ as shown by the following result, due to Holub [2].

(4.1.18) Theorem. <u>An</u> $T \in L(X, Y)$ <u>is integral if and only if there exists a</u> $\|\cdot\|_{(n)}$-<u>bounded net</u> $\{T_\lambda\}$ <u>in</u> $N_{\ell^1}(X, Y)$ <u>such that</u> T_λ <u>converges to</u> T <u>in</u> $L_\sigma(X, Y_\sigma)$, <u>namely, for any</u> $x \in X$ <u>and</u> $y' \in Y'$,

$$\lim_\lambda \langle T_\lambda x, y' \rangle = \langle Tx, y' \rangle .$$

<u>Proof</u>. <u>Necessity</u>. Clearly $p^* \otimes_\varepsilon q^* = (p \otimes_\pi q)^*|_{X' \otimes Y'}$, it follows that $X'' \tilde{\otimes}_\varepsilon Y'$ is a closed subspace of $(X' \tilde{\otimes}_\pi Y)'$, and hence that $X \tilde{\otimes}_\varepsilon Y'$ can be identified with a closed subspace of $(X' \tilde{\otimes}_\pi Y)'$ because $X \tilde{\otimes}_\varepsilon Y'$ is isometrically isomorphic to a closed subspace of $X'' \tilde{\otimes}_\varepsilon Y'$. According to the hypothesis, the linear functional f , defined by

$$f(x \otimes y') = \langle Tx, y' \rangle \quad \text{for all} \quad x \in X \quad \text{and} \quad y' \in Y' , \tag{1.35}$$

is continuous on $X \tilde{\otimes}_\varepsilon Y'$, thus the Hahn-Banach extension theorem ensures that there exists an $h \in (X' \tilde{\otimes}_\pi Y)''$ such that

$$(p^* \tilde{\otimes}_\pi q)^{**}(h) = \|T\|_{(i)} \quad \text{and} \quad h(v) = f(v) \quad \text{for all} \quad v \in X \tilde{\otimes}_\varepsilon Y' . \tag{1.36}$$

As $X' \tilde{\otimes}_\pi Y$ is weakly dense in $(X' \tilde{\otimes}_\pi Y)''$, there is a net $\{u_\lambda\}$ in $X' \tilde{\otimes}_\pi Y$ with $p^* \tilde{\otimes}_\pi q(u_\lambda) \leq \|T\|_{(i)}$ such that for any $u' \in (X' \tilde{\otimes}_\pi Y)'$

$$\langle u_\lambda, u' \rangle \to \langle h, u' \rangle .$$

In particular, by (1.35) and (1.36) we have for any $x \in X$ and $y' \in Y'$ that

$$\langle u_\lambda, x \otimes y' \rangle \to \langle h, x \otimes y' \rangle = \langle Tx, y' \rangle \tag{1.37}$$

because of $X \tilde{\otimes}_\varepsilon Y' \subset (X' \tilde{\otimes}_\pi Y)'$. Since each $u_\lambda \in X' \tilde{\otimes}_\pi Y$ defines an $T_\lambda \in N_{\ell^1}(X, Y)$ with $\|T_\lambda\|_{(n)} \leq p^* \tilde{\otimes}_\pi q(u_\lambda)$, it follows from (1.37) that the net $\{T_\lambda\}$ has the required properties.

<u>Sufficiency</u>. Let us define

$$f_\lambda(x \otimes y') = \langle T_\lambda x, y' \rangle \text{ and } f(x \otimes y') = \langle Tx, y' \rangle \text{ for all } x \in X \text{ and } y' \in Y' .$$

Then $\{f_\lambda\}$ is a bounded net in $(X \tilde{\otimes}_\varepsilon Y')'$ by the hypotheses and (4.1.17). In view of the definition of integral operators, we have only to show that

$f \in (X \otimes_\varepsilon Y')'$. Indeed, for any $\delta > 0$ and $u = \Sigma_{i=1}^n x_i \otimes y_i' \in X \otimes_\varepsilon Y'$, there is an λ such that

$$|f(u)| < |f_\lambda(u)| + \delta \leq (\sup_\lambda \|T_\lambda\|_{(i)})\, p \otimes_\varepsilon q^*(u) + \delta ,$$

hence the continuity of f follows.

It should be noted that in general one cannot replace 'net' by 'sequence' in (4.1.18). For example, let $J_\infty : L^\infty(\mu) \to L^1(\mu)$ be the embedding map for a measure space (Ω, Σ, μ) in which $J_\infty(L^\infty(\mu))$ is not separable in $L^1(\mu)$. Then J_∞ is integral by (4.1.10). If there is a sequence $\{T_n\}$ in $N_{\ell^1}(L^\infty(\mu), L^1(\mu))$ such that T_n converges to J_∞ in $L_\sigma(L^\infty(\mu), (L^1(\mu))_\sigma)$, then $T_n(L^\infty(\mu))$ is separable because each T_n is compact, hence the closed convex hull of $\cup_n T_n(L^\infty(\mu))$, denoted by $\overline{C}$, is separable. Thus there exists an $h \in L^\infty(\mu)$ such that $J_\infty(h) \notin \overline{C}$. Since $\overline{C}$ is $\sigma(L^1(\mu), L^\infty(\mu))$-closed, it follows that for any $g \in L^\infty(\mu)$, $\langle T_n h, g\rangle$ does not converge to $\langle J_\infty h, g\rangle$ which gives a contradition.

From the proof of the sufficiency of (4.1.18), we see that if $\{T_\lambda\}$ is a $\|\cdot\|_{(i)}$-bounded net in $L^i(X, Y)$ such that T_λ converges to T in $L_\sigma(X, Y_\sigma)$, then $T \in L^i(X, Y)$. On the other hand, by using the same argument given in the proof of the sufficiency of (4.1.18), it can be easily shown that if $T \in L(X, Y')$ and if there is a $\|\cdot\|_{(i)}$-bounded net $\{T_\lambda\}$ in $L^i(X, Y')$ such that T_λ converges to T in $L_\sigma(X, Y'_\sigma)$, then $T \in L^i(X, Y')$.

As $L^f(X, Y)$ is dense in $(N_{\ell^1}(X, Y), \|\cdot\|_{(n)})$, it follows from (4.1.18) that an $T \in L(X, Y)$ is integral if and only if there is a $\|\cdot\|_{(n)}$-bounded net $\{T_\lambda\}$ in $L^f(X, Y)$ such that T_λ converges to T in $L_\sigma(X, Y_\sigma)$.

4.2 The s-norm on tensor products

It is known from the previous section that the π-norms and the ε-norms have the cross-property, that is

$$p \otimes_\pi q(x \otimes y) = p(x)q(y) \quad \text{and} \quad p \otimes_\varepsilon q(x \otimes y) = p(x)q(y) ,$$

that the operator norm on $L(X, Y')$ is the dual norm of the π-norm $p \otimes_\pi q$, and that the integral norm on $L(X, Y')$ is the dual norm of the ε-norm $p \otimes_\varepsilon q$. Therefore it is natural to ask whether there is a cross-norm on $X \otimes Y$ whose Banach dual is isometrically isomorphic to $(\pi_{\ell^1}(X, Y'), \|\cdot\|_{(s)})$. This section is devoted to give the construction of such a cross-norm.

Following Schatten [1], a norm α on $X \otimes Y$ is a <u>cross-norm</u> (of p and q) if

$$\alpha(x \otimes y) = p(x)q(y) \quad \text{for all} \quad x \in X \quad \text{and} \quad y \in Y .$$

If α is a cross-norm on $X \otimes Y$, then we denote by $X \otimes_\alpha Y$ the normed space $(X \otimes Y, \alpha)$, and by $X \tilde{\otimes}_\alpha Y$ the completion of $X \otimes_\alpha Y$.

The π-norm and the ε-norm are examples of cross-norms. Moreover, the π-norm is the greatest cross-norm on $X \otimes Y$ by (4.1.3)(a). It then follows from (4.1.2) that

$$(X \tilde{\otimes}_\alpha Y)' \subset (X \tilde{\otimes}_\pi Y)' \equiv (B(X, Y), \|\cdot\|) \equiv (L(X, Y'), |||\cdot|||) ,$$

$$\|f \circ \omega\| = (p \tilde{\otimes}_\pi q)^*(f) \leq \alpha^*(f) \quad \text{for all} \quad f \in (X \tilde{\otimes}_\alpha Y)' ,$$

where $\omega : X \times Y \to X \otimes Y$ is the canonical map. Therefore we define naturally

$$B^{\alpha}(X, Y) = \{f \circ \omega \in B(X, Y) : f \in (X \tilde{\otimes}_{\alpha} Y)'\} \text{ and}$$

$$\|f \circ \omega\|_{(\alpha)} = \alpha^{*}(f) \text{ for all } f \in (X \tilde{\otimes}_{\alpha} Y)'.$$

Elements in $B^{\alpha}(X, Y)$ are called <u>bilinear forms of type</u> α (or <u>bilinear forms of finite</u> α-<u>norm</u>), and $\|\cdot\|_{(\alpha)}$ is referred to as the <u>norm of type</u> α.

When α is the ε-norm $p \otimes_{\varepsilon} q$, then

$$B^{\alpha}(X, Y) = \mathcal{J}(X, Y) \text{ and } \|\cdot\|_{(\alpha)} = \|\cdot\|_{(i)} ;$$

when α is the π-norm $p \otimes_{\pi} q$, then

$$B^{\alpha}(X, Y) = B(X, Y) \quad \text{and} \quad \|\cdot\|_{(\alpha)} = \|\cdot\| .$$

Clearly the map $f \mapsto f \circ \omega$ is an isometric isomorphism from $(X \tilde{\otimes}_{\alpha} Y)'$ onto $(B^{\alpha}(X, Y), \|\cdot\|_{(\alpha)})$, hence we identify $(B^{\alpha}(X, Y), \|\cdot\|_{(\alpha)})$ with the Banach dual $(X \tilde{\otimes}_{\alpha} Y)'$ of $X \tilde{\otimes}_{\alpha} Y$, and bilinear forms on $X \times Y$ of type α are regarded as continuous linear functionals on $X \tilde{\otimes}_{\alpha} Y$.

As $(L(X, Y'), |||\cdot|||)$ and $(B(X, Y), \|\cdot\|)$ are isometrically isomorphic under the map $T \mapsto b_T$, where b_T is the associated bilinear form of T, we define naturally

$$L^{\alpha}(X, Y') = \{T \in L(X, Y') : b_T \in B^{\alpha}(X, Y)\} \text{ and}$$

$$\|T\|_{(\alpha)} = \|b_T\|_{(\alpha)} \qquad \text{for all} \quad T \in L^{\alpha}(X, Y') .$$

Elements in $L^{\alpha}(X, Y')$ are also called <u>operators of type</u> α (or <u>operators of finite</u> α-<u>norm</u> in the terminology of Schatten [1]).

For any $T \in L^{\alpha}(X, Y')$, it is clear that

$$|||T||| \leq ||T||_{(\alpha)} = \sup\{|\Sigma_{i=1}^{n} <y_i , Tx_i>| : \alpha(\Sigma_{i=1}^{n} x_i \otimes y_i) \leq 1\} .$$

It is also obvious that the map $T \mapsto b_T$ is an isometric isomorphism from $(L^{\alpha}(X, Y'), ||\cdot||_{(\alpha)})$ onto $(B^{\alpha}(X, Y), ||\cdot||_{(\alpha)})$, $(L^{\alpha}(X, Y'), ||\cdot||_{(\alpha)})$ can be identified with the Banach dual $(X \tilde{\otimes}_{\alpha} Y)'$ of $X \tilde{\otimes}_{\alpha} Y$ by identifying any $T \in L^{\alpha}(X, Y')$ with $\ell_T \in (X \tilde{\otimes}_{\alpha} Y)'$ defined by

$$\ell_T(x \otimes y) = <y, Tx> \text{ for all } x \in X \text{ and } y \in Y .$$

A cross-norm α on $X \otimes Y$ is said to be <u>reasonable</u> if $X' \otimes Y' \subset (X \otimes_{\alpha} Y)'$ and the dual norm α^* of α is a cross-norm on $X' \otimes Y'$.

If α is a reasonable cross-norm on $X \otimes Y$, then we denote by α' the restriction to $X' \otimes Y'$ of α^* , and α' is called the <u>associated norm</u> of α . Clearly

$$\alpha'(\Sigma_{j=1}^{m} x_j' \otimes y_j') = \sup\{|\Sigma_{i=1}^{n} \Sigma_{j=1}^{m} <x_i , x_j'><y_i , y_j'>| : \alpha(\Sigma_{i=1}^{n} x_i \otimes y_i) \leq 1\} .$$

Denote by $X' \otimes_{\alpha'} Y'$ the normed space $(X' \otimes Y', \alpha')$ and by $X' \tilde{\otimes}_{\alpha'} Y$ the completion of $X' \otimes_{\alpha'} Y'$.

The π-norm $p \otimes_{\pi} q$ and the ε-norm $p \otimes_{\varepsilon} q$ are examples of reasonable cross-norms on $X \otimes Y$, and the associated norm of $p \otimes_{\pi} q$ is the ε-norm $p^* \otimes_{\varepsilon} q^*$. Furthermore, the ε-norm is the smallest reasonable cross-norm as a part of the following result, due to Schatten [1], shows.

(4.2.1) Lemma. <u>A cross-norm</u> α <u>on</u> $X \otimes Y$ <u>is reasonable if and only if</u> $p \otimes_{\varepsilon} q \leq \alpha$. <u>Consequently, the</u> ε<u>-norm is the smallest reasonable cross-</u>

<u>norm, and every reasonable cross-norm on</u> $X \otimes Y$ <u>satisfies</u>

$$p \otimes_\varepsilon q \leq \alpha \leq p \otimes_\pi q . \tag{2.1}$$

<u>Proof</u>. The necessity is clear by making use of formula (1.19) in §4.1. Conversely, if $p \otimes_\varepsilon q \leq \alpha$, then $X' \otimes Y' \subset (X \tilde{\otimes}_\varepsilon Y)' \subset (X \tilde{\otimes}_\alpha Y)'$ and

$$|\langle u, x' \otimes y'\rangle| \leq (p \otimes_\varepsilon q)^*(x' \otimes y')(p \otimes_\varepsilon q)(u)) \leq p^*(x')q^*(y')\alpha(u) \text{ for all } u \in X \otimes Y ,$$

thus $\alpha^*(x' \otimes y') \leq p^*(x')q^*(y')$. On the other hand, by the definition of the dual norm, we obtain

$$\begin{aligned} |\langle x, x'\rangle\langle y, y'\rangle| &= |\langle x \otimes y, x' \otimes y'\rangle| \leq \alpha^*(x' \otimes y')\alpha(x \otimes y) \\ &= \alpha^*(x' \otimes y')p(x)q(y) \quad \text{for all } x \in X \text{ and } y \in Y , \end{aligned}$$

which implies that

$$p^*(x')q^*(y') \leq \alpha^*(x' \otimes y') .$$

Therefore α^* is a cross-norm on $X' \otimes Y'$, and thus α is reasonable.

In view of (2.1), it is easily seen that if α is a reasonable cross-norm on $X \otimes Y$, then α' is a reasonable cross-norm on $X' \otimes Y'$.

(4.2.2) Corollary. <u>For a reasonable cross-norm</u> α <u>on</u> $X \otimes Y$, <u>we have:</u>

(a) $L^f(X, Y') \subset L^\iota(X, Y') \subset L^\alpha(X, Y')$ <u>and</u>

$$\|T\|_{(\alpha)} \leq \|T\|_{(\iota)} \quad \underline{\text{for all}} \quad T \in L^\iota(X, Y') ,$$

(b) $X' \otimes_{\alpha'} Y' = (L^f(X, Y'), \|\cdot\|_{(\alpha)})$;

(c) $X' \tilde{\otimes}_{\alpha'} Y'$ <u>can be identified with the closure of</u> $L^f(X, Y')$ <u>in</u> $(L^\alpha(X, Y'), \|\cdot\|_{(\alpha)})$, <u>and</u>

$$X' \tilde{\otimes}_{\varepsilon'} Y' \subset X' \tilde{\otimes}_{\alpha'} Y' \subset L^c(X, Y') .$$

<u>Proof.</u> Parts (b) and (c) follow from (a). To prove part (a), we notice that $\alpha^* \leq (p \otimes_\varepsilon q)^*$, so that

$$L^f(X, X') \subset L^i(X, Y') = (X \tilde{\otimes}_\varepsilon Y)' \subset (X \tilde{\otimes}_\alpha Y)' = L^\alpha(X, Y') .$$

(4.2.3) Corollary. <u>If</u> α <u>is a reasonable cross-norm on</u> $X \otimes Y'$, <u>then</u>

$$|\Sigma_{i=1}^n \langle Tx_i , y_i' \rangle| \leq \|T\|_{(i)} \alpha(\Sigma_{i=1}^n x_i \otimes y_i') \quad \text{\underline{for all}} \quad T \in L^i(X, Y) .$$

<u>Proof.</u> According to the definition of integral operators, the linear functional f , defined by

$$f(x \otimes y') = \langle Tx, y' \rangle \quad \text{for all} \quad x \in X \quad \text{and} \quad y' \in Y' ,$$

is continuous on $X \tilde{\otimes}_\varepsilon Y'$. Since α is reasonable, we conclude from (4.2.1) that

$$\begin{aligned} |\Sigma_{i=1}^n \langle Tx_i , y_i' \rangle| = |f(\Sigma_{i=1}^n x_i \otimes y_i')| &\leq \|T\|_{(i)} \; p \otimes_\varepsilon q^* (\Sigma_{i=1}^n x_i \otimes y_i') \\ &\leq \|T\|_{(i)} \alpha(\Sigma_{i=1}^n x_i \otimes y_i') . \end{aligned}$$

We shall now turn our attention to the question posed at the beginning of this section, namely: whether there is a cross-norm on $X \otimes Y$ whose dual norm is the absolutely summing norm on $\pi_{\ell^1}(X, Y')$. By means of the construction of the π-norm, we are able to construct such a norm.

Recall that an $T \in L(X, Y')$ is absolutely summing if and only if there is an $\zeta \geq 0$ such that the inequality

$$\Sigma_{i=1}^{n} q^*(Tx_i) \leq \zeta \ \sup\{\Sigma_{i=1}^{n} |<x_i , x'>| : p^*(x') \leq 1\} \tag{2.2}$$

holds for any finite subset $\{x_1, \ldots, x_n\}$ of X, and that the absolutely summing norm $\|T\|_{(s)}$ of T is defined to be the infimum of all $\zeta \geq 0$ such that (2.2) holds for any finite subset $\{x_1, \ldots, x_n\}$ of X or, equivalently,

$$\|T\|_{(s)} = \||T^{\mathbb{N}}\|| .$$

Therefore the inequality

$$\Sigma_{i=1}^{n} q^*(Tx_i) \leq \|T\|_{(s)} \ \sup\{\Sigma_{i=1}^{n} |<x_i , x'>| : p^*(x') \leq 1\} \tag{2.3}$$

holds for any finite subset $\{x_1, \ldots, x_n\}$ of X. As any element $u = \Sigma_{i=1}^{n} x_i \otimes y_i \in X \otimes Y$ is identified with the linear functional defined by

$$u(T) = \Sigma_{i=1}^{n} <y_i , Tx_i> \quad \text{for all} \quad T \in L^s(X, Y') ,$$

it follows from (2.3) that

$$|u(T)| = |\Sigma_{i=1}^{n} <y_i , Tx_i> | \leq \Sigma_{i=1}^{n} q^*(Tx_i)q(y_i) = \Sigma_{i=1}^{n} q^*(T(q(y_i)x_i))$$

$$\leq \|T\|_{(s)} \ \sup\{\Sigma_{i=1}^{n} |<x_i , x'>q(y_i)| : p^*(x') \leq 1\} ,$$

and hence that u is a continuous linear functional on $(\pi_{\ell^1}(X, Y'), \|\cdot\|_{(s)})$ such that

$$\|u\|^*_{(s)} \leq \inf\{\sup\{\Sigma_{i=1}^{n} |<x_i , x'>q(y_i)| : p^*(x') \leq 1\} : u = \Sigma_{i=1}^{n} x_i \otimes y_i\}, \tag{2.4}$$

where $\|\cdot\|^*_{(s)}$ is the dual norm of $\|\cdot\|_{(s)}$.

Denote by $p \otimes_s q$ the dual norm on $X \otimes Y$ of $\|\cdot\|_{(s)}$, that is

$$p \otimes_s q(u) = \|u\|^*_{(s)} = \sup\{|u(T)| : \|T\|_{(s)} \leq 1\} \quad \text{for all} \quad u \in X \otimes Y .$$

Then we have shown that $\pi_{\ell^1}(X, Y')$ can be identified with a vector subspace of $(X \otimes Y, p \otimes_s q)'$ and

$$(p \otimes_s q)^*(\ell_T) \leq \|T\|_{(s)} \quad \text{for all } T \in \pi_{\ell^1}(X, Y') , \tag{2.5}$$

where $\ell_T : x \otimes y \mapsto \langle y, Tx \rangle$. Furthermore, we have the following

(4.2.4) Theorem. Let $p \otimes_s q$ be the dual norm $\|\cdot\|^*_{(s)}$ on $X \otimes Y$ of the absolutely summing norm $\|\cdot\|_{(s)}$. Then $p \otimes_s q$ is a reasonable cross-norm on $X \otimes Y$,

$$p \otimes_s q(u) = \inf\{\sup\{\Sigma_{i=1}^n |\langle x_i , x'\rangle q(y_i)| : p^*(x') \leq 1\} : u = \Sigma_{i=1}^n x_i \otimes y_i\}, \tag{2.6}$$

and $(\pi_{\ell^1}(X, Y'), \|\cdot\|_{(s)})$ is isometrically isomorphic to the Banach dual $(X \otimes Y, p \otimes_s q)'$ of $(X \otimes Y, p \otimes_s q)$.

Proof. For any $u \in X \otimes Y$, let us define

$$\alpha(u) = \inf\{\sup\{\Sigma_{i=1}^n |\langle x_i , x'\rangle q(y_i)| : p^*(x') \leq 1 : u = \Sigma_{i=1}^n x_i \otimes y_i\} .$$

It is not hard to show that α is a seminorm on $X \otimes Y$. Furthermore, we claim that α is a reasonable cross-norm. To this end, it suffices to show in view of (4.2.1) that

$$p \otimes_\varepsilon q \leq \alpha \leq p \otimes_\pi q \tag{2.7}$$

In fact, for any representation $\Sigma_{i=1}^n x_i \otimes y_i$ of $u \in X \otimes Y$, we have

$$\begin{aligned} p \otimes_\varepsilon q(u) &= \sup\{q(\Sigma_{i=1}^n \langle x_i , x'\rangle y_i) : p^*(x') \leq 1\} \\ &\leq \sup\{\Sigma_{i=1}^n |\langle x_i , x'\rangle q(y_i)| : p^*(x') \leq 1\} , \end{aligned} \tag{2.8}$$

and

$$\alpha(u) \leq \sup\{\Sigma_{i=1}^{n} |\langle x_i, x'\rangle q(y_i)| : p^*(x') \leq 1\}$$

$$\leq \sup\{\Sigma_{i=1}^{n} p(x_i)p^*(x')q(y_i) : p^*(x') \leq 1\} = \Sigma_{i=1}^{n} p(x_i)q(y_i) . \qquad (2.9)$$

Combining (2.8) and (2.9), we obtain the assertion (2.7) in view of (4.1.3).

According to (2.4) and the definition of the reasonable cross-norm α, we have

$$p \otimes_s q(u) \leq \alpha(u) \quad \text{for all} \quad u \in X \otimes Y ,$$

it then follows from (2.5) that

$$\pi_{\ell^1}(X, Y') \subseteq (X \otimes Y, p \otimes_s q)' \subseteq L^{\alpha}(X, Y') \quad \text{and} \qquad (2.10)$$

$$\|T\|_{(\alpha)} \leq (p \otimes_s q)^*(\ell_T) \leq \|T\|_{(s)} \quad \text{for all} \quad T \in \pi_{\ell^1}(X, Y') . \qquad (2.11)$$

Moreover, we show that

$$L^{\alpha}(X, Y') \subseteq \pi_{\ell^1}(X, Y') \quad \text{and} \quad \|T\|_{(s)} \leq \|T\|_{(\alpha)} \quad \text{for all} \quad T \in L^{\alpha}(X, Y') . \qquad (2.12)$$

In fact, let $T \in L^{\alpha}(X, Y')$. For any finite subsets $\{x_1, ..., x_n\}$ and $\{y_1, ..., y_n\}$ of X and Y respectively, we have

$$|\Sigma_{i=1}^{n} \langle y_i, Tx_i\rangle| \leq \|T\|_{(\alpha)} \alpha(\Sigma_{i=1}^{n} x_i \otimes y_i)$$

$$\leq \|T\|_{(\alpha)} \sup\{\Sigma_{i=1}^{n} |\langle x_i, x'\rangle q(y_i)| : p^*(x') \leq 1\} .$$

Replace y_i by $(\text{sign}\langle y_i, Tx_i\rangle)y_i$ in the above inequalities, we obtain

$$\Sigma_{i=1}^{n} |\langle y_i, Tx_i\rangle| \leq \|T\|_{(\alpha)} \sup\{\Sigma_{i=1}^{n} |\langle x_i, x'\rangle q(y_i)| : p^*(x') \leq 1\} . \qquad (2.13)$$

Now for any finite set $\{u_1, ..., u_n\}$ in X and $\delta > 0$, there exists a finite set $\{v_1, ..., v_n\}$ in Y such that

$$q(v_\iota) \leq 1 \text{ and } q^*(Tu_\iota) < |<v_\iota, Tu_\iota>| + n^{-1}\delta \text{ for all } \iota = 1, \ldots, n .$$

It then follows from (2.13) that

$$\begin{aligned}\Sigma_{i=1}^{n} q^*(Tu_i) &< \Sigma_{i=1}^{n} |<v_i, Tu_i>| + \delta \\ &\leq \|T\|_{(\alpha)} \sup\{\Sigma_{i=1}^{n} |<u_i, x'>q(v_i)| : p^*(x') \leq 1\} + \delta \\ &\leq \|T\|_{(\alpha)} \sup\{\Sigma_{i=1}^{n} |<u_i, x'>| : p^*(x') \leq 1\} + \delta ,\end{aligned}$$

and hence that $T \in \pi_{\ell^1}(X, Y')$ and $\|T\|_{(s)} \leq \|T\|_{(\alpha)}$, as asserted.

Combining (2.10), (2.11) and (2.12), we obtain

$$\begin{aligned}(\pi_{\ell^1}(X, Y'), \|\cdot\|_{(s)}) &= (X \otimes Y, p \otimes_s q)' = (L^\alpha(X, Y'), \|\cdot\|_{(\alpha)}) \\ &= (X \otimes_\alpha Y)' . \end{aligned} \tag{2.14}$$

We conclude from (2.14) and a well-known result that formula (2.6) holds in view of the definition of α .

The reasonable cross-norm $p \otimes_s q$ on $X \otimes Y$, constructed by the preceding theorem, is referred to as the summing norm of p and q (or shortly s-norm) .

Denote by $X \otimes_s Y$ the normed space $(X \otimes Y, p \otimes_s q)$, by $X \tilde{\otimes}_s Y$ the completion of $X \otimes_s Y$, by $X' \otimes_{s'} Y'$ the normed space $(X' \otimes Y', (p \otimes_s q)')$ and by $X' \tilde{\otimes}_{s'} Y'$ the completion of $X' \otimes_{s'} Y'$, where $(p \otimes_s q)'$ is the associated norm of $p \otimes_s q$. Then (4.1.4) shows that $(\pi_{\ell^1}(X, Y'), \|\cdot\|_{(s)})$ can be indentified with the Banach dual of $X \tilde{\otimes}_s Y$, and that absolutely summing maps from X into Y' are precisely operators of type $p \otimes_s q$.

(4.2.5) Corollary. (a) $L^i(X, Y') \subset \pi_{\ell^1}(X, Y')$ and

$$\|T\|_{(s)} \leq \|T\|_{(i)} \quad \text{for all} \quad T \in L^i(X, Y') .$$

(b) $L^i(X, Y) \subset \pi_{\ell^1}(X, Y)$ and $\|S\|_{(s)} \leq \|S\|_{(i)}$ for all $S \in L^i(X, Y)$; consequently, the composition of two integral operators is nuclear.

(c) $X' \widetilde{\otimes}_{s'} Y'$ can be identified with the closure of $L^f(X, Y')$ in $(\pi_{\ell^1}(X, Y'), \|\cdot\|_{(s)})$ and

$$X' \widetilde{\otimes}_{\varepsilon'} Y' \subset X' \widetilde{\otimes}_{s'} Y' \subset L^c(X, Y') .$$

Proof. Parts (a) and (c) follow from (4.2.4) and (4.2.2). As Grothendiec shown (see Pietsch [1, p.66]) that the composition of two absolutely summing mappings is nuclear, part (b) is a consequence of (4.2.3) and (4.2.4).

(4.2.6) Theorem. Let X be a normed space, let (Y, q) be a Banach space and K a compact Hausdorff space. Then the following statements hold.

(a) If Y has the extension property, then

$$(\pi_{\ell^1}(X, Y), \|\cdot\|_{(s)}) = (L^i(X, Y), \|\cdot\|_{(i)})$$

(b) If Y' has the extension property, then

$$X \widetilde{\otimes}_s Y = X \widetilde{\otimes}_\varepsilon Y .$$

(c) $(\pi_{\ell^1}(C(K), Y), \|\cdot\|_{(s)}) = (L^i(C(K), Y), \|\cdot\|_{(i)})$; consequently, we have

$$C(K) \widetilde{\otimes}_s Y = C(K) \widetilde{\otimes}_\varepsilon Y = (C(K, Y), \|\cdot\|_\infty) , \qquad (2.15)$$

<u>where</u> $\|\cdot\|_\infty$ <u>is the sup-norm on</u> $C(K, Y)$.

<u>Proof</u>. Let Σ° be the closed unit ball in X' and suppose that $T \in \pi_{\ell^1}(X, Y)$. Then there is a positive Radon measure μ on Σ° such that

$$\mu(1) = \|T\|_{(s)} \quad \text{and} \quad q(Tx) \leq \int_{\Sigma^\circ} |\langle x, x'\rangle| \, d\mu(x') \quad \text{for all } x \in X .$$

Let us define $S : X \to C(\Sigma^\circ)$ by

$$(Sx)x' = \langle x, x'\rangle \quad \text{for all } x' \in \Sigma^\circ ,$$

and denote by $\|\cdot\|_1$ the norm on $L^1_\mu(\Sigma^\circ)$. Then $|||S||| \leq 1$ and

$$q(Tx) \leq \int_{\Sigma^\circ} |Sx| \, d\mu = \|J(Sx)\|_1 ,$$

where $J : C(\Sigma^\circ) \to L^1_\mu(\Sigma^\circ)$ is the canonical embedding, hence $\text{Ker}(JS) \subset \text{Ker}\, T$ since q is a norm. Therefore there exists an $Q \in L(JS(X), Y)$ such that $T = QJS$. Clearly, $|||Q||| \leq 1$. Since Y has the extension property and $JS(X) \subset L^1_\mu(\Sigma^\circ)$, there exists an $\bar{Q} \in L(L^1_\mu(\Sigma^\circ), Y)$ such that

$$|||\bar{Q}||| = |||Q||| \quad \text{and} \quad \bar{Q}(u) = Q(u) \quad \text{for all } u \in JS(X) .$$

Therefore $T = \bar{Q}JS$, hence $T \in L^i(X, Y)$ because of $J \in L^i(C(\Sigma^\circ), L^1_\mu(\Sigma^\circ))$. As

$$\|T\|_{(i)} \leq |||Q||| \, \|J\|_{(i)} \, |||S||| \leq \|J\|_{(i)} \leq \mu(1) = \|T\|_{(s)} ,$$

we obtain the desired result by making us of (4.1.5)(b).

(b) In view of (4.2.4) and part (a), we have

$$(X \tilde{\otimes}_s Y)' = (\pi_{\ell^1}(X, Y'), \|\cdot\|_{(s)}) = (L^i(X, Y'), \|\cdot\|_{(i)}) = (X \tilde{\otimes}_\varepsilon Y)' ,$$

hence part (b) follows.

(c) Suppose that $T \in \pi_{\ell^1}(C(K), Y)$. Then there is a positive Radon measure μ on the closed unit ball Σ° in $C(K)'$ such that

$$\mu(1) = \|T\|_{(s)} \quad \text{and} \quad q(Tf) \leq \int_{\Sigma^\circ} |\langle f, \psi\rangle| \, d\mu(\psi) \quad \text{for all } f \in C(K) . \tag{2.16}$$

Since K can be identified with a closed subset of Σ°, it follows that μ can be considered as a Radon measure on K. Furthermore, if $J : C(K) \to L^1_\mu(K)$ is the canonical embedding, then (2.16) shows that $\operatorname{Ker} J \subset \operatorname{Ker} T$, hence the density of $J(C(K))$ in $L^1_\mu(K)$ and the completeness of Y show that there is an $S : L^1_\mu(K) \to Y$ with $|||S||| \leq 1$ such that $T = SJ$. Consequently, $T \in L^i(C(K), Y)$ and

$$\|T\|_{(i)} \leq \|J\|_{(i)} |||S||| \leq \|T\|_{(s)}.$$

It then follows from (4.2.5) that $\|T\|_{(i)} = \|T\|_{(n)}$.

Finally, in view of (4.2.4) and (4.1.12) we obtain

$$(C(K) \tilde{\otimes}_s Y)' = (\pi_{\ell^1}(C(K), Y'), \|\cdot\|_{(s)}) = (L^i(C(K), Y'), \|\cdot\|_{(i)}) = (C(K) \tilde{\otimes}_\varepsilon Y)' ,$$

hence formula (2.15) follows.

In view of the preceding result, the following question arises naturally: Let X and Y be Banach spaces. If $X \tilde{\otimes}_s Y = X \tilde{\otimes}_\varepsilon Y$, whether X or Y' has the extension property.

By using a result of Schatten [1, p.62], it follows from (4.2.4) and (2.2.1) that the s-norm $p \otimes_s q$ on $X \otimes Y$ is uniform, that is

$$p \otimes_s q(S \otimes T(\Sigma_{i=1}^n x_i \otimes y_i)) \leq |||S||| \, |||T||| \, p \otimes_s q(\Sigma_{i=1}^n x_i \otimes y_i)$$

whenever $S \in L(X)$ and $T \in L(Y)$.

If M and N are closed subspaces of the Banach spaces (X, p) and (Y, q) respectively, and if p_M and q_N are the restrictions on M and N of p and q respectively, then Schatten [1] has shown that

$$p_M \otimes_\varepsilon q_N = p \otimes_\varepsilon q|_{M \otimes N} .$$

If $p \otimes_\lambda q$ denotes either $p \otimes_s q$ or $p \otimes_\pi q$, then it is easily seen from (4.2.4) and (4.1.3)(a) that

$$p \otimes_\lambda q(u) \leq p_M \otimes_\lambda q_M(u) \quad \text{for all} \quad u \in M \otimes N ,$$

therefore the identity map on $M \otimes N$ is a continuous linear map from $(M \otimes N, p \otimes_\lambda q|_{M \otimes N})$ onto $M \otimes_\lambda N = (M \otimes N, p_M \otimes_\lambda q_N)$. Furthermore, we have the following result, due to Holub [1].

(4.2.7) Theorem. <u>Let</u> (X, p) <u>and</u> (Y, q) <u>be Banach spaces, let</u> M <u>and</u> N <u>be closed vector subspaces of</u> X <u>and</u> Y <u>respectively, and suppose that</u> $p \otimes_\lambda q$ <u>(resp.</u> $p_M \otimes_\lambda q_N$<u>) denotes either the</u> s<u>-norm or the</u> π<u>-norm on</u> $X \otimes Y$ <u>(resp. on</u> $M \otimes N$<u>). Then there exist two continuous projections</u> $\pi_M : X \to M$ <u>and</u> $\pi_N : Y \to N$ <u>if and only if</u>

(i) $M \tilde{\otimes}_\lambda N$ <u>is topologically isomorphic to the closure</u> $\overline{M \otimes N}$ <u>of</u> $M \otimes N$ <u>in</u> $X \tilde{\otimes}_\lambda Y$.

(ii) <u>there exists a continuous projection</u> $\rho : X \tilde{\otimes}_\lambda Y \to \overline{M \otimes N}$.

<u>Proof</u>. <u>Necessity</u>. For any $u \in X \otimes Y$ and any respresentation $\Sigma_{i=1}^n x_i \otimes y_i$ of u , we have

$$p_M \otimes_\pi q_N(\pi_M \otimes \pi_N(u)) \leq \Sigma_{i=1}^{n} p_M(\pi_M(x_i))q_N(\pi_N(y_i))$$
$$\leq |||\pi_M|||\ |||\pi_N|||\ \Sigma_{i=1}^{n} p(x_i)q(y_i)$$

and

$$p_M \otimes_s q_N(\pi_M \otimes \pi_N(u)) \leq \sup\{\Sigma_{i=1}^{n} |<\pi_M x_i\, , f> q_N(\pi_N y_i)| \ :\ p_M^*(f) \leq 1,\ f \in M'\}$$
$$\leq |||\pi_M|||\ |||\pi_N|||\ \sup\{\Sigma_{i=1}^{n} |<x_i\, , x'> q(y_i)| \ :\ p^*(x') \leq 1,\ x' \in X'\}.$$

It then follows from (4.1.3) and (4.2.4) that

$$p_M \otimes_\lambda q_N(\pi_M \otimes \pi_N(u)) \leq |||\pi_M|||\ |||\pi_N|||\ p \otimes_\lambda q(u)\ ,$$

and hence that $\pi_M \tilde{\otimes}_\lambda \pi_N : X \tilde{\otimes}_\lambda Y \to M \tilde{\otimes}_\lambda N$ is continuous with

$$|||\pi_M \tilde{\otimes}_\lambda \pi_N||| \leq |||\pi_M|||\ |||\pi_N|||\ , \tag{2.17}$$

where $\pi_M \tilde{\otimes}_\lambda \pi_N$ is the unique extension of the tensor product map $\pi_M \otimes \pi_N \in L(X \otimes_\lambda Y,\ M \otimes_\lambda N)$. On the other hand, if $J_M : M \to X$ and $J_N : N \to Y$ are the embedding maps, then J_M and J_N are the restrictions to M and N of π_M and π_N respectively, and we have for any $z \in M \otimes N$ that

$$p \otimes_\pi q(J_M \otimes J_N(z)) = \inf\{\Sigma_{i=1}^{n} p(x_i)q(y_i) : \Sigma_{i=1}^{n} x_i \otimes y_i \in X \otimes Y,\ \Sigma_{i=1}^{n} x_i \otimes y_i = z\}$$
$$\leq \inf\{\Sigma_{j=1}^{m} p(u_j)q(v_j) : \Sigma_{j=1}^{m} u_j \otimes v_j \in M \otimes N,\ \Sigma_{j=1}^{m} u_j \otimes v_j = z\}$$
$$= p_M \otimes_\pi q_N(z)\ ,$$

and

$$p \otimes_s q(J_M \otimes J_N(z)) = \inf\{\sup\{\Sigma_{i=1}^{n} |<x_i\, , x'> q(y_i)| \ :\ p^*(x') \leq 1\}\ :\ z = \Sigma_{i=1}^{n} x_i \otimes y_i \text{ in } X \otimes Y\}$$
$$\leq \inf\{\sup\{\Sigma_{j=1}^{m} |<u_j\, , u'> q(v_j)| \ :\ p_M^*(u') \leq 1\}\ :\ z = \Sigma_{j=1}^{m} u_j \otimes v_j \text{ in } M \otimes N\}$$
$$= p_M \otimes_s q_N(z)\ .$$

Therefore $J_M \tilde{\otimes}_\lambda J_N : M \tilde{\otimes}_\lambda N \to X \tilde{\otimes}_\lambda Y$ is continuous and

$$p \tilde{\otimes}_\lambda q(J_M \tilde{\otimes}_\lambda J_N(z)) \leq p_M \tilde{\otimes}_\lambda q_N(z) \quad \text{for any} \quad z \in M \otimes N . \qquad (2.18)$$

Denote by I the restriction to $M \otimes N$ of $\pi_M \tilde{\otimes}_\lambda \pi_N$, then I^{-1} is the restriction to $M \otimes N$ of $J_M \tilde{\otimes}_\lambda J_N$, and hence it is an injective continuous linear map from $M \otimes_\lambda N$ into $X \otimes_\lambda Y$. Therefore I^{-1} is a topological isomorphism from $M \otimes_\lambda N$ onto $(M \otimes N, p \tilde{\otimes}_\lambda q|_{M \otimes N})$, and hence may be extended to a topological isomorphism from $M \tilde{\otimes}_\lambda N$ onto the closure $\overline{M \otimes N}$ of $M \otimes N$ in $X \tilde{\otimes}_\lambda Y$. Therefore we identify $M \tilde{\otimes}_\lambda N$ with $\overline{M \otimes N}$.

Clearly, the map $\rho = \pi_M \tilde{\otimes}_\lambda \pi_N$ is a continuous projection from $X \tilde{\otimes}_\lambda Y$ onto $\overline{M \otimes N}$ with $|||\rho||| = |||\pi_M|||\ |||\pi_N|||$ by (2.17) and (2.18).

<u>Sufficiency</u>. For any $0 \neq n_o \in N$, there exists a continuous projection Q from N onto the subspace generated by n_o, denoted by $<n_o>$, hence the necessity of this theorem shows that $M \tilde{\otimes}_\lambda <n_o>$ is a closed subspace of $M \tilde{\otimes}_\lambda N$ and $I_M \tilde{\otimes}_\lambda Q : M \tilde{\otimes}_\lambda N \to M \tilde{\otimes}_\lambda <n_o>$ is a continuous projection, where I_M is the identity map on M; consequently, $(I_M \tilde{\otimes}_\lambda Q) \circ \rho :$ $X \tilde{\otimes}_\lambda Y \to M \tilde{\otimes}_\lambda <n_o>$ is a continuous projection by condition (ii). Also $X \tilde{\otimes}_\lambda <n_o>$ is a closed subspace of $X \tilde{\otimes}_\lambda Y$, thus the restriction to $X \tilde{\otimes}_\lambda <n_o>$ of $(I_M \tilde{\otimes}_\lambda Q) \circ \rho$ is a continuous projection from $X \tilde{\otimes}_\lambda <n_o>$ onto $M \tilde{\otimes}_\lambda <n_o>$. As $p \otimes_\lambda q$ is a cross-norm, we conclude that there exists a continuous projection $\pi_M : X \to M$. A similar argument applies to Y and N to get a continuous projection $\pi_N : Y \to N$.

4.3 Linear mappings of finite rank

Throughout this section (X, p) and (Y, q) will be assumed to be normed spaces, and $e_X : X \to X''$ will denote the evaluation map.

It is known that $X' \otimes Y$ can be identified with $L^f(X, Y)$. This identification will be made in the following unless the contrary is explicitly stated. Under this identification, each $z = \Sigma_{i=1}^{n} x_i' \otimes y_i$, in $X' \otimes Y$, is a continuous linear map from X into Y with finite rank obtained from the equation

$$z(x) = \Sigma_{i=1}^{n} \langle x, x_i' \rangle y_i \quad \text{for all } x \in X .$$

Therefore the corresponding definition of the π-norm $p^* \otimes_\pi q$ can be possibly transferred to $L^f(X, Y)$, and is denoted by $\|\cdot\|_{(\pi)}$; namely, we define

$$\|\Sigma_{i=1}^{n} x_i' \otimes y_i\|_{(\pi)} = p^* \otimes_\pi q(\Sigma_{i=1}^{n} x_i' \otimes y_i) .$$

In view of the definition of the nuclear norm, the remark after (4.1.3) and (4.1.17), we have

$$\|\cdot\|_{(i)} \leq \|\cdot\|_{(n)} \leq \|\cdot\|_{(\pi)} \quad \text{on } L^f(X, Y) .$$

Therefore it is natural to ask under what conditions on X' (or Y), these norms are equal on $L^f(X, Y)$. The first purpose of this section is devoted to seeking such conditions.

On the other hand, it is known from §3.4 that

$$N^{(q)}_{\ell^1}(X, Y) \subset \pi_{\ell^1}(X, Y) \quad \text{and} \quad \|\cdot\|_{(s)} \leq \|\cdot\|_{(qn)} \quad \text{on} \quad N^{(q)}_{\ell^1}(X, Y).$$

It is natural to ask whether $\|\cdot\|_{(qn)}$ is the restriction of $\|\cdot\|_{(s)}$ to $N^{(q)}_{\ell^1}(X, Y)$. The affirmative answer will be given in this section (see (4.3.7)).

Before answering these two questions, we require the following terminology and some preliminary results: It is known from (4.1.2) that $(X' \tilde{\otimes}_\pi X'')' \equiv (B(X', X''), \|\cdot\|)$. Clearly the natural bilinear form, defined by

$$(x', x'') \mapsto \langle x', x''\rangle \quad \text{for all} \quad (x', x'') \in X' \times X'' ,$$

is continuous on $X' \times X''$ with norm 1, therefore there exists a unique continuous linear functional on $X' \tilde{\otimes}_\pi X''$, denoted by Tr, of norm 1 such that

$$Tr(u) = \Sigma^n_{i=1} \langle x'_i, x''_i\rangle \quad \text{for all} \quad u \in X' \otimes_\pi X'' , \tag{3.1}$$

where $\Sigma^n_{i=1} x'_i \otimes x''_i$ is any representation of u. As $X' \otimes X''$ is identified with a vector subspace of $(B^*(X', X''))^*$, the expression (3.1) is independent of the special representation of u. Hence the quantity $\Sigma^n_{i=1} \langle x'_i, x''_i\rangle$ is referred to as the <u>trace</u> of u.

Suppose $T \in L(Y, X'')$ and $z = \Sigma^n_{i=1} x'_i \otimes y_i \in X' \otimes Y$. Then the composition $T \circ z$, in $L^f(X, X'')$, is such that

$$T \circ z = \Sigma^n_{i=1} x'_i \otimes (Ty_i) \in X' \otimes X'' , \tag{3.2}$$

hence

$$Tr(T \circ z) = \Sigma^n_{i=1} \langle x'_i, Ty_i\rangle \quad \text{and} \quad |Tr(T \circ z)| \leq |||T||| \, \Sigma^n_{i=1} p^*(x'_i)q(y_i) . \tag{3.3}$$

It then follows that any $T \in L(Y, X'')$ associates a continuous linear functional ψ_T on $X' \tilde{\otimes}_\pi Y$ obtained from the equation

$$\psi_T(z) = Tr(T \circ z) \quad \text{for all} \quad z \in X' \otimes Y .$$

From (3.3), it is also clear that

$$(p^* \tilde{\otimes}_\pi q)^*(\psi_T) \leq |||T||| . \tag{3.4}$$

Conversely, for a given $y \in Y$ and $f \in (X' \tilde{\otimes}_\pi Y)'$, the map ϕ, defined by

$$\phi(x') = f(x' \otimes y) \quad \text{for all} \quad x' \in X' ,$$

is a continuous linear functional on X' (dependent on y and f) such that

$$|\phi(x')| \leq (p^* \tilde{\otimes}_\pi q)^*(f) q(y) p^*(x') \quad \text{for all} \quad x' \in X' , \tag{3.5}$$

hence there exists a unique $h_y \in X''$ such that

$$f(x' \otimes y) = \langle x', h_y \rangle \quad \text{for all} \quad x' \in X' .$$

Let us now define

$$T(y) = h_y \quad (y \in Y) \quad \text{if} \quad \langle x', h_y \rangle = f(x' \otimes y) . \tag{3.6}$$

Then T is obviously a linear map from Y into X'' such that

$$p^{**}(Ty) \leq (p^* \tilde{\otimes}_\pi q)^*(f) q(y) \quad \text{for all} \quad y \in Y$$

by (3.5), hence $T \in L(Y, X'')$ and

$$|||T||| \leq (p^* \tilde{\otimes}_\pi q)^*(f) \tag{3.7}$$

Furthermore, for any $z = \Sigma_{i=1}^{n} x'_i \otimes y_i \in X' \otimes Y$, we obtain from (3.6) and (3.3) that

$$f(z) = \Sigma_{i=1}^{n} f(x'_i \otimes y_i) = \Sigma_{i=1}^{n} \langle x'_i, Ty_i \rangle = Tr(T \circ z) .$$

Therefore we have proved, by using (3.4), (3.7) and (4.1.2), the following:

(4.3.1) Lemma. <u>The map</u> $T \mapsto \psi_T$, <u>defined by</u>

$$\psi_T(z) = Tr(T \circ z) \quad \underline{\text{for all}} \quad z \in X' \otimes Y \tag{3.8}$$

<u>is an isometric isomorphism from</u> $(L(Y, X''), |||\cdot|||)$ <u>onto</u> $(X' \tilde{\otimes}_\pi Y)'$ <u>(or</u> $(L^f(X, Y), \|\cdot\|_{(\pi)})')$. <u>Consequently</u>,

$$(X \tilde{\otimes}_\pi Y)' \equiv (B(X', Y), \|\cdot\|) \equiv (L(X', Y'), |||\cdot|||) \equiv (L(Y, X''), |||\cdot|||) \tag{3.9}$$

<u>and</u>

$$\|z\|_{(\pi)} = p^* \otimes_\pi q(z) = \sup\{|Tr(T \circ z)| : |||T||| \leq 1, T \in L(Y, X'')\} \quad \underline{\text{for all}} \quad z \in X' \otimes Y .$$

It should be noted that formula (3.9) show that the map $T \mapsto T' e_{X'}$ is an isometric isomorphism from $(L(Y, X''), |||\cdot|||)$ onto $(L(X', Y'), |||\cdot|||)$.

By using a similar argument, it can be shown that the map $T \mapsto \psi_T$, defined by (3.8), is an isometric isomorphism from $(L^i(Y, X''), \|\cdot\|_{(i)})$ onto $(X' \tilde{\otimes}_\varepsilon Y)'$ (or $(L^f(X, Y), |||\cdot|||)'$).

Suppose $z = \Sigma_{i=1}^{n} x'_i \otimes y_i \in X' \otimes Y$ and $S \in L(Y, X)$. Then the composition $(e_X S) \circ z$, in $L^f(X, X'')$, is such that

$$e_X \circ S \circ z = \Sigma_{i=1}^{n} x_i' \otimes (e_X(Sy_i)) \in X' \otimes X'' ,$$

hence

$$\mathrm{Tr}(e_X \circ S \circ z) = \Sigma_{i=1}^{n} \langle x_i', e_X(Sy_i)\rangle = \Sigma_{i=1}^{n} \langle Sy_i , x_i'\rangle . \tag{3.10}$$

If we denote by $L_\sigma(Y, X)$ the locally convex space of $L(Y, X)$ equipped with the topology of pointwise convergence, then (3.9) shows that the map ϕ_z, defined by

$$\phi_z(S) = e_X \circ S \circ z \quad \text{for all} \quad S \in L(Y, X) , \tag{3.11}$$

is a continuous linear functional on $L_\sigma(Y, X)$. On the other hand, since $L_\sigma(Y, X)$ is a subspace of the product space X^Y and $(X^Y)' = (X')^{(Y)}$, it follows that every $g \in (L_\sigma(Y, X))'$ is the restriction of some element in $(X')^{(Y)}$, and hence from a well-known result (see Schaefer [1, p.137]) that there exist finite subsets $\{x_1', \ldots, x_n'\}$ of X' and $\{y_1, \ldots, y_n\}$ of Y such that

$$g(S) = \Sigma_{i=1}^{n} \langle Sy_i , x_i'\rangle \quad \text{for all} \quad S \in L(Y, X) . \tag{3.12}$$

In view of (3.10), (3.11) and (3.12), we obtain the following:

(4.3.2) Lemma. <u>The map</u> $z \mapsto \phi_z$, <u>defined by</u>

$$\phi_z(S) = \mathrm{Tr}(e_X \circ S \circ z) \quad \underline{\text{for all}} \quad S \in L(Y, X) ,$$

<u>is an algebraic isomorphism from</u> $X' \otimes Y$ <u>onto</u> $(L_\sigma(Y, X))'$.

Suppose $T \in L^{i}(X, Y)$ and $u = \Sigma_{i=1}^{n} y_i' \otimes x_i'' \in Y' \otimes X''$. Then the composition $u \circ T$, in $L^{f}(X, X'')$, is such that

$$u \circ T = \Sigma_{i=1}^{n} (T'y_i') \otimes x_i'' \in X' \otimes X'' ,$$

hence

$$Tr(u \circ T) = \Sigma_{i=1}^{n} <T'y_i', x_i''> = \Sigma_{i=1}^{n} <y_i', T''x_i'> = Tr(T'' \circ u)$$

and

$$|Tr(u \circ T)| \leq |\Sigma_{i=1}^{n} <T'y_i', x_i''>| = |\Sigma_{i=1}^{n} u_{T'} (y', x_i'')|$$

$$\leq \|T'\|_{(i)} \; |||\Sigma_{i=1}^{n} y_i' \otimes x_i''||| = \|T\|_{(i)} \; |||u|||$$

by (4.1.15). This remark makes the following result clear.

(4.3.3) Lemma. *Suppose* $T \in L^{i}(X, Y)$ *and* $u = \Sigma_{i=1}^{n} y_i' \otimes x_i'' \in Y' \otimes X''$. *Then*

$$|Tr(u \circ T)| \leq \|T\|_{(i)} \, |||u||| .$$

We shall now turn our attension to the first question posed at the beginning of this section, namely: Under what conditions on X' (or Y), the norms $\|\cdot\|_{(i)}$, $\|\cdot\|_{(n)}$ and $\|\cdot\|_{(\pi)}$ on $L^{f}(X, Y)$ are the same. Before answering this question, we first notice the following facts: Denote by $L(X, Y)_1$, the closed unit ball in $(L(X, Y), |||\cdot|||)$, and by $L^{f}(X, Y)_1$ the closed unit ball in $(L^{f}(X, Y), |||\cdot|||)$; because of on $L(X, Y)_1$ the topology of precompact convergence coincides with the topology of pointwise convergence, it follows from (4.1.4) that a Banach space X has the metric approximation property if and only if, for any Banach space Y, $L^f(X, Y)_1$ is dense in $L(X, Y)_1$ for the topology of pointwise convergence.

(4.3.4) Theorem. Let X and Y be Banach spaces. If either X' or Y has the metric approximation property, then

$$\|\cdot\|_{(i)} = \|\cdot\|_{(n)} = \|\cdot\|_{(\pi)} \qquad \text{on} \qquad L^f(X, Y) .$$

Proof. For any $R = \Sigma_{i=1}^n x_i' \otimes y_i \in X' \otimes Y$, since $L^f(X, Y) \subset L^i(X, Y)$, it suffices to show that

$$\begin{aligned} \|R\|_{(\pi)} &= \sup\{|\mathrm{Tr}(TR)| : \||T\|| \leq 1 , \ T \in L(Y, X'')\} \\ &\leq \sup\{|\mathrm{Tr}(u \circ R)| : \||u\|| \leq 1, \ u \in L^f(Y, X'')\} \leq \|R\|_{(i)} \end{aligned} \qquad (3.13)$$

by virtue of (4.3.1), (4.3.3) and the fact that $\|\cdot\|_{(i)} \leq \|\cdot\|_{(n)} \leq \|\cdot\|_{(\pi)}$ on $L^f(X, Y)$.

Suppose that Y has the metric approximation property. Then $L^f(Y, X'')_1$ is dense in $L(Y, X'')_1$ for the topology on $L(Y, X'')$ of pointwise convergence. It is known from (4.3.2) that $X''' \otimes Y = (L_\sigma(Y, X''))'$ under the map $z \mapsto \phi_z$ $(z \in X''' \otimes Y)$ defined by

$$\phi_z(S) = \mathrm{Tr}(e_{X''} \circ S \circ z) \quad \text{for all} \quad S \in L(Y, X'') .$$

As $z = \Sigma_{i=1}^n (e_{X'}(x_i')) \otimes y_i \in X''' \otimes Y$, it follows that

$$\begin{aligned} &\sup\{|\mathrm{Tr}(e_{X''} \circ S \circ z)| : \||S\|| \leq 1, \ S \in L(Y, X'')\} \\ &\qquad \leq \sup\{|\mathrm{Tr}(e_{X''} \circ u \circ z)| : \||u\|| \leq 1, \ u \in L^f(Y, X'')\}. \end{aligned} \qquad (3.14)$$

On account of (3.2) and (3.3),

$$\begin{aligned}\mathrm{Tr}(e_{X''} \circ S \circ z) &= \Sigma_{i=1}^{n} \langle e_{X'}(x_i'),\ e_{X''}(Sy_i)\rangle = \Sigma_{i=1}^{n} \langle Sy_i,\ e_{X'}(x_i')\rangle \\ &= \Sigma_{i=1}^{n} \langle x_i',\ Sy_i\rangle = \mathrm{Tr}\,(SR) \qquad \text{for any} \quad S \in L(Y, X'') ,\end{aligned}$$

hence we obtain the required inequalities (3.13) by making use of (3.14).

Suppose that X' has the metric approximation property. Then $L^f(X', Y')_1$ is dense in $L(X', Y')_1$ for the topology on $L(X', Y')$ of pointwise convergence. It is known from (4.3.2) that $Y'' \otimes X' = (L_\sigma(X', Y'))'$ under the map $z \mapsto \phi_z$ $(z \in Y'' \otimes X')$ defined by

$$\phi_z(S) = \mathrm{Tr}(e_{Y'} \circ S \circ z) \quad \text{for all} \quad S \in L(X', Y') ,$$

hence for any $z \in Y'' \otimes X'$, we have

$$\begin{aligned}&\sup\{|\mathrm{Tr}(e_{Y'} \circ S \circ z)| \ :\ |||S||| \leq 1,\ S \in L(X', Y')\} \\ &\qquad \leq \sup\{|\mathrm{Tr}(e_{Y'} \circ v \circ z)| \ :\ |||v||| \leq 1,\ v \in L^f(X', Y')\} .\end{aligned}$$

Since the adjoint map R' of R belongs to $L^f(Y', X')(=Y'' \otimes X')$, it follows that

$$\begin{aligned}&\sup\{|\mathrm{Tr}(e_{Y'} \circ S \circ R')| \ :\ |||S||| \leq 1,\ S \in L(X', Y')\} \\ &\qquad \leq \sup\{|\mathrm{Tr}(e_{Y'} \circ v \circ R')| \ :\ |||v||| \leq 1,\ v \in L^f(X', Y')\}. \qquad (3.15)\end{aligned}$$

On the other hand, the formula (3.9) of (4.3.1) ensures that $(L(Y, X''), |||\cdot|||) \cong (L(X', Y'), |||\cdot|||)$ under the map $T \mapsto T' \circ e_{X'}$ $(T \in L(Y, X''))$, we then obtain from (3.15) that

$$\sup\{|\mathrm{Tr}(e_{Y'} \circ T' \circ e_{X'} \circ R')| : |||T||| \leq 1, T \in L(Y, X'')\}$$
$$\leq \sup\{|\mathrm{Tr}(e_{Y'} \circ u' \circ e_{X'} \circ R')| : |||v||| \leq 1, v \in L^f(Y, X'')\}. \quad (3.16)$$

For any $x \in X$ and $y' \in Y'$,

$$\langle x, R'y'\rangle = \langle Rx, y'\rangle = \Sigma_{i=1}^{n} \langle x, x'_i\rangle\langle y_i, y'\rangle = \langle x, \Sigma_{i=1}^{n} \langle y_i, y'\rangle x'_i\rangle$$
$$= \langle x, \Sigma_{i=1}^{n} \langle y', e_Y(y_i)\rangle x'_i\rangle ,$$

it follows that $R' = \Sigma_{i=1}^{n} (e_Y y_i) \otimes x'_i$, and hence from (3.2) and (3.3) that

$$\mathrm{Tr}(e_{Y'} \circ T' \circ e_{X'} \circ R') = \Sigma_{i=1}^{n} \langle e_Y y_i , (e_{Y'} \circ T' \circ e_{X'})x'_i\rangle = \Sigma_{i=1}^{n} \langle y_i , (T' e_{X'})x'_i\rangle$$
$$= \Sigma_{i=1}^{n} \langle Ty_i , e_{X'}(x'_i)\rangle = \Sigma_{i=1}^{n} \langle x'_i, Ty_i\rangle = \mathrm{Tr}(TR) .$$

Therefore we obtain the required inequalities (3.13) on account of (3.16), and hence the proof is complete.

(4.3.5) Corollary. Let X and Y be Banach spaces. If either X' or Y has the metric approximation property, then $X' \tilde{\otimes}_\pi Y = (N_{\ell^1}(X, Y), \|\cdot\|_{(n)})$ and

$$(N_{\ell^1}(X, Y), \|\cdot\|_{(n)})' \equiv (L(X', Y'), |||\cdot|||) \equiv (L(Y, X''), |||\cdot|||). \quad (3.17)$$

Proof. The preceding result shows that $X' \otimes_\pi Y$ is isometrically isomorphic to the subspace $(L^f(X, Y), \|\cdot\|_{(n)})$ of $(N_{\ell^1}(X, Y), \|\cdot\|_{(n)})$. As $L^f(X, Y)$ is $\|\cdot\|_{(n)}$-dense in $N_{\ell^1}(X, Y)$, it follows that $X' \tilde{\otimes}_\pi Y = (N_{\ell^1}(X, Y), \|\cdot\|_{(n)})$. Finally, (3.17) follows from (4.3.1).

(4.3.6) Corollary. Let X and Y be Banach spaces. If either X'

or Y *has the metric approximation property, then* $N_{\ell^1}(X, Y)$ *is the* $\|\cdot\|_{(i)}$-*closure of* $L^f(X, Y)$ *in* $L^i(X, Y)$ *and*

$$\|\cdot\|_{(i)} = \|\cdot\|_{(n)} \quad \text{on} \quad N_{\ell^1}(X, Y) .$$

Proof. (4.3.4) ensures that $X' \otimes_\pi Y \equiv (L^f(X, Y), \|\cdot\|_{(i)})$. As $(L^i(X, Y), \|\cdot\|_{(i)})$ is complete, it follows from (4.3.5) that $(N_{\ell^1}(X, Y), \|\cdot\|_{(n)})$ is isometrically isomorphic with the $\|\cdot\|_{(i)}$-closure of $L^f(X, Y)$ equipped with the norm induced by $\|\cdot\|_{(i)}$.

By making use of (4.3.6), we are able to answer the second question posed at the beginning of this section, namely: Whether $\|\cdot\|_{(qn)}$ is the relative norm of $\|\cdot\|_{(s)}$. The answer is affirmative as the following result shows.

(4.3.7) Theorem (Pietsch). *Let* X *and* Y *be normed spaces. Then*

$$\|\cdot\|_{(qn)} = \|\cdot\|_{(s)} \quad \text{on} \quad N^{(q)}_{\ell^1}(X, Y) .$$

Proof. It is sufficient to show that

$$\|T\|_{(qn)} \leq \|T\|_{(s)} \qquad \text{for any} \quad T \in N^{(q)}_{\ell^1}(X, Y)$$

because the inequality $\|T\|_{(s)} \leq \|T\|_{(qn)}$ always holds.

Let Λ be the closed unit ball in Y'. Then (3.4.6) shows that there exists an isometry j_Y from Y into ℓ^∞_Λ such that

$$j_Y T \in N_{\ell^1}(X, \ell^\infty_\Lambda) \quad \text{and} \quad \|T\|_{(qn)} = \|j_Y T\|_{(n)} . \tag{3.18}$$

As ℓ^∞_Λ has the metric approximation property, it follows from (4.3.6) that

$$\|j_Y T\|_{(n)} = \|j_Y T\|_{(i)} . \tag{3.19}$$

Since $\ell^\infty_\wedge$ has the extension property, it follows from (4.2.6)(a) that

$$j_Y T \in \pi_{\ell^1}(X, \ell^\infty_\wedge) \quad \text{and} \quad \|j_Y T\|_{(i)} = \|j_Y T\|_{(s)} . \tag{3.20}$$

We conclude from (3.18), (3.19) and (3.20) that

$$\|T\|_{(qn)} = \|j_Y T\|_{(s)} \leq \||j_Y\|| \, \|T\|_{(s)} = \|T\|_{(s)} ,$$

as asserted.

(4.3.8) Corollary. <u>Let</u> X <u>and</u> Y <u>be normed spaces and let</u> Y <u>be complete</u>. <u>Then</u>

<u>the</u> $\|\cdot\|_{(s)}$<u>-closure of</u> $L^f(X, Y) \subset N^{(q)}_{\ell^1}(X, Y)$.

<u>Moreover, if in addition,</u> X <u>is complete and if either</u> X' <u>or</u> Y <u>has the approximation property, then</u>

<u>the</u> $\|\cdot\|_{(s)}$<u>-closure of</u> $L^f(X, Y) = N^{(q)}_{\ell^1}(X, Y)$,

<u>hence</u> $L^f(X, Y)$ <u>is dense in</u> $(N^{(q)}_{\ell^1}(X, Y), \|\cdot\|_{(s)})$.

<u>Proof</u>. We first note from the completeness of $(\pi_{\ell^1}(X, Y), \|\cdot\|_{(s)})$ that the $\|\cdot\|_{(s)}$-closure of $L^f(X, Y)$ is contained in $\pi_{\ell^1}(X, Y)$. Let T belong to the $\|\cdot\|_{(s)}$-closure of $L^f(X, Y)$, and let $\{R_n\}$ be a sequence in $L^f(X, Y)$ such that $\lim_n \|T - R_n\|_{(s)} = 0$. Then $\{R_n\}$ is an $\|\cdot\|_{(s)}$-Cauchy sequence, and surely $\|\cdot\|_{(qn)}$-Cauchy by (4.3.7). The first part then follows from the completeness of $(N^{(q)}_{\ell^1}(X, Y), \|\cdot\|_{(qn)})$ and (4.3.7).

Finally, if either X' or Y has the approximation property, then

(3.4.4) shows that $L^f(X, Y)$ is dense in $(N^{(q)}_{\ell^1}(X, Y), \|\cdot\|_{(qn)})$, hence the second assertion follows from (4.3.7) and the first part of this corollary.

It is known from (4.2.6)(c) that if K is a compact Hausdorff space and Y is a Banach space, then $(\pi_{\ell^1}(C(K), Y), \|\cdot\|_{(s)}) \equiv (L^i(C(K), Y), \|\cdot\|_{(i)})$. The following result, which should be compared with (3.4.5), shows that $(N_{\ell^1}(C(K), Y), \|\cdot\|_{(n)}) \equiv (N^{(q)}_{\ell^1}(C(K), Y), \|\cdot\|_{(qn)})$.

(4.3.9) Corollary. Let K be a compact Hausdorff space and Y a Banach space. Then

$$(N_{\ell^1}(C(K), Y), \|\cdot\|_{(n)}) \equiv (N^{(q)}_{\ell^1}(C(K), Y), \|\cdot\|_{(qn)}) .$$

Proof. Suppose $T \in N^{(q)}_{\ell^1}(C(K), Y)$. We first notice from (4.2.6)(c) that $T \in L^i(C(K), Y)$ and $\|T\|_{(s)} = \|T\|_{(i)}$ because of $N^{(q)}_{\ell^1}(C(K), Y) \subset \pi_{\ell^1}(C(K), Y)$, hence

$$\|T\|_{(qn)} = \|T\|_{(s)} = \|T\|_{(i)} \tag{3.21}$$

by (4.3.7). On the other hand, since $C(K)'$ has the metric approximation property, it follows from (4.3.8) that there is a sequence $\{R_n\}$ in $L^f(C(K), Y)$ such that $\lim_n \|R_n - T\|_{(s)} = 0$, and hence from (4.2.7)(c) that

$$\lim_n \|R_n - T\|_{(i)} = \lim_n \|R_n - T\|_{(s)} = 0 .$$

Therefore $T \in N_{\ell^1}(C(K), Y)$ and $\|T\|_{(n)} = \|T\|_{(i)}$ by (4.3.6), and thus $\|T\|_{(n)} = \|T\|_{(qn)}$ by (3.21).

Let H_1 and H_2 be Hilbert spaces and let $\{e_i : i \in \Lambda\}$ be an orthnormal basis in H_1 . It is known from (2.2.6) that an $T \in L(H_1, H_2)$ is absolutely summing if and only if $\Sigma_i \|Te_i\|^2 < +\infty$. Therefore we define

$$\sigma(T) = (\Sigma_i \|Te_i\|^2)^{\frac{1}{2}} \quad \text{for all} \quad T \in L^s(H_1, H_2) .$$

It can be shown (see Pietsch [1, (2.5.5)]) that

$$\sigma(T) \leq \|T\|_{(s)} \leq \sqrt{3}\,\sigma(T) \quad \text{for all} \quad T \in \pi_{\ell^1}(H_1, H_2) , \qquad (3.22)$$

and that $L^f(H_1, H_2)$ is dense in $(\pi_{\ell^1}(H_1, H_2), \sigma)$ (see Pietsch [1, (2.5.3)]). Therefore $L^f(H_1, H_2)$ is dense in $(\pi_{\ell^1}(H_1, H_2), \|\cdot\|_{(s)})$ by (3.22), and thus $N^{(q)}_{\ell^1}(H_1, H_2) = \pi_{\ell^1}(H_1, H_2)$ by (4.2.8) because any Hilbert space has the approximation property. This proves the following result.

(4.3.10) Corollary. <u>For two Hilbert spaces</u> H_1 <u>and</u> H_2, $L^f(H_1, H_2)$ <u>is dense in</u> $(\pi_{\ell^1}(H_1, H_2), \|\cdot\|_{(s)})$, <u>and hence</u>

$$(N^{(q)}_{\ell^1}(H_1, H_2), \|\cdot\|_{(qn)}) \equiv (\pi_{\ell^1}(H_1, H_2), \|\cdot\|_{(s)}) .$$

4.4 Locally convex tensor products

In the present section, $(E, \mathcal{P})$ and $(F, \mathcal{J})$ will denote locally convex spaces, $\mathcal{U}_E$ and $\mathcal{U}_F$ will denote respectively neighbourhood bases at 0 for $\mathcal{P}$ and $\mathcal{J}$ consisting of closed absolutely convex sets, $B(E, F; G)$ will be the vector space of all continuous bilinear forms from $E \times F$ into another locally convex space G , and $\omega : E \times F \to E \otimes F$ will denote the canonical bilinear map.

It is easily verified that the following family

$$\mathcal{U}_\pi = \{\Gamma(V \otimes U) : V \in \mathcal{U}_E \text{ and } U \in \mathcal{U}_F\}$$

determines a unique locally convex topology on $E \otimes F$, denoted by $\mathcal{P} \otimes_\pi \mathcal{J}$, for which ω is continuous. $\mathcal{P} \otimes_\pi \mathcal{J}$ is called the π-<u>topology</u> (or <u>projective</u>

topology) of $\mathcal{P}$ and $\mathcal{J}$.

It is easily seen that the π-topology $\mathcal{P} \otimes_\pi \mathcal{J}$ is the finest locally convex topology on $E \otimes F$ for which $\omega : E \times F \to E \otimes F$ is continuous. A similar argument given in the proof of (4.1.3)(a) shows that the gauge of $\Gamma(V \otimes U)$, denoted by $p_V \otimes_\pi q_U$, is given by

$$p_V \otimes_\pi q_U(u) = \inf\{\Sigma_{i=1}^n p(x_i)q(y_i) : u = \Sigma_{i=1}^n x_i \otimes y_i\} ,$$

and has the cross-property:

$$p_V \otimes_\pi q_U(x \otimes y) = p_V(x)q_U(y) ,$$

hence $\mathcal{P} \otimes_\pi \mathcal{J}$ is determined by the family $\{p_V \otimes_\pi q_U : V \in \mathcal{U}_E, U \in \mathcal{U}_F\}$ of cross-seminorms. We shall see below that $\mathcal{P} \otimes_\pi \mathcal{J}$ is Hausdorff. Therefore we denote by $E \otimes_\pi F$ the locally convex space $(E \otimes F, \mathcal{P} \otimes_\pi \mathcal{J})$ and by $E \tilde{\otimes}_\pi F$ the completion of $E \otimes_\pi F$.

(4.4.1) Lemma. The π-topology $\mathcal{P} \otimes_\pi \mathcal{J}$ is only the locally convex topology on $E \otimes F$ with the property that for any locally convex space G, $L(E \otimes_\pi F, G)$ and $B(E, F, G)$ are algebraically isomorphic under the map $T \mapsto T \circ \omega$. Consequently, $E' \otimes F' \subset (E \otimes_\pi F)'$ and the π-topology is Hausdorff.

Proof. It is clear that the map $T \mapsto T \circ \omega$ is an algebraic isomorphism from $L(E \otimes_\pi F, G)$ onto $B(E, F, G)$. Suppose now that $\mathcal{I}$ is another locally convex topology on $E \otimes F$ with the mentioned property. Then we apply G to $E \otimes_\pi F$ to conclude that $\mathcal{P} \otimes_\pi \mathcal{J} \leq \mathcal{L}$, and we apply G to $(E \otimes F, \mathcal{L})$ to conclude that $\mathcal{L} \leq \mathcal{P} \otimes_\pi \mathcal{J}$ because $\mathcal{P} \otimes_\pi \mathcal{J}$ has the mentioned property. Therefore $\mathcal{P} \otimes_\pi \mathcal{J}$ is the only locally convex topology on $E \otimes F$ with the mentioned property.

Clearly each $x' \otimes y' \in E' \otimes F'$ defines a continuous bilinear form on $E \times F$ by the following equation

$$x' \otimes y' : (x, y) \longmapsto \langle x, x'\rangle\langle y, y'\rangle \quad \text{for all} \quad (x, y) \in E \times F .$$

Therefore $E' \otimes F'$ can be identified with a vector subspace of $B(E, F) = (E \otimes_\pi F)'$.

Finally, any non-zero $u \in E \otimes F$ can be represented in the form $u = \Sigma_{i=1}^{n} x_i \otimes y_i$, where $\{x_1, \ldots, x_n\}$ and $\{y_1, \ldots, y_n\}$ are linearly independent sets in E and F respectively. Since $\mathcal{P}$ and $\mathcal{J}$ are Hausdorff, there are $x' \in E'$ and $y' \in F'$ such that

$$\langle x_1, x'\rangle = \langle y_1, y'\rangle = 1 \quad \text{and} \quad \langle x_j, x'\rangle = \langle y_j, y'\rangle = 0 \quad \text{for all} \quad j > 1 .$$

Therefore $x' \otimes y' \in (E \otimes_\pi F)'$ is such that

$$x' \otimes y' : u \longmapsto \Sigma_{i=1}^{n} \langle x_i, x'\rangle\langle y_i, y'\rangle = 1 ,$$

thus the π-topology is Hausdorff.

It is known from Chapter 0 that $L^{eq}(E, F(\mathcal{J})')$ and $B(E, F)$ are algebraically isomorphic under the map $T \mapsto b_T$, where b_T is the associated bilinear form of T. Therefore we obtain an immediate consequence of (4.4.1) as follows.

(4.4.2) Corollary. $(E \tilde{\otimes}_\pi F)' = B(E, F) = L^{eq}(E, F(\mathcal{J})')$. Under this identification, the equicontinuous subsets of $(E \tilde{\otimes}_\pi F)'$ are the equicontinuous sets of bilinear forms on $E \times F$.

By using (4.4.2) it is easily seen that the π-topologies $\sigma(E, E') \otimes_\pi \sigma(F, F')$ and $\sigma(E', E) \otimes_\pi \sigma(F', F)$ are consistent with $\langle E \otimes F, E' \otimes F'\rangle$ because of

$$E' \otimes F' = B(E_\sigma, F_\sigma) \quad \text{and} \quad E \otimes F = B(E'_\sigma, F'_\sigma),$$

where $E_\sigma = (E, \sigma(E, E'))$ and $E'_\sigma = (E', \sigma(E', E))$; similarly for F_σ and F'_σ.

If $(E, \mathcal{P})$ and $(F, \mathcal{J})$ are metrizable, then so does $\mathcal{P} \otimes_\pi \mathcal{J}$; moreover, Grothendieck has shown (see Schaefer [1, p.94]) that each element u in the completion $E \tilde{\otimes}_\pi F$ of $E \otimes_\pi F$ can be represented as the form $u = \sum_{n=1}^{\infty} \zeta_n x_n \otimes y_n$ for some $(\zeta_n) \in \ell^1$ and two null sequences $\{x_n\}$ and $\{y_n\}$ in E and F respectively.

In order to study another important locally convex topology on $E \otimes F$, we require the following terminology and results: Clearly, $\mathfrak{S} = \{V^\circ : V \in \mathcal{U}_E\}$ is a topologizing family for $L(E'_\tau, F)$, hence the

$\mathcal{E}$ -topology is called the <u>topology of bi-equicontinuous convergence</u> on $L(E'_\tau, F)$, and the space $L(E'_\tau, F)$ equipped with this $\mathcal{E}$-topology is denoted by $L_\varepsilon(E'_\tau, F)$. It is easily seen that $\{P(V^0, U) : V \in \mathcal{U}_E, U \in \mathcal{U}_F\}$ is a neighbourhood basis at 0 for this $\mathcal{E}$-topology, and that each $\Sigma_{i=1}^n x'_i \otimes y'_i \in E' \otimes F'$ determines a continuous linear functional on $L_\varepsilon(E'_\tau, F)$, obtained by setting

$$\Sigma_{i=1}^n x'_i \otimes y'_i : T \mapsto \Sigma_{i=1}^n \langle Tx'_i, y'_i \rangle \quad \text{for all } T \in L_\varepsilon(E'_\tau, F).$$

Therefore $L_\varepsilon(E'_\tau, F)$ is Hausdorff. On the other hand, it is not hard to show that

$$S(E'_\sigma, F'_\sigma) = S(E'_\tau, F'_\sigma) = L(E'_\sigma, F_\sigma) = L(E'_\tau, F_\sigma) = L(E'_\tau, F),$$

therefore any $\mathcal{E}$-topology on $L(E'_\tau, F)$ can be transferred to $S(E'_\sigma, F'_\sigma)$. We denote by $S_{\mathcal{E}}(E'_\sigma, F'_\sigma)$ the space $S(E'_\sigma, F'_\sigma)$ equipped with the $\mathcal{E}$-topology transferred from $L_{\mathcal{E}}(E'_\tau, F)$, then $L_{\mathcal{E}}(E'_\tau, F)$ and $S_{\mathcal{E}}(E'_\sigma, F'_\sigma)$ are topologically isomorphic; in particular,

$$L_\varepsilon(E'_\tau, F) \cong S_\varepsilon(E'_\sigma, F'_\sigma).$$

Also we denote by $L^f_{\mathcal{E}}(E'_\tau, F)$ the space $L^f(E'_\tau, F)$ equipped with the relative topology induced by an $\mathcal{E}$-topology on $L(E'_\tau, F)$.

(4.4.3) Theorem. $L_\varepsilon(E'_\tau, F)$ (<u>equivalently</u>, $S_\varepsilon(E'_\sigma, F'_\sigma)$) <u>is complete if and only if both</u> E <u>and</u> F <u>are complete</u>.

<u>Proof</u>. <u>Necessity</u>. We identity $E \otimes F$ with $L^f(E'_\tau, F)$. Under this identification, for any $0 \neq y \in F$, $E \otimes y = \{x \otimes y : x \in E\}$ can be

regarded as a vector subspace of $L^f(E'_\tau, F)$. Clearly $E \otimes y$ is closed in $L_\varepsilon(E'_\tau, F)$ (and even in $L_\sigma(E'_\tau, F)$) and surely complete. The map $x \mapsto x \otimes y$ is obviously a topological isomorphism from E onto $E \otimes y$, and hence E is complete.

As $L_\varepsilon(E'_\tau, F) \cong S_\varepsilon(E'_\sigma, F'_\sigma)$, the situation is perfectly symmetric in E and F , hence the completeness of F follows from the completeness of $L_\varepsilon(E'_\tau, F)$.

<u>Sufficiency</u>. The completeness of F ensures that $L^*(E'_\tau, F)$, and surely $B^*(E'_\sigma, F'_\sigma)$, is complete under the topology of uniform convergence on equicontinuous subsets of E' (this topology need not be a vector topology, but it is clear that the notion of Cauchy net make sense for it) . Therefore it suffices to show that $S_\varepsilon(E'_\sigma, F'_\sigma)$ is closed in $B^*(E'_\sigma, F'_\sigma)$. To do this, let $\{\psi_\tau\}$ be a net in $S_\varepsilon(E'_\sigma, F'_\sigma)$, and let $\psi \in B^*(E'_\sigma, F'_\sigma)$ be such that

$$\psi_\tau \to \psi \text{ uniformly on } V^\circ \times U^\circ \quad (V \in \mathcal{U}_E \text{ and } U \in \mathcal{U}_F) .$$

Then for any $y' \in F'$, $\psi_\tau^{(y')} \to \psi^{(y')}$ uniformly on V° , where $\psi_\tau^{(y')}$ is the partial map of ψ_τ defined by $\psi_\tau^{(y')}(x') = \psi_\tau(x', y')(x' \in E')$. Clearly $\psi_\tau^{(y')} \in (E'_\sigma)'$, hence $\psi^{(y')}$ is $\sigma(E', E)$-continuous on V° , and thus $\psi^{(y')}$ is $\sigma(E', E)$-continuous on E' by Grothendieck's completeness theorem since E is complete. A similar argument applies to $\psi^{(x')}(x' \in E')$. We conclude that $\psi \in S(E'_\sigma, F'_\sigma)$.

It is clear that

$$E \otimes F = L^f(E'_\tau, F) = L^f(E'_\beta, F) = B(E'_\sigma, F'_\sigma) .$$

The topology on $L^f_\varepsilon(E'_\tau, F)$ can be transferred to $E \otimes F$, hence the topology on $E \otimes F$ transferred from $L^f_\varepsilon(E'_\tau, F)$ is called the ε-topology (or topology of bi-equicontinuous convergence) of $\mathcal{P}$ and $\mathcal{J}$, and denoted by $\mathcal{P} \otimes_\varepsilon \mathcal{J}$. Denote by $E \otimes_\varepsilon F$ the locally convex space $(E \otimes F, \mathcal{P} \otimes_\varepsilon \mathcal{J})$ and by $E \tilde{\otimes}_\varepsilon F$ the completion of $E \otimes_\varepsilon F$. Then $E \otimes_\varepsilon F$ and $L^f_\varepsilon(E'_\tau, F)$ are topologically isomorphic. If E and F are complete, then (4.4.3) shows that $E \tilde{\otimes}_\varepsilon F$ can be identified with the closure of $L^f(E'_\tau, F)$ in $L_\varepsilon(E'_\tau, F)$ (or, equivalently, the closure of $B(E'_\sigma, F'_\sigma)$ in $S_\varepsilon(E'_\sigma, F'_\sigma)$). Since

$$L^f(E'_\tau, F) \cap (P(V^o, U))^o = (V^o \otimes U^o)^o = (\Gamma(V^o \otimes U^o))^o ,$$

where the polar $(V^o \otimes U^o)^o$ is taken with $\langle E \otimes F, E' \otimes F' \rangle$, the family

$$\mathcal{U}_\varepsilon = \{(V^o \otimes U^o)^o : V \in \mathcal{U}_E \text{ and } U \in \mathcal{U}_F\}$$

is a neighbourhood basis at 0 for the (Hausdorff) ε-topology $\mathcal{P} \otimes_\varepsilon \mathcal{J}$. A similar argument given in the proof of (4.1.3) shows that the gauge of $(V^o \otimes U^o)^o$, denoted by $p_V \otimes_\varepsilon q_U$, is given by

$$p_V \otimes_\varepsilon q_U(\Sigma_{i=1}^n x_i \otimes y_i) = \sup\{|\Sigma_{i=1}^n \langle x_i, x'\rangle\langle y_i, y'\rangle| : x' \in V^o, y' \in U^o\} ,$$

and has the cross-property:

$$p_V \otimes_\varepsilon q_U(x \otimes y) = p_V(x) q_U(y) \quad \text{for all } x \in E \text{ and } y \in F ,$$

hence $\mathcal{P} \otimes_\varepsilon \mathcal{J}$ is determined by the family $\{p_V \otimes q_U : V \in \mathcal{U}_E, U \in \mathcal{U}_F\}$ of cross-seminorms. Clearly $p_V \otimes_\varepsilon q_U \leq p_V \otimes_\pi q_U$, thus

$$\mathcal{P} \otimes_\varepsilon \mathcal{J} \leq \mathcal{P} \otimes_\pi \mathcal{J} \text{ and } E' \otimes F' \subset (E \otimes_\varepsilon F)' \subset B(E, F) = L^{eq}(E, F(\mathcal{J})').$$

A similar argument given in the proof of (4.1.5)(a) yields the following more general result.

(4.4.4) <u>Example</u>. For any complete locally convex space E and a non-empty index set Λ, there are

$$\ell^1_\Lambda \tilde{\otimes}_\pi E \cong \ell^1_\pi[\Lambda, E] \quad \text{and} \quad \ell^1_\Lambda \tilde{\otimes}_\varepsilon E \cong \ell^1_\varepsilon(\Lambda, E) .$$

To consider the topological dual of $E \tilde{\otimes}_\varepsilon F$, let us define

$$\mathcal{J}(E, F) = \{f \circ \omega \in B(E, F) : f \in (E \tilde{\otimes}_\varepsilon F)'\} ,$$

where $\omega : E \times F \to E \otimes F$ is the canonical map. Elements in $\mathcal{J}(E, F)$ are called <u>integral bilinear forms</u> on $E \times F$.

Clearly, $\mathcal{J}(E, F)$ is a vector subspace of $B(E, F)$ and an $\psi \in B(E, F)$ is integral if and only if there exist $V \in \mathcal{U}_E$ and $U \in \mathcal{U}_F$ such that

$$|\Sigma_{i=1}^n \psi(x_i, y_i)| \leq p_V \otimes_\varepsilon q_U(\Sigma_{i=1}^n x_i \otimes y_i) \quad \text{for all } \Sigma_{i=1}^n x_i \otimes y_i \in E \otimes F.$$

Therefore the map $f \mapsto f \circ \omega$ is an algebraic isomorphism from $(E \tilde{\otimes}_\varepsilon F)'$ onto $\mathcal{J}(E, F)$, thus we identity $(E \tilde{\otimes}_\varepsilon F)'$ with $\mathcal{J}(E, F)$. As the ε-topology is coarser than the π-topology, we have

$$E' \otimes F' \subset (E \tilde{\otimes}_\varepsilon F)' = \mathcal{J}(E, F) \subset (E \tilde{\otimes}_\pi F)' = B(E, F) = L^{eq}(E, F(\mathfrak{I})') .$$

An $T \in L(E, F)$ is called an <u>integral linear map</u> if the bilinear form u_T, defined by

$$u_T(x, y') = \langle Tx, y' \rangle \quad \text{for all} \quad (x, y') \in E \times F' ,$$

is an integral bilinear form on $E \times F'_\beta$.

The set consisting of all integral linear maps from E into F, denoted by $L^i(E, F)$, is a vector subspace of $L(E, F)$ which is algebraically isomorphic to a vector subspace of $(E \widetilde{\otimes}_\varepsilon F'_\beta)'$. Furthermore, the continuity of u_T for $\mathscr{P} \times \beta(F', F)$ implies that $L^i(E, F) \subset L^{\ell b}(E, F)$.

In order to reduce the general case of integral bilinear forms on locally convex spaces to the case of that defined on normed spaces, we require the following terminology and results concerning with the decomposition of continuous bilinear forms.

Let $\psi \in B^*(E, F)$. If there exist $V \in \mathcal{U}_E$ and $U \in \mathcal{U}_F$ such that

$$\psi(x, y) = 0 \text{ for all } (x, y) \in ((\operatorname{Ker} p_V) \times F) \cup (E \times (\operatorname{Ker} q_U)), \qquad (4.1)$$

then there is a unique $\psi_{(V,U)} \in B^*(E_V, F_U)$ such that

$$\psi_{(V,U)}(Q_V(x), Q_U(y)) = \psi(x, y) \quad \text{for all} \quad (x, y) \in E \times F . \qquad (4.2)$$

$\psi_{(V,U)}$ (when it exists) is called the <u>induced bilinear form</u> of ψ.

It is easily seen that every <u>continuous</u> bilinear form ψ on $E \times F$ satisfies (4.1) since the continuity of ψ ensures that

$$|\psi(x, y)| \leq p_V(x) q_U(y) \quad \text{for all} \quad (x, y) \in E \times F$$

for some $V \in \mathcal{U}_E$ and $U \in \mathcal{U}_F$, hence the induced bilinear form $\psi_{(V,U)}$ of ψ exists and is continuous on $E_V \times F_U$. Conversely, if $h \in B(E_V, F_U)$, then $h \circ (Q_V \times Q_U) \in B(E, F)$, where

$$(h \circ (Q_V \times Q_U))(x, y) = h(Q_V(x), Q_U(y)) \quad \text{for all} \quad (x, y) \in E \times F,$$

thus h coincides with the induced bilinear form of $h \circ (Q_V \times Q_U)$.

Suppose $V \in \mathcal{U}_E$ and $U \in \mathcal{U}_F$. Let us define

$$P(V, U) = \{\psi \in B(E, F) : |\psi(x, y)| \leq 1 \quad \text{for all} \quad (x, y) \in V \times U\}.$$

Then $P(V, U)$ is an absolutely convex subset of $B(E, F)$, hence the vector subspace of $B(E, F)$ generated by $P(V, U)$, denoted by $B(V, U)$, is

$$B(V, U) = \bigcup_n n\, P(V, U),$$

and the gauge of $P(V, U)$ on $B(V, U)$, denoted by $p_{(V,U)}$, is

$$p_{(V,U)}(\psi) = \sup\{|\psi(x, y)| : (x, y) \in V \times U\}.$$

If $\|\cdot\|_{(V,U)}$ denotes the bilinear norm on $B(E_V, F_U)$, then the map $h \mapsto h \circ (Q_V \times Q_U)$ is an isometric isomorphism from the normed space $(B(E_V, F_U), \|\cdot\|_{(V,U)})$ onto $(B(V, U), p_{(V,U)})$ and

$$\psi_{(V,U)} \circ (Q_V \times Q_U) = \psi \text{ and } \|\psi_{(V,U)}\|_{(V,U)} = p_{(V,U)}(\psi) \quad \text{for any} \quad \psi \in B(V, U).$$

The continuity of $\psi_{(V,U)}$ ensures that it has a unique extension on $\tilde{E}_V \times \tilde{F}_U$, which is also called the <u>induced bilinear form</u> of ψ, and denoted by $\psi_{(V,U)}$.

Since the adjoint map Q_V' of Q_V is an isometric isomorphism from the Banach dual $(E_V)'$ onto the Banach space $E'(V^o)$ and since V^o is the image, under Q_V', of the closed unit ball in $(E_V)'$, it follows that

$$\hat{p}_V \otimes_\varepsilon \hat{q}_U((Q_V \otimes Q_U)u) = p_V \otimes_\varepsilon q_U(u) \quad \text{for all} \quad u \in E \otimes F. \tag{4.3}$$

Combining (4.2) and (4.3), we conclude that an $\psi \in B(E, F)$ is integral if and only if $\psi_{(V,U)} \in \mathcal{J}(E_V, F_U)$ for some $V \in \mathcal{U}_E$ and $U \in \mathcal{U}_F$.

We now summarize our results as follows:

(4.4.5) Lemma. (a) <u>If</u> $V \in \mathcal{U}_E$ <u>and</u> $U \in \mathcal{U}_F$, <u>then</u>

$$\hat{p}_V \otimes_\varepsilon \hat{q}_U((Q_V \otimes Q_U)u) = p_V \otimes_\varepsilon q_U(u) \quad \underline{\text{for all}} \quad u \in E \otimes F.$$

(b) $(B(V, U), P_{(V,U)})$ <u>and</u> $(B(E_V, F_U), \|\cdot\|_{(V,U)})$ <u>are isometrically isomorphic under the map</u> $\psi \mapsto \psi_{(V,U)}$, <u>where</u> $\psi_{(V,U)}$ <u>is the induced bilinear form of</u> ψ.

(c) $B(E, F) = \cup\{B(E_V, F_U) : V \in \mathcal{U}_E$ <u>and</u> $U \in \mathcal{U}_F\}$ <u>and</u>
$\mathcal{J}(E, F) = \cup\{\mathcal{J}(E_V, F_U) : V \in \mathcal{U}_E$ <u>and</u> $U \in \mathcal{U}_F\}$.

By means of the preceding result, the representation of integral operators from one normed space into another can be carried out for integral linear maps with domain and range to be locally convex spaces, namely:

(4.4.6) Theorem. <u>An</u> $\psi \in B(E, F)$ <u>is integral if and only if there exist</u> $V \in \mathcal{U}_E$, $U \in \mathcal{U}_F$ <u>and a positive Radon measure</u> μ <u>on</u> $V^o \times U^o$ <u>with total mass</u> ≤ 1 <u>such that</u>

$$\psi(x, y) = \int_{V^o \times U^o} \langle x, x'\rangle\langle y, y'\rangle d\mu(x', y') \quad \underline{\text{for all}} \quad (x, y) \in E \times F.$$

<u>An</u> $T \in L(E, F)$ <u>is integral if and only if there exist</u> $V \in \mathcal{U}_E$, <u>an absolutely convex bounded subset</u> B <u>of</u> F <u>and a positive Radon measure</u> μ <u>on</u> $V^o \times B^{oo}$

<u>with total mass</u> ≤ 1 <u>such that</u>

$$\langle Tx, y'\rangle = \int_{V^o \times B^{oo}} \langle x, x'\rangle\langle y, y''\rangle d\mu(x', y'') \quad \text{for all} \quad (x, y') \in E \times F',$$

<u>where</u> B^{oo} <u>is the polar of</u> B^o <u>taken in</u> F'' .

Let $T \in L(E, F'_\beta)$. If the associated bilinear form b_T , defined by

$$\psi(x, y) = b_T(x, y) = \langle y, Ty\rangle \quad \text{for all} \quad (x, y) \in E \times F,$$

is an integral bilinear form on $E \times F$, then (4.4.5) shows that there are $V \in \mathcal{U}_E$ and $U \in \mathcal{U}_F$ such that $\psi_{(V,U)} \in \mathcal{J}(E_V, F_U)$. As E_V and F_U are normed spaces, there is a unique $S \in L^i(E_V, (F_U)')$ such that

$$\langle Q_U(y), S(Q_V(x))\rangle = \psi_{(V,U)}(Q_V(x), Q_U(y)) = \langle y, Tx\rangle \quad \text{for all} \quad (x, y) \in E \times F,$$

hence

$$T = Q'_U \circ S \circ Q_V .$$

Clearly, the above equation shows that $Q'_U \circ S$ is the induced map $T_{(V,U^o)}$ of T and belongs to $L^i(E_V, F'(U^o))$, thus $T \in L^i(E, F'_\beta)$. The induced map $T_{(V,U^o)}$ of T belonging to $L^i(E_V, F'(U^o))$ characterizes the integrability of b_T on $E \times F$ as shown by the following

(4.4.7) Theorem. <u>Let</u> $T \in L^{eq}(E, F(\mathcal{J})')$. <u>Then the associated bilinear form</u> b_T <u>of</u> T <u>is integral if and only if there are</u> $V \in \mathcal{U}_E$ <u>and</u> $U \in \mathcal{U}_F$ <u>such that the induced map</u> $T_{(V,U^o)}$ <u>of</u> T <u>belongs to</u> $L^i(E_V, F'(U^o))$; <u>consequently</u>,

$$\{T \in L^{eq}(E, F(\mathcal{J})') : b_T \in \mathcal{J}(E, F)\} \subset L^i(E, F'_\beta) \tag{4.4}$$

<u>Moreover</u>, <u>if</u> F <u>is infrabarrelled</u>, <u>then</u> $L^i(E, F'_\beta) = \mathcal{J}(E, F)$.

<u>Proof</u>. The necessity and (4.4) have been proved before this theorem. To verify the sufficiency, we first notice that the bilinear form ψ , defined by

$$\psi(Q_V(x), g) = \langle T_{(V,U^o)}(Q_V(x)), g\rangle \quad \text{for all } x \in E \text{ and } g \in (F'(U^o))' ,$$

is integral on $E_V \times (F'(U^o))'$ by the hypothese. As $(F_U)'' \equiv (F'(U^o))'$, $E_V \otimes_\varepsilon F_U$ can be identified with a subspace of $E_V \otimes_\varepsilon (F'(U^o))'$, hence the restriction to $E_V \times F_U$ of ψ , denoted again by ψ , is integral on $E_V \times F_U$ and

$$\begin{aligned}\psi(Q_V(x), Q_U(y)) &= \langle T_{(V,U^o)}(Q_V(x)), e_U(Q_U(y))\rangle \\ &= \langle Q_U(y), T_{(V,U^o)}(Q_V(x))\rangle \quad \text{for all } (x, y) \in E \times F , \qquad (4.5)\end{aligned}$$

where $e_U : F_U \to (F_U)''$ is the evaluation map. Therefore $\psi \circ (Q_V \times Q_U) \in \mathcal{J}(E, F)$ by (4.4.5). When we identify $(F_U)'$ with $F'(U^o)$, the embedding map $j_{U^o} : F'(U^o) \to F'$ can be regarded as $(Q_U)'$. Therefore we conclude from (4.5) that $b_T = \psi \circ (Q_V \times Q_U)$, and hence that $b_T \in \mathcal{J}(E, F)$.

Finally, if F is infrabarrelled and if $T \in L^i(E, F'_\beta)$, then the bilinear form u_T , defined by

$$u_T(x, y'') = \langle Tx, y''\rangle \quad \text{for all } (x, y'') \in E \times F'' ,$$

is integral on $E \times F''_\beta$, where $F''_\beta = (F'', \beta(F'', F'))$. The infrabarrelledness of F implies that the evaluation $e_F : F \to F''_\beta$ is a topological isomorphism from F into F''_β , hence we identify $E \otimes_\varepsilon F$ with a subspace of $E \otimes_\varepsilon F''_\beta$ by a result of Grothendieck (see Treves [1, p.440]). Clearly, the associated bilinear form b_T of T is the restriction of u_T on $E \otimes F$, thus $b_T \in \mathcal{J}(E, F)$. This conclusion, together with (4.5) show that $L^i(E, F'_\beta)$ and $\mathcal{J}(E, F)$ are algebraically isomorphic under the map $T \mapsto b_T$.

It is known from (4.4.5) that nuclear linear maps from E into F are integral. If, in addition $(F', \beta(F', F))$ is infrabarrelled and $T' \in L^i(F'_\beta, E'_\beta)$, then (4.4.6) shows that $T \in L^i(E, F)$, which is a generalization of (4.1.14).

Another interesting topology on $E \otimes F$, which we are going to study, is the s-topology that is a generalization of the absolutely summing norm. For any $V \in \mathcal{U}_E$ and $U \in \mathcal{U}_F$, it is known from the proof of (4.2.4) that

$$p_V \otimes_s q_U(u) = \inf\{\sup\{\Sigma_{i=1}^n |\langle x_i, x'\rangle q_U(y_i)| : x' \in V^0\} : u = \Sigma_{i=1}^n x_i \otimes y_i\} \quad (4.6)$$

is a seminorm on $E \otimes F$ with the following properties:

(i) $p_V \otimes_\varepsilon q_U \leq p_V \otimes_s q_U \leq p_V \otimes_\pi q_U$; (4.7)

(ii) $p_V \otimes_s q_U(x \otimes y) = p_V(x) q_U(y)$ for all $x \in E$ and $y \in F$.

Hence the Hausdorff locally convex topology on $E \otimes F$ determined by $\{p_V \otimes_s q_U : V \in \mathcal{U}_E, U \in \mathcal{U}_F\}$ is called the s-<u>topology</u> (or <u>summing topology</u>), and denoted by $\mathcal{P} \otimes_s \mathcal{J}$. Let $E \otimes_s F$ denote the locally convex space $(E \otimes F, \mathcal{P} \otimes_s \mathcal{J})$ and let $E \tilde{\otimes}_s F$ denote the completion of $E \otimes_s F$.

Since $(E_V)' = E'(V^0)$ and since V^0 is the image, under Q'_V, of the closed unit ball in $(E_V)'$, it follows that

$$\hat{p}_V \otimes_s \hat{q}_U((Q_V \otimes Q_U)u) = p_V \otimes_s q_U(u) \quad \text{for all} \quad u \in E \otimes F .$$

On the other hand, formula (4.7) shows that

$$\mathcal{P} \otimes_\varepsilon \mathcal{J} \leq \mathcal{P} \otimes_s \mathcal{J} \leq \mathcal{P} \otimes_\pi \mathcal{J} ,$$

and hence

$$E' \otimes F' \subset (E \otimes_\varepsilon F)' \subset (E \otimes_s F)' \subset (E \otimes_\pi F)' = B(E, F) = L^{eq}(E, F(\mathcal{J})') .$$

Suppose now that $f \in (E \otimes_s F)'$ and that $T \in L^{eq}(E, F(\mathcal{J})')$ is obtained from the equation

$$\langle y, Tx \rangle = f(x \otimes y) \quad \text{for all} \quad x \in E \quad \text{and} \quad y \in F .$$

Then there exist $V \in \mathcal{U}_E$ and $U \in \mathcal{U}_F$ such that

$$|\Sigma_{i=1}^n \langle y_i , Tx_i \rangle| \leq p_V \otimes_s q_U(\Sigma_{i=1}^n x_i \otimes y_i) \quad \text{for all} \quad \Sigma_{i=1}^n x_i \otimes y_i \in E \otimes F . \tag{4.8}$$

Therefore $T(V) \subset U^o$ and the induced map $T_{(V,U^o)}$ of T is a continuous linear map from E_V into $F'(U^o)$ for which

$$T = j_{U^o} \circ T_{(V,U^o)} \circ Q_V .$$

We claim further that $T_{(V,U^o)}$ is a prenuclear-bounded linear map or, equivalently $T_{(V,U^o)} \in L^s(E_V, F'(U^o))$ in view of the normability of $F'(U^o)$. To this end, we first notice that the gauge q_{U^o} of U^o on $F'(U^o)$ is given by

$$q_{U^o}(y') = \sup\{|\langle y, y'\rangle| : y \in U\} \quad \text{for all} \quad y' \in F'(U^o) . \tag{4.9}$$

According to the definition of $p_V \otimes_s q_U$, formula (4.8) shows that the inequality

$$|\Sigma_{i=1}^n \langle y_i , Tx_i \rangle| \leq \sup\{\Sigma_{i=1}^n |\langle x_i , x'\rangle q_U(y_i)| : x' \in V^o\} \tag{4.10}$$

holds for any finite subsets $\{x_1, \dots, x_n\}$ and $\{y_1, \dots, y_n\}$ of E and F respectively. A similar argument given in the proof of (e) $\Rightarrow$ (a) of (2.2.2) together with (4.9) show that the inequality

$$\Sigma_{i=1}^n q_{U^o}(Tu_i) \leq \sup\{\Sigma_{i=1}^n |\langle u_i , x'\rangle| : x' \in V^o\} \tag{4.11}$$

holds for any finite subset $\{u_1, \dots, u_n\}$ of E . Let Σ^o be the polar, taken in $(E_V)'$, of the closed unit ball Σ in E_V . Then $Q'_V(\Sigma^o) = V^o$, hence (4.11) shows that the inequality

$$\begin{aligned} \Sigma_{i=1}^n q_{U^o}(T_{(V,U^o)}(Q_V(u_i))) &\leq \sup\{\Sigma_{i=1}^n |\langle u_i , x'\rangle| : x' \in V^o\} \\ &= \sup\{\Sigma_{i=1}^n |\langle Q_V(u_i), g\rangle| : g \in \Sigma^o\} \end{aligned}$$

holds for any finite subset $\{Q_V(x_1) , \dots, Q_V(x_n)\}$ of E_V . Therefore $T_{(V,U^o)}$ is an absolutely summing map. The induced map $T_{(V,U^o)}$ of T belonging to $L^{pn}(E_V, F'(U^o))$ characterizes the $\mathcal{P} \otimes_s \mathcal{J}$ -continuity of a linear functional induced by T as shown by the following

(4.4.8) Theorem. <u>An</u> $T \in L^{eq}(E, F(\mathcal{J})')$ <u>defines a continuous linear functional on</u> $E \otimes_s F$ <u>if and only if there exist</u> $V \in \mathcal{U}_E$ <u>and</u> $U \in \mathcal{U}_F$ <u>such that the induced map</u> $T_{(V,U^o)}$ <u>of</u> T <u>belongs to</u> $N^{(p)}_{\ell^1}(E_V, F'(U^o))$.

Moreover, if in addition, F is a normed space, then

$$(E \widetilde{\otimes}_s F)' = N^{(p)}_{\ell^1}(E, F'_\beta) = \pi_{\ell^1}(E, F'_\beta) . \qquad (4.12)$$

Proof. The necessity has been proved before this theorem. To prove the sufficiency, we first notice that $N^{(p)}_{\ell^1}(E, F'(U^o)) = L^s(E, F'(U^o))$ in view of the normability of $F'(U^o)$. Let $c \geq 0$ be such that the inequality

$$\Sigma^n_{i=1} q_{U^o}(T_{(V,U^o)}(Q_V(x_i))) \leq c \sup\{\Sigma^n_{i=1} |\langle Q_V(x_i), g\rangle| : g \in \Sigma^o\}$$

holds for any finite subset $\{Q_V(x_1), \ldots, Q_V(x_n)\}$ of E_V , where Σ^o is the closed unit ball in $(E_V)'$ and Q_{U^o} is the gauge of U^o defined on $F'(U^o)$. Then the inequality

$$\Sigma^n_{i=1} q_{U^o}(Tx_i) \leq c \sup\{\Sigma^n_{i=1} |\langle x_i, x'\rangle| : x' \in V^o\} \qquad (4.13)$$

holds for any finite subset $\{x_1, \ldots, x_n\}$ of E because of $V^o = Q'_V(\Sigma^o)$. Now for any $u \in E \otimes F$ and any representation $\Sigma^n_{i=1} x_i \otimes y_i$ of u , formula (4.13) shows that

$$|f(u)| = |\Sigma^n_{i=1} \langle y_i, Tx_i\rangle| \leq \Sigma^n_{i=1} q_U(y_i) q_{U^o}(Tx_i) = \Sigma^n_{i=1} q_{U^o}(T(q_U(y_i)x_i))$$
$$\leq c \sup\{\Sigma^n_{i=1} |\langle x_i, x'\rangle q_U(y_i)| : x' \in V^o\} ,$$

hence we obtain $|f(u)| \leq c\, p_V \otimes_s q_U(u)$, thus $f \in (E \widetilde{\otimes}_s F)'$.

Finally, if F is a normed space, then $N^{(p)}_{\ell^1}(E, F'_\beta) = \pi_{\ell^1}(E, F'_\beta)$, and U can be chosen as the closed unit ball in F in view of the proof of the first part of this theorem, thus (4.12) holds.

(4.4.9) Corollary. $(E \tilde{\otimes}_s F)'$ <u>can be identified with a vector subspace of</u> $N^{(p)}_{\ell^1}(E, F'_\beta)$ <u>and</u>

$$\{T \in L^{eq}(E, F(\mathcal{J})') : b_T \in \mathcal{J}(E, F)\} \subset N^{(p)}_{\ell^1}(E, F'_\beta) . \qquad (4.14)$$

<u>Moreover, if in addition,</u> F <u>is infrabarrelled, then</u>

$$L^i(E, F'_\beta) \subset N^{(p)}_{\ell^1}(E, F'_\beta) \quad \underline{\text{and}} \quad L^i(E, F) \subset N^{(p)}_{\ell^1}(E, F) .$$

<u>Proof</u>. The first part and (4.14) follow from (4.4.8). Suppose that F is infrabarrelled. Then we have by (4.4.7) and the first part that

$$L^i(E, F'_\beta) = (E \tilde{\otimes}_\varepsilon F)' \subset (E \tilde{\otimes}_s F)' \subset N^{(p)}_{\ell^1}(E, F'_\beta) .$$

To prove that $L^i(E, F) \subset N^{(p)}_{\ell^1}(E, F)$, we first notice from the infrabarrelledness of F that the evaluation map $e_F : F \to (F'', \beta(F'', F'))$ is an into topological isomorphism, hence every continuous seminorm q'' on $(F'', \beta(F'', F'))$ is the extension of some continuous seminorm q on F. Now for any $T \in L^i(E, F)$, the linear functional f, defined by

$$f(x \otimes y') = \langle Tx, y'\rangle \quad \text{for all} \quad x \in E \quad \text{and} \quad y' \in F' ,$$

is continuous on $E \otimes_\varepsilon F'_\beta$ and surely on $E \otimes_s F'_\beta$, therefore $T \in N^{(p)}_{\ell^1}(E, F''_\beta)$ by (4.14). On the other hand, (3.3.5) shows that there exists an $V \in \mathcal{U}_E$ such that for any continuous seminorm q'' on F''_β it is possible to find an $c(q'') \geq 0$ such that the inequality

$$\Sigma_{i=1}^{n} q''(Tx_i) \leq c(q'') \sup\{\Sigma_{i=1}^{n} |\langle x_i, x'\rangle| : x' \in V^o\} \qquad (4.15)$$

holds for any finite subset $\{x_1, \ldots, x_n\}$ of E. As $T(E) \subset F$, it follows from the infrabarrelledness of F and (4.15) that $T \in N^{(p)}_{\ell^1}(E, F)$.

As a consequence of (4.4.9) and (3.3.6) we obtain the following:

(4.4.10) Corollary. Let F and G be infrabarrelled spaces. If $T_1 \in L^i(E, F)$ and $T_2 \in L^i(F, G)$, then $T_2 \circ T_1 \in N_{\ell^1}(E, G)$.

Since normed spaces are infrabarrelled, the preceding result is a generalization of (4.2.5)(b).

4.5 Tensor product mappings

Let E_j and F_j be locally convex spaces, and suppose that $T_j \in L(E_j, F_j)$ $(j = 1, 2)$. Then the tensor product map $T_1 \otimes T_2$ belongs to $L(E_1 \otimes_\varepsilon E_2, F_1 \otimes_\varepsilon F_2)$ as well as to $L(E_1 \otimes_\pi E_2, F_1 \otimes_\pi F_2)$, thus it can be uniquely extended to a continuous linear map from $E_1 \tilde{\otimes}_\varepsilon E_2$ into $F_1 \tilde{\otimes}_\varepsilon F_2$ (resp. from $E_1 \tilde{\otimes}_\pi E_2$ into $F_1 \tilde{\otimes}_\pi F_2$), which is denoted by $T_1 \tilde{\otimes}_\varepsilon T_2$ (resp. $T_1 \tilde{\otimes}_\pi T_2$). Furthermore, Grothendieck has shown (see Treves [1, pp.440-441]) that if T_j is a topological homomorphism from E_j onto a dense subspace of F_j $(j = 1, 2)$, then $T_1 \tilde{\otimes}_\pi T_2$ is a topological homomorphism from $E_1 \tilde{\otimes}_\pi E_2$ onto a dense subspace of $F_1 \tilde{\otimes}_\pi F_2$, and that if T_j is a topological isomorphism from E_j into F_j $(j = 1, 2)$, then $T_1 \tilde{\otimes}_\varepsilon T_2$ is a topological isomorphism from $E_1 \tilde{\otimes}_\varepsilon E_2$ into $F_1 \tilde{\otimes}_\varepsilon F_2$. This section is devoted to a study of some special classes of linear maps which are preserved for some locally convex tensor product topologies.

(4.5.1) Lemma. *If* $T_j \in L(E_j, F_j)$ $(j = 1, 2)$, *then* $T_1 \otimes T_2 \in L(E_1 \otimes_s E_2, F_1 \otimes_s F_2)$.

Proof. Let U_j be an absolutely convex o-neighbourhood in F_j and q_j the gauge of U_j $(j = 1, 2)$. Then $V_j = T_j^{-1}(U_j)$ is an absolutely convex o-neighbourhood in E_j and

$$q_j(T_j x^{(j)}) \leq p_j(x^{(j)}) \quad \text{for all} \quad x^{(j)} \in E_j , \tag{5.1}$$

where p_j is the gauge of V_j $(j = 1, 2)$. For any $u \in E_1 \otimes_s E_2$ and any representation $\Sigma_{i=1}^n x_i^{(1)} \otimes x_i^{(2)}$ of u , where $x_i^{(j)} \in E_j$ $(i = 1, \ldots, n$ and $j = 1, 2)$, we have by (5.1) that

$$\begin{aligned} q_1 \otimes_s q_2(T_1 \otimes T_2(u)) &= q_1 \otimes_s q_2(\Sigma_{i=1}^n (T_1 x_i^{(1)}) \otimes (T_2 x_i^{(2)})) \\ &\leq \sup\{\Sigma_{i=1}^n |\langle T_1 x_i^{(1)}, g\rangle q_2(T_2 x_i^{(2)})| : g \in F_1' , g \in U_1^o\} \\ &\leq \sup\{\Sigma_{i=1}^n |\langle x_i^{(1)}, w'\rangle p_2(x_i^{(2)})| : w' \in E_1', w' \in V_1^o\} , \end{aligned}$$

hence

$$q_1 \otimes_s q_2(T_1 \otimes T_2(u)) \leq p_1 \otimes_s p_2(u)$$

since the representation $\Sigma_{i=1}^n x_i^{(1)} \otimes x_i^{(2)}$ of u was arbitrary, and thus $T_1 \otimes T_2 \in L(E_1 \otimes_s E_2, F_1 \otimes_s F_2)$.

Because of the preceding result, $T_1 \otimes T_2$ can be uniquely extended to a continuous linear map from $E_1 \tilde{\otimes}_s E_2$ into $F_1 \tilde{\otimes}_s F_2$, which is denoted by $T_1 \tilde{\otimes}_s T_2$.

In the remainder of this section, we denote by $\mathcal{I}$ either the π-topology or the s-topology or the ε-topology, by $E_1 \otimes_{\mathcal{I}} E_2$ the locally convex space $(E_1 \otimes E_2, \mathcal{I})$, by $E_1 \tilde{\otimes}_{\mathcal{I}} E_2$ the completion of $E_1 \otimes_{\mathcal{I}} E_2$, and by $T_1 \tilde{\otimes}_{\mathcal{I}} T_2$ either $T_1 \tilde{\otimes}_{\pi} T_2$ or $T_1 \tilde{\otimes}_s T_2$ or $T_1 \tilde{\otimes}_{\varepsilon} T_2$.

By an <u>ideal</u> we mean a set A consisting of continuous linear maps such that if $T \in A \cap L(E, F)$ then

$$Q \circ T \circ S \in A \quad \text{whenever} \quad S \in L(G, E) \quad \text{and} \quad Q \in L(F, N),$$

where G, E, F and N are locally convex spaces.

For example, the set of all nuclear linear maps, the set of all absolutely summing maps, the set of all quasi-nuclear-bounded linear maps, and the set of all prenuclear-bounded linear maps are all ideals.

Suppose now that A is an ideal. Then we write

$$A(E, F) = A \cap L(E, F).$$

The following result will be useful in showing the converse of a number of our later results.

(4.5.2) Lemma. <u>Let</u> A <u>be an ideal</u>. <u>If</u>

$$T_1 \tilde{\otimes}_{\mathcal{I}} T_2 \in A(E_1 \tilde{\otimes}_{\mathcal{I}} E_2, F_1 \tilde{\otimes}_{\mathcal{I}} F_2) \quad \underline{\text{and}} \quad T_1 \neq 0, \quad \underline{\text{then}} \quad T_2 \in A(E_2, F_2).$$

<u>Proof</u>. There exists an $u_1 \in E_1$ such that $T_1 u_1 \neq 0$, hence there are

an absolutely convex o-neighbourhood U_1 in F_1 and $h_1 \in U_1^o$ such that

$$q_1(T_1u_1) = \langle T_1u_1, h_1\rangle = 1 ,$$

where q_1 is the gauge of U_1. Let us define $S : E_2 \to E_1 \tilde{\otimes}_{\mathcal{I}} E_2$ and $Q : F_1 \tilde{\otimes}_{\mathcal{I}} F_2 \to F_2$ by setting

$$S(x^{(2)}) = u_1 \otimes x^{(2)} \quad \text{for all} \quad x^{(2)} \in E_2 ,$$

$$Q(\Sigma_{i=1}^n y_i^{(1)} \otimes y_i^{(2)}) = \Sigma_{i=1}^n \langle y_i^{(1)}, h_1\rangle y_i^{(2)} \quad \text{for all} \quad \Sigma_{i=1}^n y_i^{(1)} \otimes y_i^{(2)} \in F_1 \otimes F_2 .$$

Then for any continuous seminorm p_i on E_i $(i = 1, 2)$,

$$p_1 \otimes_\pi p_2(Sx^{(2)}) = p_1 \otimes_\varepsilon p_2(Sx^{(2)}) = p_1 \otimes_s p_2(Sx^{(2)}) = p_1(u_1)p_2(x^{(2)}) .$$

If q_2 is any continuous seminorm on F_2, then

$$\begin{aligned} q_2(Q(\Sigma_{i=1}^n y_i^{(1)} \otimes y_i^{(2)})) &= q_2(\Sigma_{i=1}^n \langle y_i^{(1)}, h_1\rangle y_i^{(2)}) \\ &\leq \sup\{q_2(\Sigma_{i=1}^n \langle y_i^{(1)}, f_1\rangle y_i^{(2)}) : f_1 \in U_1^o\} \\ &= q_1 \otimes_\varepsilon q_2(\Sigma_{i=1}^n y_i^{(1)} \otimes y_i^{(2)}) . \end{aligned}$$

Therefore $S \in L(E_2, E_1 \tilde{\otimes}_{\mathcal{I}} E_2)$ and $Q \in L(F_1 \tilde{\otimes}_{\mathcal{I}} F_2, F_2)$ because of

$$q_1 \otimes_\varepsilon q_2 \leq q_1 \otimes_s q_2 \leq q_1 \otimes_\pi q_2 \quad \text{on} \quad F_1 \otimes F_2 .$$

Clearly $T_2 = Q \circ (T_1 \tilde{\otimes} T_2) \circ S$, we conclude that $T_2 \in A(E_2, F_2)$ because A is an ideal and $T_1 \tilde{\otimes}_{\mathcal{I}} T_2 \in A(E_1 \tilde{\otimes}_{\mathcal{I}} E_2, F_1 \tilde{\otimes}_{\mathcal{I}} F_2)$.

The preceding result was proved by Holub [1] in the case when E_j and F_j are Banach spaces.

In order to show that the nuclearity of linear maps is completely preserved for any locally convex tensor product topology, we need the following result, due to Holub [1].

(4.5.3) Theorem. Let (X_j, p_j) and (Y_j, q_j) be normed vector spaces, $T_j \in L(X_j, Y_j)$ $(j = 1, 2)$ and suppose that α is a reasonable cross-norm. Then $T_1 \otimes T_2 \in N_{\ell^1}(X_1 \otimes_\alpha X_2, Y_1 \otimes_\alpha Y_2)$ if and only if $T_j \in N_{\ell^1}(X_j, Y_j)$ $(j = 1, 2)$.

Proof. The necessity follows from (4.5.2). To prove the sufficiency, let $\{f_n^{(j)}\}$ in X_j' and $\{y_n^{(j)}\}$ in Y_j be such that $\Sigma_{n=1}^\infty p_j^*(f_n^{(j)})q_j(y_n^{(j)}) < \infty$ and

$$T_j(x^{(j)}) = \Sigma_{n=1}^\infty f_n^{(j)}(x_n^{(j)})y_n^{(j)} \quad \text{for all} \quad x^{(j)} \in X_j \quad (j = 1, 2) .$$

For any $u = \Sigma_{i=1}^n x_i^{(1)} \otimes x_i^{(2)} \in X_1 \otimes_\alpha X_2$,

$$\begin{aligned} T_1 \otimes T_2(u) &= \Sigma_{i=1}^n (T_1 x_i^{(1)}) \otimes (T_2 x_i^{(2)}) \\ &= \Sigma_{i=1}^n [\Sigma_{n,k} f_n^{(1)}(x_i^{(1)}) f_k^{(2)}(x_i^{(2)}) y_n^{(1)} \otimes y_n^{(2)}] \\ &= \Sigma_{n,k} \langle \Sigma_{i=1}^n x_i^{(1)} \otimes x_i^{(2)}, f_n^{(1)} \otimes f_k^{(2)} \rangle y_n^{(1)} \otimes y_k^{(2)} . \end{aligned} \tag{5.2}$$

Since α is a reasonable cross-norm, it follows that

$$\begin{aligned} &\Sigma_{n,k} \alpha^*(f_n^{(1)} \otimes f_k^{(2)})\, \alpha(y_n^{(1)} \otimes y_k^{(2)}) \\ &\quad = \Sigma_{n,k} p_1^*(f_n^{(1)}) p_2^*(f_k^{(2)}) q_1(y_n^{(1)}) q_2(y_k^{(2)}) < \infty . \end{aligned} \tag{5.3}$$

Clearly

$$h_{n,k} = f_n^{(1)} \otimes f_k^{(2)} \in X_1' \otimes X_2' \subset (X_1 \otimes_\alpha X_2)' \quad \text{and}$$

$$z_{n,k} = y_n^{(1)} \otimes y_k^{(2)} \in Y_1 \otimes Y_2 .$$

We obtain from (5.2) and (5.3) that

$$\Sigma_{n,k}\alpha^*(h_{n,k})\alpha(z_{n,k}) < \infty \quad \text{and} \quad T_1 \otimes T_2(u) = \Sigma_{n,k}\langle u, h_{n,k}\rangle z_{n,k} ,$$

which shows that $T_1 \otimes T_2 \in N_{\ell^1}(X_1 \otimes_\alpha X_2, Y_1 \otimes_\alpha Y_2)$.

(4.5.4) Corollary. <u>Let</u> $T_j \in L(E_j, F_j)$ $(j = 1, 2)$. <u>Then</u> $T_1 \tilde{\otimes}_{\mathcal{I}} T_2$ <u>belongs to</u> $N_{\ell^1}(E_1 \tilde{\otimes}_{\mathcal{I}} E_2, F_1 \tilde{\otimes}_{\mathcal{I}} F_2)$ <u>if and only if</u> $T_j \in N_{\ell^1}(E_j, F_j)$ $(j = 1, 2)$.

<u>Proof</u>. The necessity follows from (4.5.2). To prove the sufficiency, it suffices to show that $T_1 \otimes T_2 \in N_{\ell^1}(E_1 \otimes_{\mathcal{I}} E_2, F_1 \tilde{\otimes}_{\mathcal{I}} F_2)$. To do this, we first notice from the hypotheses and a well-known result that each T_j is the composition of the following continuous linear maps

$$E_j \xrightarrow{P_j} X_j \xrightarrow{S_j} Y_j \xrightarrow{Q_j} F_j ,$$

where X_j and Y_j are normed spaces and $S_j \in N_{\ell^1}(X_j, Y_j)$. Then $S_1 \otimes S_2$ belongs to $N_{\ell^1}(X_1 \otimes_\pi X_2, Y_1 \otimes_\pi Y_2)$ by (4.5.3), and $P_1 \otimes P_2 \in L(E_1 \otimes_{\mathcal{I}} E_2, Y_1 \otimes_\pi Y_2)$ and $Q_1 \otimes Q_2 \in L(Y_1 \otimes_\pi Y_2, F_1 \tilde{\otimes}_{\mathcal{I}} F_2)$. It is easily seen that

$$T_1 \otimes T_2 = (Q_1 \otimes Q_2) \circ (S_1 \otimes S_2) \circ (P_1 \otimes P_2) ,$$

we conclude that $T_1 \otimes T_2 \in N_{\ell^1}(E_1 \otimes_{\mathcal{I}} E_2, F_1 \tilde{\otimes}_{\mathcal{I}} F_2)$.

(4.5.5) Theorem. <u>Let</u> $T_j \in L(E_j, F_j)$ $(j = 1, 2)$. <u>Then</u> $T_1 \otimes T_2$ <u>belongs to</u> $N^{(q)}_{\ell^1}(E_1 \otimes_\varepsilon E_2, F_1 \otimes_\varepsilon F_2)$ <u>if and only if</u> $T_j \in N^{(q)}_{\ell^1}(E_j, F_j)$ $(j = 1, 2)$.

<u>Proof</u>. The necessity follows from (4.5.2). To prove the sufficiency,

let $(\zeta_n^{(j)}) \in \ell^1$ and let two equicontinuous sequences $\{f_n^{(j)}\}$ in E_j' be such that for any continuous seminorm q_j on F_j it is possible to find an $c_j \geq 0$ such that

$$q_j(T_j x^{(j)}) \leq c_j \Sigma_{n=1}^{\infty} |\zeta_n^{(j)} < x^{(j)}, f_n^{(j)}>| \quad \text{for all } x^{(j)} \in E_j \quad (j = 1, 2). \quad (5.4)$$

Suppose $u = \Sigma_{i=1}^{m} x_i^{(1)} \otimes x_i^{(2)} \in E_1 \otimes_\varepsilon E_2$. Then

$$\Sigma_{i=1}^{m} x_i^{(1)} \otimes (T_2 x_i^{(2)}) : g^{(2)} \mapsto \Sigma_{i=1}^{m} <T_2 x^{(2)}, g^{(2)}> x_i^{(1)} \in E_1 \quad (g^{(2)} \in F_2'),$$

and hence

$$q_1(T_1(\Sigma_{i=1}^{m} <T_2 x_i^{(2)}, g^{(2)}> x_i^{(1)})) \leq c_1 \Sigma_{n=1}^{\infty} |\zeta_n^{(1)} \Sigma_{i=1}^{m} <T_2 x_i^{(2)}, g^{(2)}><x_i^{(1)}, f_n^{(1)}>|. \quad (5.5)$$

whenever $g^{(2)} \in E_2'$ by (5.4). If $U_j = \{y^{(j)} \in F_j : q_j(y^{(j)}) \leq 1\}$ $(j = 1, 2)$, then

$$\begin{aligned} & q_1 \otimes_\varepsilon q_2(T_1 \otimes T_2(\Sigma_{i=1}^{m} x_i^{(1)} \otimes x_i^{(2)})) \\ & = \sup\{q_1(\Sigma_{i=1}^{m} <T_2 x_i^{(2)}, h^{(2)}> T_1 x_i^{(1)}) : h^{(2)} \in U_2^o\} \\ & = \sup\{q_1(T_1(\Sigma_{i=1}^{m} <T_2 x_i^{(2)}, h^{(2)}> x_i^{(1)})) : h^{(2)} \in U_2^o\} \end{aligned}$$

and

$$q_2(\Sigma_{i=1}^{m} <x_i^{(1)}, f_n^{(1)}> T_2 x_i^{(2)}) = \sup\{|\Sigma_{i=1}^{m} <x_i^{(1)}, f_n^{(1)}><T_2 x_i^{(2)}, g^{(2)}>| : g^{(2)} \in U_2^o\}.$$

It follows from (5.5) that

$$q_1 \otimes_\varepsilon q_2(T_1 \otimes T_2(\Sigma_{i=1}^{m} x_i^{(1)} \otimes x_i^{(2)})) \leq c_1 \Sigma_{n=1}^{\infty} |\zeta_n^{(1)}| q_2(\Sigma_{i=1}^{m} <x_i^{(1)}, f_n^{(1)}> T_2 x_i^{(2)}). \quad (5.6)$$

On the other hand, since $\Sigma_{i=1}^{m} <x_i^{(1)}, f_n^{(1)}> x_i^{(2)} \in E_2$ for all n, it follows from (5.4) that

$$q_2(T_2(\Sigma_{i=1}^{m} <x_i^{(1)}, f_n^{(1)}> x_i^{(2)})) \leq c_2 \Sigma_{k=1}^{\infty} |\zeta_k^{(2)} \Sigma_{i=1}^{m} <x_i^{(1)}, f_n^{(1)}><x_i^{(2)}, f_k^{(2)}>|.$$

Substituting this into (5.6) we obtain

$$\begin{aligned} q_1 \otimes_\varepsilon q_2(T_1 \otimes T_2(\Sigma_{i=1}^m x_i^{(1)} \otimes x_i^{(2)})) \\ &\leq c_1 \Sigma_{n=1}^\infty |\zeta_n^{(1)}| (c_2(\Sigma_{k=1}^\infty |\zeta_k^{(2)} \Sigma_{i=1}^m <x_i^{(1)}, f_n^{(1)}><x_i^{(2)}, f_k^{(2)}>|) \\ &= c_1 c_2 \Sigma_{n,k} |\zeta_n^{(1)} \zeta_k^{(2)} <\Sigma_{i=1}^m x_i^{(1)} \otimes x_i^{(2)}, f_n^{(1)} \otimes f_k^{(2)}>| . \end{aligned} \tag{5.7}$$

Since $\{f_n^{(j)}\}$ is an equicontinuous sequence in E_j' $(j = 1, 2)$, there exists an absolutely convex o-neighbourhood V_j in E_j such that $f_n^{(j)} \in V_j^o$ for all n $(j = 1, 2)$. Hence

$$f_n^{(1)} \otimes f_k^{(2)} \in V_1^o \otimes V_2^o \text{ for all } n \text{ and } k . \tag{5.8}$$

Let us define h_r and σ_r by

$$h_{(n-1)(n-1)+t} = f_t^{(1)} \otimes f_n^{(2)} \text{ and } \sigma_{(n-1)(n-1)+t} = \zeta_t^{(1)} \zeta_n^{(2)} \qquad \text{for } 1 \leq t \leq n$$

$$h_{(n-1)(n-1)+t+n} = f_n^{(1)} \otimes f_{n-t}^{(2)} \text{ and } \sigma_{(n-1)(n-1)+t+n} = \zeta_n^{(1)} \zeta_{n-t}^{(2)} \text{ for } 1 \leq t < n .$$

Then $\{h_r\}$ is an equicontinuous sequence in $(E_1 \otimes_\varepsilon E_2)'$ by (5.8), and $(\sigma_r) \in \ell^1$ since

$$\Sigma_{r=1}^\infty |\sigma_r| = (\Sigma_{n=1}^\infty |\zeta_n^{(1)}|)(\Sigma_{k=1}^\infty |\zeta_k^{(2)}|) < \infty .$$

In view of (5.7), we obtain

$$q_1 \otimes_\varepsilon q_2(T_1 \otimes T_2(\Sigma_{i=1}^m x_i^{(1)} \otimes x_i^{(2)})) \leq c_1 c_2 \Sigma_{r=1}^\infty |\sigma_r <\Sigma_{i=1}^m x_i^{(1)} \otimes x_i^{(2)}, h_r>|$$

which shows that $T_1 \otimes T_2 \in N_{\ell^1}^{(q)}(E_1 \otimes_\varepsilon E_2, F_1 \otimes_\varepsilon F_2)$.

The preceding theorem was proved by Holub [1] in the case when E_j and F_j are Banach spaces. He also gave an example to show that there are

$T_j \in N^{(q)}_{\ell^1}(E_j, F_j)$ such that $T_1 \otimes T_2 \notin N^{(q)}_{\ell^1}(E_1 \otimes_\pi E_2, F_1 \otimes_\pi F_2)$. For instance, let e_n be the sequence with 1 in the n-th place and zero elsewhere, and let $(a_n) \in c_o$ be such that $(a_n) \notin \ell^4$. Then the map $T \in L(\ell^1, \ell^2)$, defined by

$$T(\Sigma_{n=1}^{\infty} b_n e_n) = \Sigma_{n=1}^{\infty} a_n b_n e_n \quad \text{for all} \quad (b_n) \in \ell^1 ,$$

is quasi-nuclear-bounded (see the proof of (3.4.8)) such that $T \otimes T \notin N^{(q)}_{\ell^1}(\ell^1 \otimes_\pi \ell^1, \ell^2 \otimes_\pi \ell^2)$.

To prove that prenuclear-bounded linear maps are preserved for the ε-topology, we need the following result, due to Holub [1].

(4.5.6) Theorem. <u>Let</u> (X_j, p_j) <u>and</u> (Y_j, q_j) <u>be normed spaces</u>. <u>If</u> $T_j \in \pi_{\ell^1}(X_j, Y_j)$, <u>then</u> $T_1 \otimes T_2 \in \pi_{\ell^1}(X_1 \otimes_\varepsilon X_2, Y_1 \otimes_\varepsilon Y_2)$.

<u>Proof</u>. By (2.2.2), there exist a $\sigma(X_j', X_j)$-closed equicontinuous subset B_j of X_j' and a positive Radon measure μ_j on B_j such that

$$q_j(T_j x^{(j)}) \leq \int_{B_j} |\langle x^{(j)}, f^{(j)}\rangle| d\mu_j(f^{(j)}) \quad \text{for all} \quad x^{(j)} \in X_j \ (j = 1, 2) . \tag{5.9}$$

As X_j is a normed space, we can assume that B_j is the polar Σ_j^o of the closed unit ball Σ_j in X_j . Let $u = \Sigma_{i=1}^m x_i^{(1)} \otimes x_i^{(2)} \in X_1 \otimes_\varepsilon X_2$. Then

$$\Sigma_{i=1}^m (T_1 x_i^{(1)}) \otimes x_i^{(2)} : g^{(1)} \mapsto \Sigma_{i=1}^m \langle T_1 x_i^{(1)}, g^{(1)}\rangle x_i^{(2)} \in Y_2 \quad \text{for all} \quad g^{(1)} \in Y_1' ,$$

hence we have by (5.9) that

$$q_2(T_2(\Sigma_{i=1}^m \langle T_1 x_i^{(1)}, g^{(1)}\rangle x_i^{(2)}) \leq \int_{\Sigma_2^o} |\Sigma_{i=1}^m \langle T_1 x_i^{(1)}, g^{(1)}\rangle \langle x_i^{(2)}, f^{(2)}\rangle| d\mu_2(f^{(2)})$$

$$\leq \int_{\Sigma_2^o} q_1(\Sigma_{i=1}^m \langle x_i^{(2)}, f^{(2)}\rangle T_1 x_i^{(1)}) d\mu_2(f^{(2)}) \quad \text{for all} \quad g^{(1)} \in U_1^o , \tag{5.10}$$

where U_1^o is the polar of the closed unit ball U_1 in Y_1. Since

$$\Sigma_{i=1}^m x_i^{(1)} \otimes x_i^{(2)} : f^{(2)} \mapsto \Sigma_{i=1}^m \langle x_i^{(2)}, f^{(2)} \rangle x_i^{(1)} \in X_1 \quad \text{for all} \quad f^{(2)} \in \Sigma_2^o ,$$

it follows from (5.9) that

$$q_1(\Sigma_{i=1}^m \langle x_i^{(2)}, f^{(2)} \rangle T_1 x_i^{(1)}) \leq \int_{\Sigma_1^o} |\Sigma_{i=1}^m \langle x_i^{(2)}, f^{(2)} \rangle \langle x_i^{(1)}, f^{(1)} \rangle| \, d\mu_1(f^{(1)}) .$$

Substituting this into (5.10) and applying Fubini's theorem, we obtain

$$\begin{aligned} q_2(\Sigma_{i=1}^m &\langle T_1 x_i^{(1)}, g^{(1)} \rangle T_2 x_i^{(2)}) \\ &\leq \int_{\Sigma_2^o} \left[\int_{\Sigma_1^o} |\Sigma_{i=1}^m \langle x_i^{(2)}, f^{(2)} \rangle \langle x_i^{(1)}, f^{(1)} \rangle| \, d\mu_1(f^{(1)}) \right] d\mu_2(f^{(2)}) \\ &= \int_{\Sigma_1^o \times \Sigma_2^o} |\Sigma_{i=1}^m \langle x_i^{(1)}, f^{(1)} \rangle \langle x_i^{(2)}, f^{(2)} \rangle| \, d(\mu_1 \times \mu_2) \quad \text{for all} \quad g^{(1)} \in U_1^o , \end{aligned}$$

hence

$$q_1 \otimes_\varepsilon q_2(T_1 \otimes T_2(u)) \leq \int_{\Sigma_1^o \times \Sigma_2^o} |\Sigma_{i=1}^m \langle x_i^{(1)}, f^{(1)} \rangle \langle x_i^{(2)}, f^{(2)} \rangle| \, d(\mu_1 \times \mu_2) \qquad (5.11)$$

by virtue of (4.1.3).

If Σ_ε^o is the closed unit ball in $(X_1 \otimes_\varepsilon X_2)'$, then the $\sigma((X_1 \otimes_\varepsilon X_2)', X_1 \otimes_\varepsilon X_2)$-closure of $\Sigma_1^o \otimes \Sigma_2^o$ in $(X_1 \otimes_\varepsilon X_2)'$, denoted by B, is contained in Σ_ε^o, hence B, equipped with the relative topology induced by $\sigma((X_1 \otimes_\varepsilon X_2)', X_1 \otimes_\varepsilon X_2)$, is a compact Hausdorff space, and thus $X_1 \otimes_\varepsilon X_2$ is isometrically isomorphic to a subspace of $C(B)$. Clearly, on $\Sigma_1^o \otimes \Sigma_2^o$ we have

$$\sigma((X_1 \otimes_\varepsilon X_2)', X_1 \otimes_\varepsilon X_2) \leq \text{the } (p_1 \otimes_\varepsilon p_2)'\text{-topology}$$
$$\leq \text{the } p_1^* \otimes_\pi p_2^*\text{-topology} .$$

It follows that the restriction of ω to $\Sigma_1^o \times \Sigma_2^o$, denoted by ω again, is continuous from $\Sigma_1^o \times \Sigma_2^o$ into $\Sigma_1^o \otimes \Sigma_2^o$, and hence that the map $S : C(B) \to C(\Sigma_1^o \times \Sigma_2^o)$, defined by

$$S(\psi) = \psi \circ \omega \quad \text{for all} \quad \psi \in C(B) ,$$

is an (into) isometry because of

$$\|\psi\| = \sup\{|\psi(z')| : z' \in B\} = \sup\{|\psi(f^{(1)} \otimes f^{(2)})| : f^{(1)} \otimes f^{(2)} \in \Sigma_1^o \otimes \Sigma_2^o\}$$
$$= \sup\{|\psi \circ \omega(f^{(1)}, f^{(2)})| : (f^{(1)}, f^{(2)}) \in \Sigma_1^o \times \Sigma_2^o\} = \|S(\psi)\| .$$

Therefore $\nu = S'(\mu_1 \times \mu_2)$ is a positive Radon measure on B such that

$$\int_B \psi(z')d\nu(z') = \nu(\psi) = (\mu_1 \times \mu_2)(S\psi)$$
$$= \int_{\Sigma_1^o \times \Sigma_2^o} \psi \circ \omega(f^{(1)}, f^{(2)})d(\mu_1 \times \mu_2) \quad \text{for all} \quad \psi \in C(B) . \quad (5.12)$$

As $u = \Sigma_{i=1}^m x_i^{(1)} \otimes x_i^{(2)} \in C(B)$, we obtain from (5.11) and (5.12) that

$$q_1 \otimes_\varepsilon q_2(T_1 \otimes T_2(u)) \leq \int_{\Sigma_1^o \times \Sigma_2^o} |\langle u, z'\rangle| d\nu(z') \quad \text{for all} \quad u \in X_1 \otimes_\varepsilon X_2 ,$$

which shows that $T_1 \otimes T_2 \in \pi_{\ell^1}(X_1 \otimes_\varepsilon X_2, Y_1 \otimes_\varepsilon Y_2)$ by (2.2.2).

It is worthwhile to note that a similar argument given in the proof of the preceding result shows that if X_j and Y_j are Banach spaces and if $T_j \in \pi_{\ell^1}(X_j, Y_j)$ $(j = 1, 2)$, then $T_1 \tilde{\otimes}_\varepsilon T_2 \in \pi_{\ell^1}(X_1 \tilde{\otimes}_\varepsilon X_2, Y_1 \tilde{\otimes}_\varepsilon Y_2)$. Also Holub [1] has given an example to show that if $T_j \in \pi_{\ell^1}(X_j, Y_j)$, then

$T_1 \tilde{\otimes}_\pi T_2 \notin \pi_{\ell^1}(X_1 \tilde{\otimes}_\pi X_2, Y_1 \tilde{\otimes}_\pi Y_2)$. For instance, the embedding map $J : \ell^1 \to \ell^2$ is such a map.

By using (3.3.5), a similar argument given in the proof of (4.5.4) yields the following

(4.5.7) Corollary. <u>Suppose</u> $T_j \in L(E_j, F_j)$ $(j = 1, 2)$. <u>Then</u> $T_1 \otimes T_2$ <u>belongs to</u> $N^{(p)}_{\ell^1}(E_1 \otimes_\varepsilon E_2, F_1 \otimes_\varepsilon F_2)$ <u>if and only if</u> $T_j \in N^{(p)}_{\ell^1}(E_j, F_j)$ $(j = 1, 2)$.

By using a result of characterizations of precompact-bounded linear maps, we are able to show that precompact-bounded linear maps are preserved for the ε-topology as follows.

(4.5.8) Theorem. <u>Suppose</u> $T_j \in T(E_j, F_j)$ $(j = 1, 2)$. <u>Then</u> $T_1 \otimes T_2$ <u>belongs</u> to $L^{pb}(E_1 \otimes_\varepsilon E_2, F_1 \otimes_\varepsilon F_2)$ <u>if and only if</u> $T_j \in L^{pb}(E_j, F_j)$ $(j = 1, 2)$.

<u>Proof</u>. The necessity follows from (4.5.2). To prove the sufficiency, it is known from (1.3.4) that there exist $(\zeta_n^{(i)}) \in c_0$ and equicontinuous sequences $\{f_n^{(i)}\}$ in E_i' such that for arbitrary continuous seminorms q_i on E_i it is possible to find $\mu_i \geq 0$ such that

$$q_i(T_i x^{(i)}) \leq \mu_i \sup_n |\zeta_n^{(i)} \langle x^{(i)}, f_n^{(i)} \rangle| \quad \text{for all } x^{(i)} \in E_i \ (i = 1, 2). \tag{5.13}$$

Suppose $u = \Sigma_{j=1}^m x_j^{(1)} \otimes x_j^{(2)} \in E_1 \otimes_\varepsilon E_2$. Then we have for any $g^{(2)} \in F_2'$ that

$$\Sigma_{j=1}^m x_j^{(1)} \otimes T_2 x_j^{(2)} : g^{(2)} \mapsto \Sigma_{j=1}^m \langle T_2 x_j^{(2)}, g^{(2)} \rangle x_j^{(1)} \in E_1 ,$$

hence

$$q_1(T_1(\Sigma_{j=1}^{m} <T_2x_j^{(2)}, g^{(2)}>x_j^{(1)})) \leq \mu_1 \sup_n |\zeta_n^{(1)} \Sigma_{j=1}^{m} <T_2x_j^{(2)}, g^{(2)}><x_j^{(1)}, f_n^{(1)}>| \quad (5.$$

by (5.13). If $U_j = \{y^{(j)} \in F_j : q_j(y^{(j)}) \leq 1\}$, then (5.13) and (5.14) show that

$$\begin{aligned}
q_1 \otimes_\varepsilon q_2(T_1 \otimes T_2(u)) &= \sup\{q_1(\Sigma_{j=1}^{m} <T_2x_j^{(2)}, g^{(2)}>T_1x_j^{(1)}) : g^{(2)} \in U_2^o\} \\
&\leq \sup\{\mu_1 \sup_n |\zeta_n^{(1)} \Sigma_{j=1}^{m} <T_2x_j^{(2)}, g^{(2)}><x_j^{(1)}, f_n^{(1)}>|\} : g^{(2)} \in U_2^o\} \\
&\leq \mu_1 \sup_n\{|\zeta_n^{(1)}| \sup\{|\Sigma_{j=1}^{m} <T_2x_j^{(2)}, g^{(2)}><x_j^{(1)}, f_n^{(1)}>|\} : g^{(2)} \in U_2^o\} \\
&= \mu_1 \sup_n\{|\zeta_n^{(1)}| q_2(\Sigma_{j=1}^{m} <x_j^{(1)}, f_n^{(1)}>T_2x_j^{(2)})\} \\
&\leq \mu_1\mu_2 \sup_n\{|\zeta_n^{(1)}| \sup_k |\zeta_k^{(2)} \Sigma_{j=1}^{m} <x_j^{(1)}, f_n^{(1)}><x_j^{(2)}, f_k^{(2)}>|\} \\
&= \mu_1\mu_2 \sup_{n,k}\{|\zeta_n^{(1)}\zeta_k^{(2)} <\Sigma_{j=1}^{m} x_j^{(1)} \otimes x_j^{(2)}, f_n^{(1)} \otimes f_k^{(2)}>|\} . \qquad (5.15)
\end{aligned}$$

Let V_j be absolutely convex o-neighbourhoods in E_j such that $f_n^{(j)} \in V_j^o$ for all n $(j = 1, 2)$. Then

$$f_n^{(1)} \otimes f_k^{(2)} \in V_1^o \otimes V_2^o \text{ for all } n \text{ and } k . \qquad (5.16)$$

Let us define two sequence $\{\sigma_r\}$ and $\{h_r\}$ by

$$\sigma_{(n-1)(n-1)+t} = \zeta_t^{(1)}\zeta_n^{(2)} \text{ and } h_{(n-1)(n-1)+t} = f_t^{(1)} \otimes f_n^{(2)} \quad \text{if } 1 \leq t \leq n ,$$

$$\sigma_{(n-1)(n-1)+t+n} = \zeta_n^{(1)}\zeta_{n-t}^{(2)} \text{ and } h_{(n-1)(n-1)+t+n} = f_n^{(1)} \otimes f_{n-t}^{(2)} \quad \text{if } 1 \leq t < n .$$

Then $(\sigma_r) \in c_o$ and $\{h_r\}$ is an equicontinuous sequence in $(E_1 \otimes_\varepsilon E_2)'$ by (5.16). In view of (5.15), we have

$$q_1 \otimes_\varepsilon q_2(T_1 \otimes T_2(u)) \leq \mu_1 \mu_2 \sup_n |\sigma_r \langle u, h_r \rangle| ,$$

which shows that $T_1 \otimes T_2 \in L^{pb}(E_1 \otimes_\varepsilon E_2, F_1 \otimes_\varepsilon F_2)$ by (1.3.4).

4.6 Criteria for nuclearity

We have known from §4.4 that the ε-topology is, in general, coarser than the π-topology on a tensor product $E \otimes F$. Therefore, it is natural to ask under what condition on E (or F), the ε-topology coincides with the π-topology. The equality of these two topologies characterizes the nuclearity of E as shown by the following important result.

(4.6.1) Theorem (Grothendieck). The following statements are equivalent:

(a) $(E, \mathcal{P})$ is a nuclear space.

(b) $E \tilde{\otimes}_\pi F \cong E \tilde{\otimes}_\varepsilon F$ for any locally convex space F.

(c) $E \tilde{\otimes}_\pi Y \cong E \tilde{\otimes}_\varepsilon Y$ for any Banach space Y.

(d) For any locally convex space F, $E \otimes_\pi F$ is topologically isomorphic to a dense subspace of $L_\varepsilon(E'_\tau, F)$.

Proof. The implications (d) $\Rightarrow$ (b) $\Rightarrow$ (c) are obvious. It is known that E is nuclear if and only if its completion $\tilde{E}$ is nuclear; the implication (c) $\Rightarrow$ (a) then follows from (3.2.4)(f) and (4.4.4). Therefore we complete the proof by showing the implications (a) $\Rightarrow$ (b) $\Rightarrow$ (d).

(a) $\Rightarrow$ (b): It suffices to show that $\mathcal{P} \otimes_\pi \mathcal{J} \leq \mathcal{P} \otimes_\varepsilon \mathcal{J}$. To this end, we first notice that

$$\hat{p}_V \otimes q_U((Q_V \otimes I)u) = p_V \otimes_\pi q_U(u) \quad \text{for all} \quad u \in E \otimes F , \tag{6.1}$$

where I is the identity map on F. For any $V \in \mathcal{U}_E$, the nuclearity of $Q_V : E \to E_V$ ensures that Q_V is of the form

$$Q_V(x) = \Sigma_{n=1}^{\infty} \zeta_n f_n(x) Q_V(u_n) \quad \text{for all} \quad x \in E , \tag{6.2}$$

where $(\zeta_n) \in \ell^1$, $\{f_n\}$ is an equicontinuous sequence in E' and $\{Q_V(u_n)\}$ is a bounded sequence in E_V, therefore we may assume without loss of generality that

$$f_n \in V^0 \quad \text{and} \quad \hat{p}_V(Q_V(u_n)) \leq 1 \quad \text{for all} \quad n \geq 1 .$$

On the other hand, for any fixed $y \in F$, the map $z \mapsto z \otimes y$ $(z \in E_V)$ is a continuous linear map from E_V into $E_V \otimes_\pi F$, therefore we have from (6.2) that

$$\begin{aligned}(Q_V \otimes I)(x \otimes y) = Q_V(x) \otimes y &= \Sigma_{n=1}^{\infty} \zeta_n f_n(x)(Q_V(u_n) \otimes y) \\ &= \Sigma_{n=1}^{\infty} \zeta_n f_n(x)(Q_V \otimes I)(x \otimes y) .\end{aligned} \tag{6.3}$$

For any $U \in \mathcal{U}_F$, we claim that there is a constant $c \geq 0$ such that

$$p_V \otimes_\pi q_U(w_0) \leq c p_V \otimes_\varepsilon q_U(w_0) \quad \text{for all} \quad w_0 \in E \otimes F . \tag{6.4}$$

In fact, for any $w_0 = \Sigma_{j=1}^{m} x_j \otimes y_j \in E \otimes F$, since $z_0 = (Q_V \otimes I)w_0 \in E_V \otimes F$, the Hahn-Banach extension theorem shows that there is an $\tilde{g} \in (E_V \otimes_\pi F)'$ such that

$$\tilde{g}(z_0) = \hat{p}_V \otimes_\pi q_U(z_0) \text{ and } |\tilde{g}(z)| \leq \hat{p}_V \otimes_\pi q_U(z) \text{ for all } z \in E_V \otimes_\pi F . \tag{6.5}$$

In particular, we have from (6.3) that

$$\tilde{g}(Q_V(x) \otimes y) = \Sigma_{n=1}^{\infty} \zeta_n f_n(x) \tilde{g}(Q_V(u_n) \otimes y) .$$

From this it follows by (6.5) that

$$\begin{aligned}
\hat{p}_V \otimes_\pi q_U(z_o) &= \tilde{g}(z_o) = \Sigma_{j=1}^{m} \tilde{g}(Q_V(x_j) \otimes y_j) = \Sigma_{j=1}^{m} \Sigma_{n=1}^{\infty} \zeta_n f_n(x_j) \tilde{g}(Q_V(u_n) \otimes y_j) \\
&= \Sigma_{n=1}^{\infty} \tilde{g}(Q_V(u_n) \otimes (\Sigma_{j=1}^{m} \zeta_n f_n(x_j) y_j)) \\
&\leq \Sigma_{n=1}^{\infty} \hat{p}_V \otimes_\pi q_U(Q_V(u_n) \otimes (\Sigma_{j=1}^{m} \zeta_n f_n(x_j) y_j)) \\
&= \Sigma_{n=1}^{\infty} \hat{p}_V(Q_V(u_n)) q_U(\Sigma_{j=1}^{m} \zeta_n f_n(x_j) y_j)) \\
&\leq \Sigma_{n=1}^{\infty} q_U(\Sigma_{j=1}^{m} \zeta_n f_n(x_j) y_j) . \qquad (6.6)
\end{aligned}$$

As $f_n \in V^o$, we obtain from the expression of $p_V \otimes_\varepsilon q_U$ that

$$\begin{aligned}
q_U(\Sigma_{j=1}^{m} \zeta_n f_n(x_j) y_j) &= \sup\{|\Sigma_{j=1}^{m} \zeta_n f_n(x_j) \langle y_j, y' \rangle| : y' \in U^o\} \\
&\leq |\zeta_n| p_V \otimes_\varepsilon q_U(\Sigma_{j=1}^{m} x_j \otimes y_j) = |\zeta_n| p_V \otimes_\varepsilon q_U(w_o) .
\end{aligned}$$

In view of (6.1) and $z_o = (Q_V \otimes I) w_o$, we obtain from (6.6) that

$$p_V \otimes_\pi q_U(w_o) = \hat{p}_V \otimes_\pi q_U(z_o) \leq (\Sigma_{n=1}^{\infty} |\zeta_n|) p_V \otimes_\varepsilon q_U(w_o) ,$$

hence formula (6.4) holds by taking $c = \Sigma_{n=1}^{\infty} |\zeta_n|$.

(c) $\Rightarrow$ (d): In view of the statement (b), we have

$$E \otimes_\pi F \cong E \otimes_\varepsilon F \cong L_\varepsilon^f(F'_\tau, E) ,$$

hence we have to show that $L^f(F'_\tau, E)$ is dense in $L_\varepsilon(F'_\tau, E)$. To this end, let $T \in L_\varepsilon(F'_\tau, E)$, $U \in \mathcal{U}_E$ and $V \in \mathcal{U}_E$. Then $T(U^o)$ is a $\sigma(E, E')$-compact absolutely convex subset of E since $T \in L(F'_\sigma, E_\sigma)$, hence $T(U^o)$ is $\mathcal{P}$-precompact by the nuclearity of E . On the other hand, since nuclear spaces have the approximation property, there are $x'_i \in E'$ and $x_i \in E$ $(i = 1, \ldots, n)$ such that

$$\Sigma_{i=1}^{n} \langle u, x'_i \rangle x_i - u \in V \quad \text{for all} \quad u \in T(U^o) ;$$

consequently,

$$\Sigma_{i=1}^{n} \langle Ty', x'_i \rangle x_i - Ty' \in V \quad \text{for all} \quad y' \in U^o . \tag{6.7}$$

Let $y_i = T'x'_i$ and define T_n by

$$T_n(y') = \Sigma_{i=1}^{n} \langle y_i , y' \rangle x_i \quad \text{for all} \quad y' \in F' .$$

Then $y_i \in F$ and $T_n \in L^f(E'_\tau, F)$ is such that

$$T_n(y') - T(y') \in V \quad \text{for all} \quad y' \in U^o$$

by (6.7). Therefore $T_n - T \in P(U^o, V)$, and thus $L^f(F'_\tau, E)$ is dense in $L_\varepsilon(F'_\tau, E)$.

If E is nuclear, then (4.6.1), (4.4.2) and the definition of integral bilinear forms show that every continuous bilinear form on $E \times F$ is integral; also (4.6.1) and the definition of the s-topology imply that

$$E \tilde{\otimes}_\pi F \cong E \tilde{\otimes}_s F \cong E \tilde{\otimes}_\varepsilon F$$

for any locally convex space F.

Theorem (4.6.1) has many important application; we mention a few below.

(4.6.2) Corollary. Let $(E, \mathcal{P})$ be a nuclear space. Then

$$E \tilde{\otimes}_{\pi} F \cong E \tilde{\otimes}_{\varepsilon} F \cong \tilde{L}_{\varepsilon}(E'_{\tau}, F) \cong \tilde{S}_{\varepsilon}(E'_{\sigma}, F'_{\sigma}) . \tag{6.8}$$

Furthermore, if in addition, E is quasi-complete, then $E \otimes_{\pi} F$ is topologically isomorphic to a dense subspace of $L_{\varepsilon}(E'_{\beta}, F)$, and hence $L^{f}(E'_{\beta}, F)$ is dense in $L_{\varepsilon}(E'_{\beta}, F)$.

Proof. Furmula (6.8) follows from (4.6.1)(d). If, in addition, E is quasi-complete, then E is semi-reflexive, hence $\beta(E', E) = \tau(E', E)$, and thus the result follows from (4.6.1)(d).

As an immediate consequence of (4.4.3) and (4.6.2) we obtain the following:

(4.6.3) Corollary. Let E and F be complete. If E is nuclear, then

$$E \tilde{\otimes}_{\pi} F \cong E \tilde{\otimes}_{\varepsilon} F \cong L_{\varepsilon}(E'_{\tau}, F) \cong L_{\varepsilon}(E'_{\beta}, F) \cong S_{\varepsilon}(E'_{\sigma}, F'_{\sigma}) ,$$

namely, $L_{\varepsilon}(E'_{\tau}, F)$ (or equivalently, $S_{\varepsilon}(E'_{\sigma}, F'_{\sigma})$) is the completion of $E \otimes_{\pi} F$ as well as of $E \otimes_{\varepsilon} F$.

(4.6.4) Proposition. The following statements are equivalent.

(a) E and F are nuclear spaces.

(b) $L_{\varepsilon}(E'_{\tau}, F)$ is a nuclear space.

(c) $E \otimes_{\varepsilon} F$ is a nuclear space.

Moreover, if one of the statements (a), (b) and (c) holds, then $E \otimes_{\pi} F \cong E \otimes_{\varepsilon} F$ and all $E \otimes_{\pi} F$ and $E \tilde{\otimes}_{\pi} F$ are nuclear spaces.

<u>Proof</u>. The implication (a) ⇒ (b) is a consequence of (3.5.1), and the implication (b) ⇒ (c) is obvious because subspaces of a nuclear space are nuclear. To prove the implication (c) ⇒ (a), we have known that $E \otimes_{\varepsilon} F \cong L_{\varepsilon}^{f}(E'_{\tau}, F)$, so that E (resp. F) is topologically isomorphic to a subspace of $E \otimes_{\varepsilon} F$ (see the proof of necessity of (4.4.3)). Therefore E and F are nuclear spaces.

Finally, if one of the statements (a), (b) and (c), say (c), holds, then $E \otimes_{\pi} F \cong E \otimes_{\varepsilon} F$ by (4.6.1), thus $E \otimes_{\pi} F$ and $E \tilde{\otimes}_{\pi} F$ are nuclear spaces since the completion of a nuclear space is nuclear.

By using (1.4.4), it can be shown that $E \otimes_{\varepsilon} F$ is a Schwartz (resp. s-nuclear) space if and only if E and F are Schwartz (resp. s-nuclear) spaces.

(4.6.5) Lemma. (a) <u>If</u> E <u>is semi-reflexive</u>, <u>then</u> $L_{\beta}(E, F) \cong L_{\varepsilon}(E, F)$, <u>hence</u> $L_{\beta}(E, F)$ <u>is topologically isomorphic to a subspace of</u> $L_{\varepsilon}(E_{\tau}, F)$.

(b) <u>If</u> E <u>is a complete barrelled space for which its strong dual</u> $(E', \beta(E', E))$ <u>is semi-reflexive</u>, <u>then</u> E <u>is reflexive</u>, <u>hence</u>

$$L_{\beta}(E, F) \cong L_{\varepsilon}(E_{\tau}, F) . \tag{6.9}$$

<u>Proof</u>. (a) It is not hard to show that

$$S(E_{\sigma}, F'_{\sigma}) = L(E_{\sigma}, F_{\sigma}) = L(E_{\tau}, F) \quad \text{and} \quad L(E, F) \subset L(E_{\tau}, F) .$$

Clearly, the semi-reflexivity of E ensures that bounded subsets of E are identical with $\beta(E', E)$-equicontinuous subsets of $E'' = E$, hence $L_\varepsilon(E, F) \cong L_\beta(E, F)$; consequently, $L_\beta(E, F)$ is topologically isomorphic to the subspace $L(E, F)$ of the locally convex space $L_\varepsilon(E_\tau, F)$.

(b) The completeness and the barrelledness of E imply that E is topologically isomorphic to a closed subspace of $(E'', \beta(E'', E'))$, while the semi-reflexivity of $(E', \beta(E', E))$ shows that E and $(E'', \beta(E'', E'))$ have the same topological dual E' . In view of the strong separation theorem, the closedness of E in $(E'', \beta(E'', E'))$ implies that $E = E''$, thus E is reflexive. Consequently, (6.9) follows from part (a) since E is a Mackey space.

The following result should be compared with (4.6.2) and (4.6.3).

(4.6.6) Proposition. Let E be a co-nuclear semi-reflexive space. Then

$$\tilde{L}_\varepsilon(E_\tau, F) \cong E'_\beta \tilde{\otimes}_\pi F \cong E'_\beta \tilde{\otimes}_\varepsilon F . \qquad (6.10)$$

Furthermore, if in addition, F is nuclear, then all $L_\varepsilon(E, F)$, $L_\varepsilon(E_\tau, F)$, $B_\varepsilon(E_\sigma, F'_\sigma)$ and $S_\varepsilon(E_\sigma, F'_\sigma)$ are nuclear spaces.

Proof. Formula (6.10) follows from (4.6.2). If F is nuclear, then $L_\beta(E, F)$ is nuclear by (3.5.1), hence $L_\varepsilon(E, F)$ is nuclear by (4.6.5)(a). On the other hand, (4.6.4) shows that $E'_\beta \tilde{\otimes}_\pi F$ is nuclear since E'_β is nuclear, hence $L_\varepsilon(E_\tau, F)$ is nuclear by (6.10) because a locally convex space is nuclear if and only if its completion is nuclear.

As $S_\varepsilon(E_\sigma, F'_\sigma) \cong L_\varepsilon(E_\tau, F)$ and $B_\varepsilon(E_\sigma, F'_\sigma)$ is a subspace of $S_\varepsilon(E_\sigma, F'_\sigma)$, we conclude that $S_\varepsilon(E_\sigma, F'_\sigma)$ and $B_\varepsilon(E_\sigma, F'_\sigma)$ are nuclear spaces.

(4.6.7) Corollary. <u>Let</u> E <u>be a complete barrelled space, and let</u> E'_β <u>be a complete nuclear space. Then</u> E <u>is reflexive and</u>

$$\tilde{L}_\beta(E, F) \cong E'_\beta \tilde{\otimes}_\pi F \cong E'_\beta \tilde{\otimes}_\varepsilon F . \tag{6.11}$$

<u>Furthermore, if in addition,</u> F <u>is complete, then</u> $L_\beta(E, F)$ <u>is complete, hence</u>

$$L_\beta(E, F) \cong E'_\beta \tilde{\otimes}_\pi F \cong E'_\beta \tilde{\otimes}_\varepsilon F . \tag{6.12}$$

<u>Proof</u>. The completeness and the nuclearity of E'_β imply that E'_β is semi-reflexive, hence E is reflexive and $L_\beta(E, F) \cong L_\varepsilon(E_\tau, F)$ by (4.6.5)(b), thus (6.11) holds by (4.6.6). Finally, if F is complete, then $L_\varepsilon((E'')_\tau, F)$ is complete by (4.4.3), hence $L_\beta(E, F)$ is complete by the reflexivity of E and (4.6.5).

(4.6.8) Theorem (Grothendieck). <u>A metrizable locally convex space</u> E <u>is nuclear if and only if it is a co-nuclear space</u>.

<u>Proof</u>. It is known from §3.5 that a locally convex space is nuclear if and only if its completion is nuclear. Hence we may assume that E is a Fréchet space.

If E is nuclear, then E is a Montel space by (3.2.5) and (1.2.6)(b), hence E'_β is a Montel space and $\beta(E', E) = \tau(E', E)$. To prove the nuclearity of E'_β , it suffices to show by (3.2.4)(j) that

$$L(E'_\tau, Y) \subseteq N_{\ell^1}(E'_\tau, Y) \quad \text{for any Banach space } Y . \tag{6.11}$$

In fact, we first notice from (4.6.3) that $E \tilde{\otimes}_\pi Y \cong L_\varepsilon(E'_\tau, Y)$ since E is complete and nuclear, so that $L_\varepsilon(E'_\tau, Y)$ is the completion of $E \otimes_\pi Y$. As E and Y being metrizable, it follows from a well-known result (see Schaefer [1, p.94]) that each $T \in L(E'_\tau, Y)$ is of the form

$$T(x') = \sum_{n=1}^{\infty} \zeta_n \langle x_n, x'\rangle y_n \quad \text{for all } x' \in E' ,$$

where $(\zeta_n) \in \ell^1$, $\{x_n\}$ and $\{y_n\}$ are null sequences in E and Y respectively. Since E is metrizable and reflexive, it follows that $\{x_n\}$ is an $\tau(E', E)$-equicontinuous sequence in E'' , and hence that $T \in N_{\ell^1}(E'_\tau, Y)$. Therefore (6.11) holds.

Conversely, if E'_β is nuclear, then E'_β is semi-reflexive since E'_β is complete, hence E is reflexive by (4.6.5)(b); consequently $\beta(E', E) = \tau(E', E)$. On the other hand, for proving the nuclearity of E , it suffices by (4.6.1) to show that for any Banach space Y , the canonical embedding map $J : E \tilde{\otimes}_\pi Y \to E \tilde{\otimes}_\varepsilon Y$ is a topological isomorphism. As $E \tilde{\otimes}_\pi Y$ and $E \tilde{\otimes}_\varepsilon Y$ being Fréchet spaces, we have only to show by making use of the Banach open mapping that J is bijective since J is always continuous.

We first claim that J is surjective. Indeed, (4.4.3) shows that $E \tilde{\otimes}_\varepsilon Y$ is topologically isomorphic with a subspace of $L_\varepsilon(E'_\beta, Y)$, and the nuclearity of E'_β implies that $L(E'_\beta, Y) = N_{\ell^1}(E'_\beta, Y)$ by (3.2.4)(j).

Hence each $u \in E \tilde{\otimes}_{\varepsilon} Y$ can be identified with a nuclear linear map from E'_{β} into Y, thus u is of the form

$$u(x') = \sum_{n=1}^{\infty} \zeta_n \langle x', x''_n \rangle y_n \quad \text{for all } x' \in E' ,$$

where $(\zeta_n) \in \ell^1$, $\{y_n\}$ is a null sequence in Y and $\{x''_n\}$ is a sequence in $E'' = E$ which converges to 0 uniformly on a suitable o-neighbourhood in E'_{β}. Now the reflexivity of E implies that $\{x''_n\}$ is a null sequence in E, hence $u \in E \tilde{\otimes}_{\pi} Y$ since E and Y are complete and metrizable (see Schaefer [1, p.94]). Therefore J is surjective.

Finally, to prove the injectivity of J, it suffices to show that $\mathcal{J}(E, Y)$ is weakly dense in $B(E, Y)$ because of

$$(E \tilde{\otimes}_{\varepsilon} Y)' = \mathcal{J}(E, Y) \quad \text{and} \quad (E \tilde{\otimes}_{\pi} Y)' = B(E, Y) .$$

On the other hand, we notice that (see Schaefer [1, p.94]) each $u \in E \tilde{\otimes}_{\pi} Y$ belongs to $\bar{\Gamma}(K_E \otimes K_Y)$ for some compact subsets K_E and K_Y of E and Y respectively since any null sequence in E (or Y) forms a relatively compact set, and that

$$E' \otimes Y' = L^f(Y, E'_{\beta}) \quad \text{and} \quad B(E, Y) = L^{eq}(Y, E(\tau)') \subset L(Y, E'_{\beta}) .$$

Therefore, for proving the weak density of $\mathcal{J}(E, Y)$ in $B(E, Y)$, it suffices to verify that $L^f(Y, E'_{\beta})$ is dense in $L^{eq}(Y, E(\tau)')$ for the topology of compact convergence.

To this end, let $T \in L^{eq}(Y, E(\tau)')$, let K_Y be a compact subset of Y and M a o-neighbourhood in E'_{β}. Then $T(K_Y)$ is a compact subset of E'_β, hence there exists an $\sum_{i=1}^{n} x_i \otimes x'_i \in E \otimes E'$ (note that $E'' = E$) such that

$$\sum_{i=1}^{n} \langle x_i , Ty \rangle x'_i - Ty \in M \quad \text{for all} \quad y \in K_Y$$

since E'_β has the approximation property. Let $y'_i = T'x_i$ $(i = 1, 2, \ldots, n)$. Then $y'_i \in Y'$ $(i = 1, \ldots, n)$ and $\sum_{i=1}^{n} x'_i \otimes y'_i \in E' \otimes Y'$ is such that

$$\sum_{i=1}^{n} \langle y'_i , y \rangle x'_i - Ty \in M \quad \text{for all} \quad y \in K_Y ,$$

therefore $L^f(Y, E'_\beta)$ is dense in $L^{eq}(Y, E(\tau)')$ for the topology of compact convergence.

(4.6.9) Corollary. A (DF)-space E is nuclear if and only if it is co-nuclear.

Proof. We first notice that the strong dual of a (DF)-space is a Fréchet space, and from (1.4.10) that a (DF)-space which is either nuclear or co-nuclear is infrabarrelled.

Therefore E'_β is nuclear if and only if $(E'', \beta(E'', E'))$ is nuclear by (4.6.8), and this is the case if and only if E is nuclear by the remark before (3.5.3) since E is always infrabarrelled. This completes the proof.

Suppose now that F is complete and that E is a nuclear Fréchet space. Then (4.6.3), (4.6.7) and (4.6.8) imply that

(i) $E \tilde{\otimes}_\pi F \cong E \tilde{\otimes}_\varepsilon F \cong L_\varepsilon(E'_\beta, F) \cong L(E'_\beta, F)$

(ii) $E'_\beta \tilde{\otimes}_\pi F \cong E'_\beta \tilde{\otimes}_\varepsilon F \cong L_\beta(E, F)$;

moreover, if in addition, F is a Fréchet space, then Grothendieck has shown (see Schaefer [1]) that

$$\text{(iii)} \quad E'_\beta \,\tilde{\otimes}_\pi\, F'_\beta \cong E'_\beta \,\tilde{\otimes}_\varepsilon\, F'_\beta \cong (E \,\tilde{\otimes}_\pi\, F)'_\beta \cong (L_\beta(E'_\beta, F))'_\beta ,$$

$$\text{(iv)} \quad E \,\tilde{\otimes}_\pi\, F''_\beta \cong E \,\tilde{\otimes}_\varepsilon\, F''_\beta \cong (E \,\tilde{\otimes}_\pi\, F)''_\beta .$$

On the other hand, each (DF)-space which is either nuclear or co-nuclear is separable (since bounded sets are separable), and hence infrabarrelled. As a (DF)-space is nuclear if and only if it is a co-nuclear space, it follows that the conclusions (i) and (ii) still hold whenever E is a quasi-complete nuclear (DF)-space.

CHAPTER 5. TENSOR PRODUCTS OF ORDERED CONVEX SPACES

5.1 The ℓ-norm on tensor products

Throughout this chapter the scalar field for vector spaces is assumed to be the field $\mathbb{R}$ of real numbers. By a (positive) cone in a vector space E is meant a non-empty convex subset E_+ of E satisfying $\lambda E_+ \subseteq E_+$ for all $\lambda \geq 0$. A cone E_+ is proper if $E_+ \cap (-E_+) = \{0\}$. A cone E_+ in E determines a transitive and reflexive relation "$\leq$" by

$$x \leq y \quad \text{if} \quad y - x \in E_+ ;$$

moreover, this relation is compatible with the vector structure. The pair (E, E_+) (or $(E, \leq)$) is called an ordered vector space. A locally convex space with a cone is called an ordered convex space, and a normed space with a cone is called an ordered normed space.

For any subset V of an ordered vector space (E, E_+), we define

$$F(V) = (V + E_+) \cap (V - E_+) ;$$
$$D(V) = \{x \in V : x = \lambda x_1 - (1 - \lambda)x_2, \lambda \in [0, 1], x_1, x_2 \in V \cap E_+\} ;$$
$$S(V) = \cup\{[-u, u] : u \in V \cap E_+\} .$$

Then $F(V)$ is called the order-convex hull of V, while $D(V)$ is called the decomposable kernel of V. Furthermore, V is said to be

(a) order-convex if $V = F(V)$;

(b) o-convex if it is both order-convex and convex;

(c) decomposable if $V = D(V)$;

(d) <u>absolutely order-convex</u> if $S(V) \subseteq V$;

(e) <u>absolutely dominated</u> if $V \subseteq S(V)$;

(f) <u>solid</u> if $V = S(V)$.

Let (E, E_+) be an ordered vector space and p a seminorm on E . Then we say that p is

(i) <u>strongly monotone</u> if

$$y \leq x \leq z \Rightarrow p(x) \leq \max\{p(y), p(z)\} ;$$

(ii) <u>absolutely monotone</u> if

$$-u \leq x \leq u \Rightarrow p(x) \leq p(u) ;$$

(iii) <u>monotone</u> if

$$0 \leq u \leq w \Rightarrow p(u) \leq p(w) ;$$

(iv) <u>decomposable</u> if

$$p(x) = \inf\{p(u) + p(w) : u, w \in E_+ \text{ and } u - w = x\} \text{ for any } x \in E ;$$

(v) <u>a Riesz seminorm</u> (or <u>a regular seminorm</u>) if

$$p(x) = \inf\{p(u) : u \in E_+ \text{ and } -u \leq x \leq u\} \text{ for any } x \in E .$$

Let p be a seminorm on an ordered vector space (E, E_+) and

$$V = \{x \in E : p(x) < 1\} .$$

It is known from Wong [1] that p is strongly monotone (resp. absolutely monotone) if and only if V is O-convex (resp. absolutely order-convex), that

p is a decomposable if and only if V is decomposable, and that p is a Riesz seminorm if and only if V is solid. It is also easily seen that if V is o-convex and decomposable, then V is solid.

If p is a seminorm on an ordered vector space (E, E_+) and

$$V = \{x \in E : p(x) < 1\} ,$$

then the gauge of $F(V)$, denoted by p_F , is given by

$$p_F(x) = \sup\{|f(x)| : V^\pi \cap E_+^*\} \quad \text{for any } x \in E ,$$

where V^π is the polar in E^* of V . If, in addition, E_+ is generating, then the gauges of $D(V)$ and $S(V)$, denoted p_D and p_S respectively, are given by

$$p_D(x) = \inf\{p(x_1) + p(x_2) : x_1, x_2 \in E_+ \text{ and } x = x_1 - x_2\} \quad \text{for any } x \in E$$

and

$$p_S(x) = \inf\{p(u) : -u \leq x \leq u\} \quad \text{for any } x \in E .$$

An ordered convex space $(E, E_+, \mathcal{P})$ is said to be <u>locally decomposable</u> (resp. <u>locally o-convex</u>, <u>locally solid</u>) if $\mathcal{P}$ admits a neighbourhood basis at 0 consisting of absolutely convex and decomposable (resp. O-convex and circled, convex and solid) sets in E .

Let $(E, E_+, \mathcal{P})$ be an ordered convex space, let $\mathcal{U}_E$ be a neighbourhood basis at 0 for $\mathcal{P}$ consisting of closed absolutely convex sets in E , and let

$$F(\mathcal{U}_E) = \{F(V) : V \in \mathcal{U}_E\}$$
$$D(\mathcal{U}_E) = \{D(V) : V \in \mathcal{U}_E\}$$
$$S(\mathcal{U}_E) = \{S(V) : V \in \mathcal{U}_E\} .$$

Then $F(\mathcal{U}_E)$ determines a unique locally o-convex topology, denoted by $\mathcal{P}_F$, which is the least upper bound of all locally o-convex topologies coarser than $\mathcal{P}$. $\mathcal{P}_F$ is referred to as the <u>locally o-convex topology associated with</u> $\mathcal{P}$. If E_+ is generating (i.e., $E = E_+ - E_+$), then $D(\mathcal{U}_E)$ (resp. $S(\mathcal{U}_E)$) determines a unique locally decomposable (resp. locally solid) topology, denoted by $\mathcal{P}_D$ (resp. $\mathcal{P}_S$). The topology $\mathcal{P}_D$ (resp. $\mathcal{P}_S$) is called the <u>locally decomposable</u> (resp. <u>locally solid</u>) <u>topology associated with</u> $\mathcal{P}$.

It is known from Wong [1, (1.3.15)] that if $(E, E_+, \mathcal{P})$ is a locally solid space, then the locally solid topology $\sigma_S(E, E')$ associated with $\sigma(E, E')$ coincides with the topology $o(E, E')$ of uniform convergence on all order-intervals in E'.

Let $(E, E_+, \mathcal{P})$ be an ordered convex space and $(F, \mathcal{J})$ a locally convex space. Recall that an $T \in L^*(E, F)$ is said to be <u>cone-absolutely summing</u> if for any $\mathcal{J}$-continuous seminorm q on F there exists a $\mathcal{P}$-continuous seminorm p on E such that the inequality

$$\Sigma_{i=1}^{n} q(Tu_i) \leq p(\Sigma_{i=1}^{n} u_i)$$

holds for any finite subset $\{u_1, \ldots, u_n\}$ of E_+. It is known from Wong [1, (3.1.1)] that if $(E, E_+, \mathcal{P})$ is a locally solid space, then an $T \in L^*(E, F)$ is cone-absolutely summing if and only if $T \in L(E(\sigma_S), F)$, and this is the case if and only if $T^\wedge$ is a continuous linear map from $(\ell^1\langle\Lambda, E\rangle, C_\varepsilon(\Lambda, E), \mathcal{P}_{\varepsilon D})$ into $(\ell^1[\Lambda, F], \mathcal{J}_\pi)$ for any non-empty index set Λ, where

$$C_\varepsilon(\Lambda, E) = E_+^\Lambda \cap \ell^1(\Lambda, E) \quad \text{and} \quad \ell^1\langle\Lambda, E\rangle = C_\varepsilon(\Lambda, E) - C_\varepsilon(\Lambda, E).$$

Therefore the notion of cone-absolutely summing does not depend upon the locally solid topology $\mathcal{P}$ on E, but only on the dual pair $\langle E, E' \rangle$. The set consisting of all cone-absolutely summing mappings, denoted by $L^{\ell}(E, F)$, is a vector subspace of $L(E, F)$ whenever $(E, E_+, \mathcal{P})$ is a locally decomposable space, and contains $\pi_{\ell^1}(E, F)$. There are some alternative characterizations for mappings to be cone-absolutely summing which are anologus to those of absolutely summing mappings (see Wong [1, (3.1.1)] and (2.2.2)). Also it is known from Wong [1, (3.2.12) and (3.3.1)] that the notion of cone-absolutely summing mappings is very useful for establishing some characterizations of $\sigma_S(E, E')$ which are anologues to the nuclearity of a locally convex space.

Throughout this section, (X, X_+, p) will be assumed to be an ordered normed space such that the unit ball

$$\Sigma_X = \{x \in X : p(x) < 1\}$$

is 0-convex and decomposable, and (Y, q) will be a normed space whose dual norm is denoted by q^*. Then p is a Riesz norm, thus the norm topology is locally solid. In view of the definition of cone-absolutely summing mappings, we see that an $T \in L(X, Y')$ is cone-absolutely summing if and only if there is an $\zeta \geq 0$ such that the inequality

$$\Sigma_{i=1}^{n} q^*(Tu_i) \leq \zeta p(\Sigma_{i=1}^{n} u_i) \tag{1.1}$$

holds for any finite subset $\{u_1, \ldots, u_n\}$ of X_+. Thus we define for any $T \in L^{\ell}(X, Y')$ that

$$\|T\|_{(\ell)} = \inf\{\zeta \geq 0 : (1.1) \text{ holds for any finite set in } X_+\}. \tag{1.2}$$

It is not difficult to show that

$$\|T\|_{(\ell)} = |||T^{\mathbb{N}}||| = \sup\{q^*_{\pi}(T^{\mathbb{N}}(x_n, \mathbb{N}]) : p_{\varepsilon D}([x_n, \mathbb{N}]) \leq 1\} \tag{1.3}$$

whenever $T \in L^{\ell}(X, Y')$. Therefore $\|\cdot\|_{(\ell)}$ is a norm on $L^{\ell}(X, Y')$ and

$$\|T\|_{(\ell)} \leq \|T\|_{(s)} \quad \text{for all} \quad T \in \pi_{\ell^1}(X, Y') .$$

As Σ_X is decomposable, it follows from Wong [1, (2.2.6)] that for any embedding $J_i : (X, X_+, p) \to (\ell^1\langle \mathbb{N}, X\rangle, C_\varepsilon(\mathbb{N}, X), p_{\varepsilon D})$ we have

$$|||J_i||| \leq 1 ,$$

and hence from (1.3) that

$$|||T||| \leq \|T\|_{(\ell)} \quad \text{for any} \quad T \in L^{\ell}(X, Y') .$$

Therefore, by a standard argument it is easily shown that $(L^{\ell}(X, Y'), \|\cdot\|_{(\ell)})$ is a Banach space. Furthermore, in view of the definition of $\|\cdot\|_{(\ell)}$, it is easily verified that

$$\|T\|_{(\ell)} = \sup\{\Sigma_i q^*(Tu_i) : \{u_i\} \text{ finite } \subset X_+ \text{ and } p(\Sigma_i u_i) \leq 1\} . \qquad (1.2)^*$$

When X is a Banach lattice, $\|T\|_{(\ell)}$ is the ℓ-norm of T defined by Schaefer [2, p.247].

Now the question arises naturally that whether there is a cross-norm on $X \otimes Y$ whose Banach dual is isometrically isomorphic to $(L^{\ell}(X, Y'), \|\cdot\|_{(\ell)})$. By making use of the same idea for the constraction of the s-norm, we are able to construct such a norm. Before doing this, we first notice that $X = X_+ - X_+$ since Σ_X is assumed to be decomposable; thus each $z \in X \otimes Y$ can be represented as the form

$$z = \Sigma_{i=1}^{m} u_i \otimes y_i \text{ with } u_i \in X_+ \text{ and } y_i \in Y \ (i = 1, 2, \dots, n) \qquad (1.4)$$

On the other hand, for any $T \in L^{\ell}(X, Y')$, we obtain from (1.2) that

$$|z(T)| = |\Sigma_{i=1}^{m} \langle y_i , Tu_i\rangle| \leq \Sigma_{i=1}^{m} q^*(Tu_i)q(y_i) = \Sigma_{i=1}^{m} q^*(T(q(y_i)u_i))$$
$$\leq \|T\|_{(\ell)} p(\Sigma_{i=1}^{m} q(y_i)u_i) ,$$

thus each $z \in X \otimes Y$ can be identified with a $\|\cdot\|_{(\ell)}$-continuous linear functional on $L^{\ell}(X, Y')$ such that

$$\|z\|^*_{(\ell)} \leq \inf\{p(\Sigma_{i=1}^{m} q(y_i)u_i) : z = \Sigma_{i=1}^{m} u_i \otimes y_i , u_i \in X_+\} , \qquad (1.5)$$

where $\|\cdot\|^*_{(\ell)}$ is the dual norm of $\|\cdot\|_{(\ell)}$.

Denote by $p \otimes_{\ell} q$ the dual norm on $X \otimes Y$ of the norm $\|\cdot\|_{(\ell)}$, that is

$$p \otimes_{\ell} q(z) = \|z\|^*_{(\ell)} = \sup\{|z(T)| : \|T\|_{(\ell)} \leq 1\} \quad (z \in X \otimes Y) .$$

Then (1.5) shows that $L^{\ell}(X, Y')$ can be identified with a vector subspace of $(X \otimes Y, p \otimes_{\ell} q)'$ and

$$(p \otimes_{\ell} q)^*(\ell_T) \leq \|T\|_{\ell} \quad \text{for all } T \in L^{\ell}(X, Y') , \qquad (1.6)$$

where $\ell_T : x \otimes y \to \langle y, Tx\rangle$. Furthermore, we have the following result.

(5.1.1) Theorem. $p \otimes_{\ell} q$ <u>is a reasonable cross-norm on</u> $X \otimes Y$,

$$p \otimes_{\ell} q(z) = \inf\{p(\Sigma_{i=1}^{m} q(y_i)u_i) : z = \Sigma_{i=1}^{m} u_i \otimes y_i \text{ with } u_i \in X_+\} , \qquad (1.7)$$

$(L^{\ell}(X, Y'), \|\cdot\|_{(\ell)})$ <u>is isometrically isomorphic to the Banach dual</u> $(X \otimes Y, p \otimes_{\ell} q)'$ <u>of</u> $(X \otimes Y, p \otimes_{\ell} q)$, <u>and</u>

$$p \otimes_{\pi} q(z) = \inf\{\Sigma_{i=1}^{m} p(u_i)q(y_i) : z = \Sigma_{i=1}^{m} u_i \otimes y_i \text{ with } u_i \in X_+\} \qquad (1.8)$$

<u>Proof</u>. We first verify the formula (1.8). Indeed, (4.1.3)(a) shows that

$$p \otimes_\pi q(z) \leq \inf\{\Sigma_{i=1}^{m} p(u_i)q(y_i) : z = \Sigma_{i=1}^{m} u_i \otimes y_i \text{ with } u_i \in X_+\}. \quad (1.9)$$

For any $\delta > 0$, there exist finite subsets $\{x_1, \ldots, x_n\}$ and $\{y_1, \ldots, y_n\}$ of X and Y respectively such that

$$z = \Sigma_{i=1}^{n} x_i \otimes y_i \quad \text{and} \quad \Sigma_{i=1}^{n} p(x_i)q(y_i) < p \otimes_\pi q(z) + \delta . \quad (1.10)$$

Since Σ_X is decomposable, there exist $u_i^{(1)}, u_i^{(2)} \in X_+$ such that

$$x_i = u_i^{(1)} - u_i^{(2)} \quad \text{and} \quad p(u_i^{(1)}) + p(u_i^{(2)}) < p(x_i) + \delta \quad (i = 1, \ldots, n). \quad (1.11)$$

Let us define

$$w_i = \begin{cases} u_i^{(1)} & \text{if } i = 1, 2, \ldots, n \\ u_{i-n}^{(2)} & \text{if } i = n+1, \ldots, 2n \end{cases} \quad \text{and} \quad \bar{y}_i = \begin{cases} y_i & \text{if } i = 1, \ldots, n \\ -y_{i-n} & \text{if } i = n+1, \ldots, 2n . \end{cases}$$

Then (1.10) and (1.11) show that

$$z = \Sigma_{i=1}^{2n} w_i \otimes \bar{y}_i \quad \text{with} \quad w_i \in X_+ \ (i = 1, \ldots, 2n)$$

and

$$\begin{aligned} \Sigma_{i=1}^{2n} p(w_i)q(\bar{y}_i) &= \Sigma_{i=1}^{n} (p(u_i^{(1)}) + p(u_i^{(2)}))q(y_i) \\ &< \Sigma_{i=1}^{n} p(x_i)q(y_i) + \delta\, \Sigma_{i=1}^{n} q(y_i) \\ &< p \otimes_\pi q(z) + \delta\ (1 + \Sigma_{i=1}^{n} q(y_i)) . \end{aligned}$$

As δ was arbitrary, we obtain

$$\inf\{\Sigma_{i=1}^{m} p(u_i)q(y_i) : z = \Sigma_{i=1}^{m} u_i \otimes y_i \text{ with } u_i \in X_+\} \leq p \otimes_\pi q(z) .$$

Combing (1.9), we obtain our assertion.

For any $z \in X \otimes Y$, let us define

$$\alpha(z) = \inf\{p(\Sigma_{i=1}^{m} q(y_i)u_i) : z = \Sigma_{i=1}^{m} u_i \otimes y_i \text{ with } u_i \in X_+\}$$

Then α is clearly a seminorm on $X \otimes Y$ such that

$$\alpha(z) \leq \Sigma_{i=1}^{m} q(y_i)p(u_i)$$

whenever $z = \Sigma_{i=1}^{m} u_i \otimes y_i$ with $u_i \in X_+$, thus

$$\alpha(z) \leq p \otimes_\pi q(z)$$

by (1.8). To prove that

$$p \otimes_\varepsilon q(z) \leq \alpha(z) , \tag{1.12}$$

let $z = \Sigma_{i=1}^{m} u_i \otimes y_i$ with $u_i \in X_+$. Then

$$p \otimes_\varepsilon q(z) = \sup\{p(\Sigma_{i=1}^{m} <y_i , y'>u_i) : q^*(y') \leq 1\} .$$

For any $y' \in Y'$ with $q^*(y') \leq 1$, we have

$$- \Sigma_{i=1}^{m} q(y_i)u_i \leq \Sigma_{i=1}^{m} <y_i , y'>u_i \leq \Sigma_{i=1}^{m} q(y_i)u_i$$

since $u_i \in X_+$. As Σ_X is 0-convex, it follows that p is absolutely monotone, and hence that

$$p(\Sigma_{i=1}^{m} <y_i , y'>u_i) \leq p(\Sigma_{i=1}^{m} q(y_i)u_i) .$$

Therefore we obtain (1.12) since $q^*(y') \leq 1$ was arbitrary. As $p \otimes_\varepsilon q$ and $p \otimes_\pi q$ are norms on $X \otimes Y$, it follows that α is a reasonable cross-norm.

In view of the definition of the reasonable cross-norm α and (1.5), we see that

$$p \otimes_\ell q(z) \leq \alpha(z) \quad \text{for any } z \in X \otimes Y ;$$

it then follows from (1.6) that

$$L^{\ell}(X, Y') \subseteq (X \otimes Y, p \otimes_{\ell} q)' \subseteq L^{\alpha}(X, Y') \qquad (1.13)$$

and

$$\|T\|_{(\alpha)} \leq (p \otimes_{\ell} q)^*(\ell_T) \leq \|T\|_{(\ell)} \quad \text{for any} \quad T \in L^{\ell}(X, Y') . \qquad (1.14)$$

We complete the proof by showing that

$$L^{\alpha}(X, Y') \subseteq L^{\ell}(X, Y') \text{ and } \|T\|_{(\ell)} \leq \|T\|_{(\alpha)} \text{ for all } T \in L^{\alpha}(X, Y'). \qquad (1.15)$$

In fact, let $T \in L^{\alpha}(X, Y')$. For any finite subsets $\{u_1, \ldots, u_m\}$ and $\{y_1, \ldots, y_m\}$ of X_+ and Y respectively, we have

$$|\Sigma_{i=1}^{m} \langle y_i , Tu_i \rangle| \leq \|T\|_{(\alpha)} \alpha(\Sigma_{i=1}^{m} u_i \otimes y_i) \leq \|T\|_{(\alpha)} p(\Sigma_{i=1}^{m} q(y_i)u_i) .$$

Replace y_i by $(\text{sign} \langle y_i , Tu_i \rangle)y_i$ in the above inequalities, we obtain

$$\Sigma_{i=1}^{m} |\langle y_i , Tu_i \rangle| \leq \|T\|_{(\alpha)} p(\Sigma_{i=1}^{m} q(y_i)u_i) . \qquad (1.16)$$

For any $\delta > 0$ there exists a finite subset $\{v_1, \ldots, v_m\}$ of Y such that

$$q(v_i) \leq 1 \text{ and } q^*(Tu_i) < |\langle v_i , Tu \rangle| + m^{-1}\delta \quad \text{for all} \quad i = 1, \ldots, m .$$

It then follows from (1.16) that

$$\Sigma_{i=1}^{m} q^*(Tu_i) < \Sigma_{i=1}^{m} |\langle v_i , Tu_i \rangle| + \delta \leq \|T\|_{(\alpha)} p(\Sigma_{i=1}^{m} q(v_i)u_i) + \delta . \qquad (1.17)$$

Since $q(v_i) \leq 1$ and $u_i \in X_+$, it follows that

$$0 \leq \Sigma_{i=1}^{m} q(v_i)u_i \leq \Sigma_{i=1}^{m} u_i ,$$

and hence that

$$p(\Sigma_{i=1}^{m} q(v_i)u_i) \leq p(\Sigma_{i=1} u_i)$$

since p is strongly monotone; consequently

$$\Sigma_{i=1}^{m} q^*(Tu_i) < \|T\|_\alpha \, p(\Sigma_{i=1}^{m} u_i) + \delta \ ,$$

which implies that

$$T \in L^{\ell}(X, Y') \quad \text{and} \quad \|T\|_{(\ell)} \leq \|T\|_{(\alpha)} \ .$$

(5.1.2) Corollary. $p \otimes_s q(z) \leq p \otimes_\ell q(z)$ <u>for all</u> $z \in X \otimes Y$.

<u>Proof</u>. This follows from (5.1.1) and (4.2.4) since $\pi_{\ell^1}(X, Y() \subseteq L^{\ell}(X, Y')$ and $\|T\|_{(\ell)} \leq \|T\|_{(s)}$ for all $T \in \pi_{\ell^1}(X, Y')$.

If p is additive on X_+ , that is

$$p(u + v) = p(u) + p(v) \quad \text{for all} \quad u, v \in X_+ \ ,$$

then the topology on X induced by the norm p coincides with $\sigma_S(X, X')$, hence $L^{\ell}(X, Y) = L(X, Y)$ for any normed space (Y, q) (see Wong [1, (3.2.12)]). Furthermore, by the idea using in Schaefer [2, (3.7), chap.IV], one can prove the following more general result.

(5.1.3) Proposition. <u>Let</u> (X, X_+, p) <u>be an ordered normed space such that the unit ball</u> Σ_X <u>is</u> $\mathcal{O}$-<u>convex and decomposable</u>. <u>Then</u> p <u>is additive if and only if for any normed space</u> (Y, q) <u>there are</u> $L(X, Y) = L^{\ell}(X, Y)$ <u>and</u>

$$|||T||| = \|T\|_{(\ell)} \quad \underline{\text{for all}} \quad T \in L(X, Y) \ . \tag{1.18}$$

<u>Proof</u>. <u>Necessity</u>. It is known that $L(X, Y) = L^{\ell}(X, Y)$. For any $T \in L(X, Y)$ and any finite subset $\{u_1, \ldots, u_n\}$ of X_+ with $p(\Sigma_{i=1}^{n} u_i) \leq 1$, we have

$$\Sigma_{i=1}^{n} q(Tu_i) \leq |||T|||\Sigma_{i=1}^{n} p(u_i) = |||T|||\Sigma_{i=1}^{n} p(u_i) \leq |||T||| ,$$

hence (1.2)* implies that

$$||T||_{(\ell)} \leq |||T||| .$$

Consequently, (1.18) holds since $|||T||| \leq ||T||_{(\ell)}$ is always true.

<u>Sufficiency</u>. The identity map I_X on X is cone-absolutely summing and $||I_X||_{(\ell)} = |||I_X||| = 1$. For any $u, v \in X_+$, we have

$$p(u) + p(v) \leq ||I_X||_{(\ell)} p(u + v) = p(u + v) ,$$

thus p is additive on X_+ .

(5.1.4) Corollary. <u>If</u> p <u>is additive on</u> X_+ , <u>then</u>

$$p \otimes_\ell q = p \otimes_\pi q \quad \underline{\text{for any normed space}} \quad (Y, q) .$$

<u>Proof</u>. This follows from (5.1.1), (5.1.3) and (4.1.3)(a).

In the sequel we assume that $(E, E_+, \mathcal{P})$ is a locally solid space which has a neighbourhood base $\mathcal{U}_E$ at 0 consisting of 0-convex, decomposable sets, and that $(F, \mathcal{T})$ is a locally convex space which has a neighbourhood base $\mathcal{U}_F$ at 0 consisting of absolutely convex sets. For any $V \in \mathcal{U}_E$ and $U \in \mathcal{U}_F$, it is known from (5.1.1) and (5.1.2) that

$$p_V \otimes_\ell q_U(z) = \inf\{p_V(\Sigma_{i=1}^{m} q_U(y_i)u_i) : z = \Sigma_{i=1}^{m} u_i \otimes y_i \quad \text{with } u_i \in E_+\} \qquad (1.19)$$

is a seminorm on $E \otimes F$ such that

$$p_V \otimes_\varepsilon q_U \leq p_V \otimes_s q_U \leq p_V \otimes_\ell q_U \quad \text{and} \tag{1.20}$$

$$p_V \otimes_\ell q_U(x \otimes y) = p_V(x) q_U(y) \quad \text{for all} \quad x \in E \quad \text{and} \quad y \in F .$$

Hence the Hausdorff locally convex topology on $E \otimes F$ determined by $\{p_V \otimes_\ell q_U : V \in \mathcal{U}_E, U \in \mathcal{U}_F\}$ is called the ℓ-topology of $\mathcal{P}$ and $\mathcal{J}$, and denoted by $\mathcal{P} \otimes_\ell \mathcal{J}$. As usual, we denote by $E \otimes_\ell F$ the locally convex space $(E \otimes F, \mathcal{P} \otimes_\ell \mathcal{J})$, and by $E \tilde{\otimes}_\ell F$ the completion of $E \otimes_\ell F$.

It is clear that

$$\mathcal{P} \otimes_\varepsilon \mathcal{J} \leq \mathcal{P} \otimes_s \mathcal{J} \leq \mathcal{P} \otimes_\ell \mathcal{J} \leq \mathcal{P} \otimes_\pi \mathcal{J} . \tag{1.21}$$

For any o-convex, decomposable $\mathcal{P}$-neighbourhood of 0 in E , $p_V^{-1}(0)$ is order-convex, and $Q_V(V)$ is a decomposable subset of the ordered vector space $(E_V, (E_V)_+)$, where $(E_V)_+$ is the canonical cone in the quotient space E_V, that is

$$(E_V)_+ = E_V \cap Q_V(E_+) . \tag{1.22}$$

(5.1.5) Lemma. <u>Let</u> $f \in (E \tilde{\otimes}_\ell F)'$ <u>and let</u> T <u>be defined by</u>

$$\langle y, Tx\rangle = f(x \otimes y) \quad \underline{\text{for all}} \quad x \in E \quad \underline{\text{and}} \quad y \in F . \tag{1.23}$$

<u>Then there exist</u> $V \in \mathcal{U}_E$ <u>and</u> $U \in \mathcal{U}_F$ <u>such that</u> $T(V) \subset U^o$ <u>and the induced map</u> $T_{(V^o, U)}$ <u>of</u> T <u>is a cone-absolutely summing map from</u> $(E_V, (E_V)_+, p_V)$ <u>into</u> $F'(U^o)$.

<u>Proof</u>. Let $V \in \mathcal{U}_E$ and $U \in \mathcal{U}_F$ be such that

$$|f(z)| \leq p_V \otimes_\ell q_U(z) \quad \text{for all} \quad z \in E \otimes F .$$

For any representation $z = \sum_{i=1}^n u_i \otimes y_i$ with $u_i \in E_+$ and $y_i \in F$, we have

$$|\Sigma_{i=1}^{n} <y_i , Tu_i>| \leq p_V \otimes_\ell q_U(z) \leq p_V(\Sigma_{i=1}^{n} q_U(y_i)u_i) , \tag{1.24}$$

hence $T(V \cap E_+) \subset U^o$, and thus $T(V) \subset U^o$ since V is decomposable and U is convex. The boundedness of U^o in F'_β implies that $\operatorname{Ker} p_V \subset \operatorname{Ker} T$, hence the induced map $T_{(V, U^o)}$ of T is a continuous linear map from $(E_V, (E_V)_+, p_V)$ into $F'(U^o)$ such that

$$T = j_{U^o} \circ T_{(V,U^o)} \circ Q_V .$$

When we replace y_i by (sign $<y_i , Tu_i>)y_i$ in (1.24), we obtain

$$\Sigma_{i=1}^{n} |<y_i , Tu_i>| \leq p_V(\Sigma_{i=1}^{n} q_U(y_i)u_i) \tag{1.25}$$

whenever $\{u_1, \ldots, u_n\} \subset E_+$ and $\{y_1, \ldots, y_n\} \subset F$. As the gauge of U^o denoted by q_{U^o} , is given by

$$q_{U^o}(y') = \sup\{|<y, y'>| : y \in U\} \quad \text{for all } y' \in F'(U^o) ,$$

it follows from (1.25) that the inequality

$$\Sigma_{i=1}^{n} q_{U^o}(Tu_i) \leq p_V(\Sigma_{i=1}^{n} u_i) \tag{1.26}$$

holds for any finite subset $\{u_1, \ldots, u_n\}$ of E_+ since p_V is monotone.

Now for any finite subset $\{Q_V(\overline{u}_i) : i = 1, \ldots, n\}$ of $(E_V)_+$, we have

$$\overline{u}_i = u_i + n_i \quad (i = 1, \ldots, n) ,$$

for some $u_i \in E_+$ and $n_i \in p_V^{-1}(0)$ $(i = 1, \ldots, n)$. Note that

$$\begin{aligned}\Sigma_{i=1}^{n} q_{U^o}(T_{(V,U^o)}(Q_V(\overline{u}_i))) &= \Sigma_{i=1}^{n} q_{U^o}(T_{(V,U^o)}(Q_V(u_i))) \\ &= \Sigma_{i=1}^{n} q_{U^o}(Tu_i)\end{aligned}$$

and that

$$\hat{p}_V(\Sigma_{i=1}^n Q_V(\bar{u}_i)) = \hat{p}_V(\Sigma_{i=1}^n Q_V(u_i)) = \hat{p}_V(Q_V(\Sigma_{i=1}^n u_i))$$
$$= p_V(\Sigma_{i=1}^n u_i) .$$

It then follows from (1.26) that

$$\Sigma_{i=1}^n q_{U^o}(T_{(V,U^o)}(Q_V(\bar{u}_i))) \leq \hat{p}_V(\Sigma_{i=1}^n Q_V(\bar{u}_i))$$

whenever $\{Q_V(\bar{u}_i) : i = 1, \ldots, n\} \subset (E_V)_+$, and hence that $T_{(V,U^o)}$ is cone-absolutely summing.

(5.1.6) Theorem. <u>An</u> $T \in L^{eq}(E, F(\mathcal{J})')$ <u>defines a continuous linear functional on</u> $E \otimes_\ell F$ <u>if and only if there exist</u> $V \in \mathcal{U}_E$ <u>and</u> $U \in \mathcal{U}_F$ <u>such that the induced map</u> $T_{(V,U^o)}$ <u>of</u> T <u>is a cone-absolutely summing map from</u> $(E_V, (E_V)_+, p_V)$ <u>into</u> $F'(U^o)$. <u>Furthermore</u>, <u>if</u> F <u>is a normed space</u>, <u>then</u>

$$(E \tilde{\otimes}_\ell F)' = L^\ell(E, F) . \tag{1.27}$$

<u>Proof</u>. The necessity is a restatement of (5.1.6). To prove the sufficiency, let $\zeta \geq 0$ be such that the inequality

$$\Sigma_{i=1}^n q_{U^o}(T_{(V,U^o)}(Q_V(\bar{u}_i))) \leq \zeta \hat{p}_V(\Sigma_{i=1}^n Q_V(\bar{u}_o)) \tag{1.28}$$

holds for any finite subset $\{Q_V(\bar{u}_1), \ldots, Q_V(\bar{u}_n)\}$ of $(E_V)_+$. For any $z \in E \otimes F$ and any representation $z = \Sigma_{i=1}^n u_i \otimes y_i$ with $u_i \in E_+$ and $y_i \in F$, we have $\{Q_V(u_1), \ldots, Q_V(u_n)\} \subset (E_V)_+$, thus

$$|f(z)| = |\Sigma_{i=1}^n \langle y_i, Tu_i\rangle| \leq \Sigma_{i=1}^n q_U(y_i) q_{U^o}(Tu_i)$$
$$= \Sigma_{i=1}^n q_U(y_i) q_{U^o}(T_{(V,U^o)}(Q_V(u_i)))$$

$$= \Sigma_{i=1}^{n} q_{U^o}(T_{(V,U^o)}(q_U(y_i) q_V(u_i)))$$

$$\leq \zeta \hat{p}_V(\Sigma_{i=1}^{n} q_U(y_i) q_V(u_i)) = \zeta p_V(\Sigma_{i=1}^{n} q_U(y_i) u_i)$$

by (1.28). As the representation of z is arbitrary, it follows that

$$|f(z)| \leq \zeta p_V \otimes_\ell q_U(z) ,$$

and hence that $f \in (E \tilde{\otimes}_\ell F)'$.

Finally, if F is a normed space, then U can be chosen as the closed unit ball in F in view of the proof of (5.1.6), thus (1.27) holds.

(5.1.7) Corollary. <u>If</u> $(E, E_+, \mathcal{P})$ <u>is metrizable and if</u> $\mathcal{P} = \sigma_S(E, E')$, <u>then for any normed space</u> (Y, q) <u>we have</u>

$$\mathcal{P} \otimes_\ell \mathcal{J} = \mathcal{P} \otimes_\pi \mathcal{J} ,$$

<u>where</u> $\mathcal{J}$ <u>is the norm-topology on</u> Y <u>induced by</u> q .

<u>Proof</u>. In view of Wong [1, (3.2.12)], $L^\ell(E, Y') = L(E, Y')$. Clearly $\mathcal{P} \otimes_\ell \mathcal{J}$ and $\mathcal{P} \otimes_\pi \mathcal{J}$ are metrizable; it then follows (5.1.7), (4.4.4) and Mackey-Aren's theorem $\mathcal{P} \otimes_\ell \mathcal{J} = \mathcal{P} \otimes_\pi \mathcal{J}$.

5.2 <u>The projective cone and biprojective cone</u>

Let $(E, E_+, \mathcal{P})$ and $(F, F_+, \mathcal{J})$ be ordered convex spaces whose ordered strong dual spaces are denoted by (E'_β, E'_+) and (F'_β, F'_+) respectively, where E'_+ and F'_+ are the dual cones of E_+ and F_+ respectively. It is easily seen that the set, defined by

$$B_+(E, F) = \{\psi \in B(E, F) : \psi(u, v) \geq 0 \text{ for all } u \in E_+ \text{ and } v \in F_+\},$$

is always a cone, which is called the <u>natural cone</u> in $B(E, F)$. Furthermore, if $E_+ - E_+$ is dense in $(E, \mathcal{P})$, then $B_+(E, F)$ is proper.

It is known from (4.4.1) that $B(E, F)$ can be identified with the topological dual $(E \otimes_\pi F)'$ of $E \otimes_\pi F$ by identifying each $\psi \in B(E, F)$ with a continuous linear functional on $E \otimes_\pi F$ obtained from the equation

$$\psi(\Sigma_{i=1}^n x_i \otimes y_i) = \Sigma_{i=1}^n \psi(x_i, y_i).$$

It is natural to ask whether there exists a cone C_p on $E \otimes F$ such that its dual cone is the usual cone $B_+(E, F)$. The answer is affirmative, in fact, the cone C_π, defined by

$$C_\pi = co(E_+ \otimes F_+) = \{\Sigma_{i=1}^n u_i \otimes v_i : u_i \in E_+, y_i \in F_+\}, \tag{2.1}$$

has the required property. We call C_π the <u>projective cone of</u> E_+ <u>and</u> F_+.

If $\omega : E \times F \to E \otimes F$ is the canonical bilinear map, then for any ordered convex space $(G, G_+, \mathcal{L})$, the map $T \rightsquigarrow T \circ \omega$ is an order isomorphism from $(L(E \otimes_\pi F, G), L_+(E \otimes_\pi F, G))$ onto $(B(E, F; G), B_+(E, F; G))$, where

$$L_+(E \otimes_\pi F, G) = \{T \in L(E \otimes_\pi F, G) : T(C_\pi) \subset G_+\}.$$

Moreover, if (X, X_+, p) and (Y, Y_+, q) are ordered normed spaces, then the map $T \rightsquigarrow b_T$ is an isometrically order isomorphism from $(L(X, Y'), L_+(X, Y'), |||\cdot|||)$

onto $(B(X, Y), B_+(X, Y), \|\cdot\|)$, where b_T is the associated bilinear map of T and

$$L_+(X, Y') = \{T \in L(X, Y') : T(X_+) \subseteq Y'_+\} ,$$

also the map $T \rightsquigarrow T \circ \omega$ is an isometrically order isomorphism from the ordered Banach space $((X \otimes Y)', C'_\pi)$ onto $(B(X, Y), B_+(X, Y), \|\cdot\|)$.

On the other hand, it is known that $E \otimes_\varepsilon F$ can be identified with a subspace of $S_\varepsilon(E'_\sigma, F'_\sigma)$; therefore we define

$$C_i = \{\Sigma_{i=1}^n u_i \otimes v_i : \Sigma_{i=1}^n \langle u_i, x'\rangle\langle v_i, y'\rangle \geq 0 \text{ for all } x' \in E'_+ \text{ and } y' \in F'_+\} .$$

It is clear that C_i is a cone in $E \otimes F$ such that

$$C_\pi \subseteq C_i$$

C_i is called the <u>biprojective cone</u> of E_+ and F_+ . As $\langle E \otimes F, E' \otimes F'\rangle$ is a dual pair, it follows that

$$C_i = -(E'_+ \otimes F'_+)^\circ = -(co(E'_+ \otimes F'_+))^\circ ; \qquad (2.2)$$

in other words, C_i is the dual cone of the projective cone C_π, of E'_+ and F'_+ . Therefore C_i is closed for any locally convex topology on $E \otimes F$ which is finer than $\sigma(E \otimes F, E' \otimes F')$. Furthermore, if F_+ is closed, then

$$C_i = (E \otimes F) \cap S_+(E'_\sigma, F'_\sigma) ,$$

where

$$S_+(E'_\sigma, F'_\sigma) = \{\psi \in S(E'_\sigma, F'_\sigma) : \psi(u', v') \geq 0 \text{ for all } u' \in E'_+, v' \in F'_+\} ;$$

thus $(E \otimes_\varepsilon F, C_i)$ is topologically isomorphic and order isomorphic to a subspace of $(S_\varepsilon(E'_\sigma, F'_\sigma), S_+(E'_\sigma, F'_\sigma))$.

We denote by $\leq_\pi$ the vector ordering in $E \otimes F$ induced by the projective cone C_π, that is

$$t \leq_\pi z \quad \text{if} \quad z - t \in C_\pi ,$$

and call $\leq_\pi$ the *projective ordering*. Similarly, we denote by $\leq_i$ the vector ordering in $E \otimes F$ induced by the biprojective cone C_i, and call $\leq_i$ the *biprojective ordering*.

Also we write $C_{\pi'}$ for the projective cone of E'_+ and F'_+, that is

$$C_{\pi'} = \operatorname{co}(E'_+ \otimes F'_+) .$$

As $\langle E \otimes F, E' \otimes F' \rangle$ is a dual pair, we see from (2.2) that the biprojective cone C_i of E_+ and F_+ is the dual cone of the projective cone in $E' \otimes F'$ of E'_+ and F'_+.

(5.2.1) Lemma. *If E_+ and F_+ are generating cones in E and F respectively, then so do C_π and C_i.*

Proof. It suffices to show that C_π is generating since $C_\pi \subseteq C_i$. For this, let $z = \Sigma_{i=1}^n x_i \otimes y_i$, and let $u_i^{(1)}, u_i^{(2)}$ in E_+ and $v_i^{(1)}, v_i^{(2)}$ in F_+ be such that

$$x_i = u_i^{(1)} - u_i^{(2)} \quad \text{and} \quad y_i = v_i^{(1)} - v_i^{(2)} \quad (i = 1, 2, \ldots, n) .$$

Then

$$\begin{aligned} z &= \Sigma_{i=1}^n (u_i^{(1)} - u_i^{(2)}) \otimes (v_i^{(1)} - v_i^{(2)}) \\ &= \Sigma_{i=1}^n (u_i^{(1)} \otimes v_i^{(1)} + u_i^{(2)} \otimes v_i^{(2)}) - \Sigma_{i=1}^n (u_i^{(1)} \otimes v_i^{(2)} + u_i^{(2)} \otimes v_i^{(1)}) \in C_p - C_p \end{aligned}$$

which obtains our assertion.

(5.2.2) **Lemma.** <u>The biprojective ordering</u> $\leq_i$ <u>in</u> $E \otimes F$ <u>is always</u> <u>Archimedean. Furthermore, if</u> E'_+ <u>and</u> F'_+ <u>are total in</u> $(E', \sigma(E', E))$ <u>and</u> $(F', \sigma(F', F))$ <u>respectively, then</u> $\leq_i$ <u>is almost-Archimedean.</u>

Proof. Let $z = \Sigma_{i=1}^{m} x_i \otimes y_i$ in $E \otimes F$ and $t = \Sigma_{j=1}^{n} u_j \otimes v_i$ in C_i be such that $kz \leq_i t$ for all $k > 0$. For any $u \in E'_+$ and $v' \in F'_+$

$$\Sigma_{j=1}^{n} \langle u_j, u' \rangle \langle v_j, v' \rangle - k \Sigma_{i=1}^{m} \langle x_i, u' \rangle \langle y_i, v' \rangle \geq 0 .$$

As the usual ordering in $\mathbb{R}$ is Archimedean, we have

$$\Sigma_{i=1}^{m} \langle x_i, u' \rangle \langle y_i, v' \rangle \leq 0 ,$$

thus $z = \Sigma_{i=1}^{m} x_i \otimes y_i \leq_i 0$.

To prove the second part, let $t \pm kz \in C_i$ for all positive integer k. As the usual ordering in $\mathbb{R}$ is almost-Archimedean, it follows that

$$\Sigma_{i=1}^{m} \langle x_i, u' \rangle \langle y_i, v' \rangle = 0 \text{ for all } u' \in E'_+ \text{ and } v' \in F'_+ ,$$

and hence from the totalities of E'_+ and F'_+ in E' and F' respectively that $z = \Sigma_{i=1}^{m} x_i \otimes y_i = 0$. Therefore the biprojective ordering $\leq_i$ is almost-Archimedean.

The remainder of this section is going to verify that if $(E, E_+, \mathscr{P})$ and $(F, F_+, \mathscr{T})$ are locally decomposable (resp. locally o-convex) spaces, then $(E \otimes F, C_\pi, \mathscr{P} \otimes_\pi \mathscr{T})$ is locally decomposable (resp. $(E \otimes F, C_i, \mathscr{P} \otimes_\varepsilon \mathscr{T})$ is locally o-convex). To this end, let us define for any subset A of $E \otimes F$ that

$$D_\pi(A) = \{ z \in A : z = \lambda z_1 - (1 - \lambda) z_2, \lambda \in [0, 1], z_1, z_2 \in A \cap C_\pi \} ,$$

$$D_i(A) = \{z \in A : z = \lambda z_1 - (1-\lambda) z_2, \lambda \in [0, 1], z_1, z_2 \in A \cap C_i\} ,$$
$$F_\pi(A) = (A + C_\pi) \cap (A - C_\pi) ,$$
$$F_i(A) = (A + C_i) \cap (A - C_i) ,$$
$$S_\pi(A) = \cup\{[-t, t] : t \in A \cap C_\pi\} ,$$
$$S_i(A) = \cup\{[-t, t] : t \in A \cap C_i\} .$$

As $C_\pi \subset C_i$, we have

$$D_\pi(A) \subseteq D_i(A) , \quad F_\pi(A) \subseteq F_i(A) \quad \text{and} \quad S_\pi(A) \subseteq S_i(A) .$$

If A is absolutely convex, then

$$D_\pi(A) = co\{-(A \cap C_\pi) \cup (A \cap C_\pi)\} \quad \text{and} \quad D_i(A) = co\{-(A \cap C_i) \cup (A \cap C_i)\} .$$

(5.2.3) Lemma. <u>If</u> V <u>and</u> W <u>are absolutely convex subsets of</u> E <u>and</u> F <u>respectively, then</u>

$$S(V) \otimes S(W) \subseteq S_\pi(V \otimes W) \tag{2.3}$$

<u>and</u>

$$D(V) \otimes D(W) \subseteq D_\pi(co(V \otimes W)) . \tag{2.4}$$

<u>In particular, if</u> V <u>and</u> W <u>are absolutely dominated, then</u> $V \otimes W$ <u>is absolutely dominated with respect to the projective cone</u> C_π ; <u>if</u> $V \subseteq \alpha D(V)$ <u>and</u> $W \subseteq \beta D(W)$ <u>for some</u> $\alpha \geq 1$ <u>and</u> $\beta \geq 1$, <u>then</u>

$$co(V \otimes W) \subseteq \alpha\beta D_\pi(co(V \otimes W)) . \tag{2.5}$$

<u>Proof</u>. Let $x \in S(V)$ and $y \in S(W)$. There exist $u \in V \cap E_+$ and $v \in W \cap F_+$ such that $u \pm x \in E_+$ and $v \pm y \in F_+$, hence $(u \pm x) \otimes (v \pm y) \in C_\pi$. Consequently,

$$2(u \otimes v + x \otimes y) = (u + x) \otimes (v + y) + (u - x) \otimes (v - y) \in C_\pi$$

and

$$2(u \otimes v - x \otimes y) = (u + x) \otimes (v - y) + (u - x) \otimes (v - y) \in C_\pi \ ,$$

thus $u \otimes v \pm x \otimes y \in C_\pi$. As $u \otimes v \in (V \otimes W) \cap C_\pi$, we conclude that $x \otimes y \in S_\pi(V \otimes W)$; thus (2.3) holds.

Clearly, if V and W are absolutely dominated, then (2.3) shows that $V \otimes W$ is absolutely dominated with respect to the projective cone C_π .

To prove (2.4), let z , in $D(V) \otimes D(W)$, be such that

$$z = (\lambda u_1 - \lambda_2 u_2) \otimes (\zeta_1 v_1 - \zeta_2 v_2) \ ,$$

where $u_1, u_2 \in E_+$, $v_1, v_2 \in F_+$, and $\lambda_1, \lambda_2, \zeta_1$ and ζ_2 are positive numbers satisfying

$$\lambda_1 + \lambda_2 = \zeta_1 + \zeta_2 = 1 \ .$$

Then

$$\begin{aligned} z &= (\lambda_1\zeta_1 u_1 \otimes v_1 + \lambda_2\zeta_2 u_2 \otimes v_2) - (\lambda_1\zeta_2 u_1 \otimes v_2 + \lambda_2\zeta_1 u_2 \otimes v_1) \\ &= \beta_1\left(\frac{\lambda_1\zeta_1}{\beta_1} u_1 \otimes v_1 + \frac{\lambda_2\zeta_2}{\beta_1} u_2 \otimes v_2\right) - \beta_2\left(\frac{\lambda_1\zeta_2}{\beta_2} u_1 \otimes v_2 + \frac{\lambda_2\zeta_1}{\beta_2} u_2 \otimes v_1\right) \ , \end{aligned}$$

where

$$\beta_1 = \lambda_1\zeta_1 + \lambda_2\zeta_2 \quad \text{and} \quad \beta_2 = \lambda_1\zeta_2 + \lambda_2\zeta_1 \ .$$

Clearly β_1 and β_2 are positive numbers and satisfy $\beta_1 + \beta_2 = 1$. Let

$$z_1 = \frac{\lambda_1\zeta_1}{\beta_1} u_1 \otimes v_1 + \frac{\lambda_2\zeta_2}{\beta_1} u_2 \otimes v_2$$

and

$$z_2 = \frac{\lambda_1\zeta_2}{\beta_2} u_1 \otimes v_2 + \frac{\lambda_2\zeta_1}{\beta_1} u_2 \otimes v_1 .$$

Then $z_1, z_2 \in (co(V \otimes W)) \cap C_\pi$ and

$$z = \beta_1 z_1 - \beta_2 z_2 \in D_\pi(co(V \otimes W)) ,$$

which proves the formula (2.4).

Note that $D_\pi(co(V \otimes W))$ is absolutely convex, the formula (2.5) follows from (2.4).

(5.2.4) Lemma. Let V be an absolutely convex $\mathcal{P}$-neighbourhood of 0 in E and W an absolutely convex $\mathcal{J}$-neighbourhood of 0 in F. Then

$$S_i((V^o \otimes W^o)^o) \subseteq (S_{\pi'}(\overline{co}(V^o \otimes W^o)))^o \tag{2.6}$$

and

$$F_i((V^o \otimes W^o)^o) \subseteq (D_{\pi'}(\overline{co}(V^o \otimes W^o)))^o , \tag{2.7}$$

where the bar is the $\sigma((E \otimes_\varepsilon F)', E \otimes F)$-closure,

$$S_{\pi'}(\overline{co}(V^o \otimes W^o)) = \{g \in (E \otimes_\varepsilon F)' : h \pm g \in C_{\pi'}, \text{ for some } h \in (\overline{co}(V^o \otimes W^o)) \cap C_{\pi'}\} \tag{2.8}$$

and

$$D_{\pi'}(\overline{co}(V^o \otimes W^o)) = \Gamma((\overline{co}(V^o \otimes W^o)) \cap C_{\pi'}) . \tag{2.9}$$

In particular, if V and W are absolutely order-convex, then $(V^o \otimes W^o)^o$ is absolutely order-convex with respect to the biprojective cone C_i ; if $F(V) \subseteq \alpha V$ and $F(W) \subseteq \beta W$ for some $\alpha, \beta \geq 1$, then

$$F_i((V^o \otimes W^o)^o) \subseteq \alpha\beta(V^o \otimes W^o)^o . \tag{2.10}$$

<u>Proof</u>. For any subset B of $E \otimes F$, we denote by B^{o} the polar of B taken in $(E \otimes_{\varepsilon} F)'$. For any absolutely convex subset N of $(E \otimes_{\varepsilon} F)'$, let

$$S^{\varepsilon}(N) = \{g \in (E \otimes_{\varepsilon} F)' : h \pm g \in C_i' \text{ for some } h \in N \cap C_i'\}$$

and

$$D^{\varepsilon}(N) = \Gamma(N \cap C_i'),$$

where C_i' is the dual cone of C_i in $(E \otimes_{\varepsilon} F)'$. As $E' \otimes F' \subseteq (E \otimes_{\varepsilon} F)'$ and C_i is the dual cone of $C_{\pi'} = co(E_+' \otimes F_+')$, it follows that $C_{\pi'} \subseteq C_i'$, and hence from (2.8) and (2.9) that

$$S_{\pi'}(N) \subseteq S^{\varepsilon}(N) \quad \text{and} \quad D_{\pi'}(N) \subseteq D^{\varepsilon}(N) \tag{2.11}$$

whenever N is any absolutely convex subset of $(E \otimes_{\varepsilon} F)'$.

As $(V^{o} \otimes W^{o})^{o}$ is an absolutely convex $\mathcal{P} \otimes_{\varepsilon} \mathcal{J}$-neighbourhood of 0 in $E \otimes F$, we have, by Wong [1, (1.1.9) and (1.1.7)] and the bipolar theorem, that

$$\begin{aligned} S_i((V^{o} \otimes W^{o})^{o}) &\subseteq (S_i((V^{o} \otimes W^{o})))^{oo} = (S^{\varepsilon}((V^{o} \otimes W^{o})^{oo}))^{o} \\ &= (S^{\varepsilon}(\overline{co}(V^{o} \otimes W^{o})))^{o} \end{aligned}$$

and

$$\begin{aligned} F_i((V^{o} \otimes W^{o})^{o}) &\subseteq (F_i((V^{o} \otimes W^{o})))^{oo} = (D^{\varepsilon}((V^{o} \otimes W^{o})^{oo}))^{o} \\ &= (D^{\varepsilon}(\overline{co}(V^{o} \otimes W^{o})))^{o}, \end{aligned}$$

thus formulae (2.6) and (2.7) follow from (2.11).

We first notice from (2.8) and (2.9) that

$$S_{\pi'}(N) \subseteq E' \otimes F' \quad \text{and} \quad D_{\pi'}(N) \subseteq E' \otimes F'$$

whenever N is any subset of $(E \otimes_\varepsilon F)'$. If V and W are absolutely order-convex, then

$$V^o \subseteq (S(V))^o = S(V^o) \quad \text{and} \quad W^o \subseteq S(W^o) \ ,$$

by a result of Jamesion (see Wong [1, (1.1.9)]). In view of (2.3) of (5.2.3),

$$V^o \otimes W^o \subseteq S(V^o) \otimes S(W^o) \subseteq S_{\pi'}(V^o \otimes W^o) \subseteq S_{\pi'}(\overline{\mathrm{co}}(V^o \otimes W^o)) \ ;$$

it then follows from (2.6) that

$$S_i((V^o \otimes W^o)^o) \subseteq (V^o \otimes W^o)^o \ .$$

Therefore $(V^o \otimes W^o)^o$ is absolutely order-convex.

To prove (2.10), it is known from Wong [1, (1.1.7)] that

$$V^o \subseteq \alpha D(V^o) \quad \text{and} \quad W^o \subseteq \beta D(W^o) \ .$$

In view of (2.4) of (5.2.3),

$$V^o \otimes W^o \subseteq \alpha\beta D(V^o) \otimes D(W^o) \subseteq \alpha\beta D_{\pi'}(\mathrm{co}(V^o \otimes W^o))$$
$$\subseteq \alpha\beta D_{\pi'}(\overline{\mathrm{co}}(V^o \otimes W^o)) \ ;$$

it then follows from (2.7) that

$$F_i((V^o \otimes W^o)^o) \subseteq \alpha\beta(V^o \otimes W^o)^o \ ,$$

which obtains (2.10).

(5.2.5) Theorem. (a) If $(E, E_+, \mathcal{P})$ and $(F, F_+, \mathcal{T})$ are locally decomposable spaces, then $(E \otimes F, C_\pi, \mathcal{P} \otimes_\pi \mathcal{T})$ is locally decomposable.

(b) <u>If</u> $(E, E_+, \mathcal{P})$ <u>and</u> $(F, F_+, \mathcal{J})$ <u>are locally o-convex spaces, then</u> $(E \otimes F, C_i, \mathcal{P} \otimes_\varepsilon \mathcal{J})$ <u>is locally o-convex.</u>

<u>Proof</u>. (a) Let $\mathcal{U}_E$ and $\mathcal{U}_F$ be respectively neighbourhood bases at 0 for $\mathcal{P}$ and $\mathcal{J}$ consisting of absolutely convex, decomposable sets. As

$$\{\Gamma(V \otimes W) : V \in \mathcal{U}_E, W \in \mathcal{U}_F\}$$

being a neighbourhood base at 0 for $\mathcal{P} \otimes_\pi \mathcal{J}$, it follows from (5.2.3) and the absolute convexity of $D_\pi(\Gamma(V \otimes W))$ that

$$\Gamma(V \otimes W) \subseteq D_\pi(\Gamma(V \otimes W)) ;$$

in other words, each $\Gamma(V \otimes W)$ is absolutely convex and decomposable. Therefore $(E \otimes F, C_\pi, \mathcal{P} \otimes_\pi \mathcal{J})$ is a locally decomposable space.

(b) Let $\mathcal{U}_E$ and $\mathcal{U}_F$ be respectively neighbourhood bases at 0 for $\mathcal{P}$ and $\mathcal{J}$ consisting of o-convex and circled sets. As

$$\{(V^o \otimes W^o)^o : V \in \mathcal{U}_E, W \in \mathcal{U}_F\}$$

is a neighbourhood base at 0 for $\mathcal{P} \otimes_\varepsilon \mathcal{J}$, it follows from (2.7) of (5.2.4) that

$$F_i((V^o \otimes W^o)^o) \subseteq (V^o \otimes W^o)^o$$

since $V = F(V)$ and $W = F(W)$; in other words, each $(V^o \otimes W^o)^o$ is o-convex. Therefore $(E \otimes F, C_i, \mathcal{P} \otimes_\varepsilon \mathcal{J})$ is a locally o-convex space.

It is natural to ask if $(E, E_+, \mathcal{P})$ and $(F, F_+, \mathcal{J})$ are locally o-convex spaces (resp. locally decomposable spaces), is $(E \otimes_\pi F, C_\pi)$ locally o-convex (resp. $(E \otimes_{\varepsilon} F, C_i)$ locally decomposable)? The answer is negative as shown by the following example.

(5.2.6) Example. Let ℓ^2 be the sequence space of real numbers equipped with the usual norm $\|x\| = (\Sigma_{i=1}^{\infty} |x_i|^2)^{\frac{1}{2}}$ and the usual ordering. Then ℓ^2 is a Hilbert space which is also a Banach lattice. As ℓ^2 is reflexive, it follows that the ordered Banach dual space $(\ell^2 \otimes_{\pi} \ell^2, C_{\pi})'$ of $(\ell^2 \otimes_{\pi} \ell^2, C_{\pi})$ can be identified with $(L(\ell^2, \ell^2), L_+(\ell^2, \ell^2), |||\cdot|||)$. As the cone $L_+(\ell^2, \ell^2)$ is not generating (see Peressini [1, p.171]), it follows from the Grosberg-Krein theorem (see Wong and Ng [1, (5.15)]) that C_{π} is not normal in $\ell^2 \otimes_{\pi} \ell^2$ since $L_+(\ell^2, \ell^2)$ is the dual cone of C_{π}. Therefore $(\ell^2 \otimes_{\pi} \ell^2, C_{\pi})$ is not locally o-convex. On the other hand, since ℓ^2 is the ordered Banach dual of ℓ^2 and C_i is the dual cone of the projective cone $C_{\pi'}$, it follows from the Grosberg-Krein theorem that C_i is not generating. Therefore $(\ell^2 \otimes_{\varepsilon} \ell^2, C_i)$ is not locally decomposable.

Let (X, X_+, p) be an ordered normed space, let Σ_X be the closed unit ball in X and let $\alpha \geq 1$. We say that X_+ is

(a) α-normal in (X, p) if $F(\Sigma_X) \subseteq \alpha\Sigma_X$;

(b) α-generating in (X, p) if $\Sigma_X \subseteq \alpha D(\Sigma_X)$;

(c) nearly α-generating in (X, p) if $\Sigma_X \subseteq \alpha \overline{D(\Sigma_X)}$;

(d) approximately α-generating in (X, p) if X_+ is $(\alpha + \delta)$-generating for any $\delta > 0$.

The Groberg-Krein duality theorem says that X_+ is α-normal if and only if X'_+ is α-generating in (X', p^*) (see Wong and Ng [1, (5.15)]). If X_+ is complete in (X, p), then the Andô-Ellis theorem says that X'_+ is α-normal in (X', p^*) if and only if X_+ is nearly α-generating in (X, p), and this is the case if and only if X_+ is approximately α-generating in (X, p) (see Wong and Ng [1, (5.16)]).

Combining (4.1.4), (2.5) of (5.2.3) and (2.10) of (5.2.4), we obtain immediately the following result.

(5.2.7) Proposition. Let (X, X_+, p) and (Y, Y_+, q) be ordered normed spaces. Then the following statements hold

(a) If X_+ is α-generating in (X, p) and Y_+ is β-generating in (Y, q), then C_π is $\alpha\beta$-generating in $(X \otimes Y, p \otimes_\pi q)$.

(b) If X_+ is α-normal in (X, p) and Y_+ is β-normal in (Y, q) then C_i is $\alpha\beta$-normal in $(X \otimes Y, p \otimes_\varepsilon q)$.

The following result, due to Wickstead [1], is slightly a generalizatio of (5.2.7).

(5.2.8) Proposition. Let (X, X_+, p) and (Y, Y_+, q) be ordered normed spaces.

(a) If X_+ is α-generating in (X, p) and Y_+ is β-normal in (Y, q), then $L_+(X, Y)$ is $\alpha\beta$-normal in $(L(X, Y), |||\cdot|||)$.

(b) Let X_+ be closed and proper, and let $L_+(X, Y)$ be γ-normal in $(L(X, Y), |||\cdot|||)$. Then X'_+ is γ-normal in (X', p^*) and Y_+ is γ-normal in (Y, q). If, in addition, X_+ is complete, then X_+ is nearly γ-generating in (X, p) and Y_+ is γ-normal in (Y, q).

(c) <u>Let</u> X_+ <u>and</u> Y_+ <u>be closed and proper. If</u> $L_+(X, Y)$ <u>is</u> γ-<u>generating in</u> $(L(X, Y), |||\cdot|||)$, <u>then</u> X_+ <u>is</u> γ-<u>normal in</u> (X, p) <u>and</u> Y_+ <u>is</u> γ-<u>generating in</u> (Y, q) .

<u>Proof</u>. (a) Let $T \in L(X, Y)$ and let A and B , in $L(X, Y)$, be such that

$$A \leqslant T \leqslant B \quad \text{and} \quad |||A||| \leqslant 1 \quad \text{and} \quad |||B||| \leqslant 1 . \tag{2.12}$$

Since X_+ is α-generating, it follows that $\Sigma_X \subseteq \alpha D(\Sigma_X) = \alpha\Gamma(\Sigma_X \cap X_+)$, and hence that

$$|||T||| \leqslant \alpha \sup\{q(Tu) : u \in \Sigma_X \cap X_+\} . \tag{2.13}$$

As $|||A||| \leqslant 1$ and $|||B||| \leqslant 1$, it follows that

$$A(\Sigma_X \cap X_+),\ B(\Sigma_X \cap X_+) \subseteq \Sigma_Y ,$$

and hence from (2.12) and the β-normality of Y_+ that

$$T(\Sigma_X \cap X_+) \subseteq F(\Sigma_Y) \subseteq \beta\Sigma_Y .$$

Therefore we obtain from (2.13) that

$$|||T||| \leqslant \alpha\beta ;$$

in other words, $L_+(X, Y)$ is $\alpha\beta$-normal in $(L(X, Y), |||\cdot|||)$.

(b) We first show that X'_+ is γ-normal in (X', p^*) . Let $f \in X'$ and let $g, h \in \Sigma_X^o$ be such that $g \leqslant f \leqslant h$. Choose $v \in Y_+$ with $g(v) = 1$, and define

$$T_f(x) = f(x)v,\ T_g(x) = g(x)v \ \text{ and } \ T_h(x) = h(x)v \ \text{ for all } \ x \in X .$$

Clearly T_f, T_g and T_h are continuous linear maps from X into Y and satisfy

$$T_g \leq T_f \leq T_h, \ |||T_g||| = p^*(g), \ |||T_f||| = p^*(f) \text{ and } |||T_h||| = p^*(h) .$$

As $L_+(X, Y)$ being γ-normal in $(L(X, Y), |||\cdot|||)$, we see that

$$p^*(f) = |||T_f||| \leq \gamma \max\{|||T_g|||, |||T_h|||\} \leq \gamma ,$$

thus X'_+ is γ-normal in (X', p^*) .

To prove the γ-normality of Y_+ in (Y, q) , let $y \in F(\Sigma_Y)$ and let s, z, in Σ_Y , be such that $s \leq y \leq z$. Suppose $0 \neq u_o \in X_+$. Then the closedness of X_+ ensures that there exists an $f_o \in X'_+$ such that

$$p^*(f_o) = 1 \text{ and } f_o(u_o) \neq 0 .$$

Let us define

$$T_s(x) = f_o(x)s, \ T_y(x) = f_o(x)y \text{ and } T_z(x) = f_o(x)z \text{ for all } x \in X .$$

Then T_s, T_y and T_z are continuous linear maps from X into Y , and satisfy

$$T_s \leq T_y \leq T_z, \ |||T_s||| = q(s), \ |||T_z||| = q(z) \text{ and } |||T_y||| = q(y) .$$

As $L_+(X, Y)$ being γ-normal in $(L(X, Y), |||\cdot|||)$, we see that

$$q(y) = |||T_y||| \leq \gamma \max\{|||T_s|||, |||T_z|||\} \leq \gamma ,$$

thus Y_+ is γ-normal in (Y, q) .

Finally, if X_+ is assumed to be complete in (X, p) , then Y_+ is γ-normal in (Y, q) and X_+ is nearly γ-generating by Andô-Ellis theorem.

(c) We first show that X_+ is γ-normal in (X, p). To this end, it suffices to show by Grosberg-Krein's theorem that X'_+ is γ-generating in (X', p^*). Indeed, let v_o, in Y_+, be such that $q(v_o) = 1$. Since Y_+ is closed and proper, there exists an $g \in Y'_+$ such that

$$q^*(g) = 1 \quad \text{and} \quad g(v_o) = 1 .$$

For any $f \in X'$, let

$$T_f(x) = f(x)v_o \quad \text{for all} \quad x \in X .$$

Then $T_f \in L(X, Y)$ and $|||T_f||| = p^*(f)$. As $L_+(X, Y)$ is γ-generating in $(L(X, Y), |||\cdot|||)$, there exist $T_1, T_2 \in L_+(X, Y)$ such that

$$T_f = T_1 - T_2 \quad \text{and} \quad |||T_1||| + |||T_2||| \leq \gamma|||T_f||| .$$

Let us define

$$f_i(x) = g(T_i(x)) \quad \text{for all} \quad x \in X \text{ and } i = 1, 2 .$$

Then $f_i \in X'_+$ since g and T_i are positive,

$$\begin{aligned} f_1(x) - f_2(x) &= g(T_1(x)) - g(T_2(x)) = g(T_f(x)) \\ &= f(x)g(v_o) = f(x) \quad \text{for all} \quad x \in X , \end{aligned}$$

and

$$\begin{aligned} p^*(f_1) + p^*(f_2) &\leq q^*(g)(|||T_1||| + |||T_2|||) \\ &\leq \gamma|||T_f||| = \gamma p^*(f) . \end{aligned}$$

Therefore, X'_+ is γ-generating in (X', p^*).

Finally, by a similar argument, it can be shown that Y_+ is γ-generating in (Y, q).

5.3 The ordered projective topology and the ordered biprojective topology

It is known from (5.2.6) that if $(E, E_+, \mathcal{P})$ and $(F, F_+, \mathcal{J})$ are locally o-convex spaces, then $(E \otimes F, C_\pi, \mathcal{P} \otimes_\pi \mathcal{J})$ is, in general, not locally o-convex, and that if $(E, E_+, \mathcal{P})$ and $(F, F_+, \mathcal{J})$ are locally decomposable spaces, then $(E \otimes F, C_i, \mathcal{P} \otimes_\varepsilon \mathcal{J})$ is, in general, not locally decomposable. But we can associate the locally o-convex of $\mathcal{P} \otimes_\pi \mathcal{J}$, and the locally decomposable topology of $\mathcal{P} \otimes_\varepsilon \mathcal{J}$, and then show that these two topologies are finer than the ε-topology as well as coarser than the π-topology.

(5.3.1) Lemma. Let $(E, E_+, \mathcal{P})$ be an ordered convex space and V an absolutely convex $\mathcal{P}$-neighbourhood of C. Suppose further that p is the gauge of V and that p_F is the gauge of $F(V) = (V + E_+) \cap (V - E_+)$. Then

$$p_F(x) = \sup\{|f(x)| : f \in V^o \cap E'_+\} \text{ for all } x \in E. \tag{3.1}$$

Proof. As $V^o \cap E'_+ \subseteq \Gamma(V^o \cap E'_+) = (F(V))^o$, it follows from

$$p_F(x) = \sup\{|g(x)| : g \in (F(V))^o\}$$

that

$$p_F(x) \geq \sup\{|f(x)| : f \in V^o \cap E'_+\} .$$

To prove the opposite inequality, let $\lambda > 0$ be such that $x \in \lambda F(V)$. Then there exist $a, b \in V$ such that $\lambda a \leq x \leq \lambda b$, hence

$$-\lambda \leq \lambda f(a) \leq f(x) \leq \lambda f(b) \leq \lambda \quad \text{for all } f \in V^o \cap E'_+ ,$$

and thus

$$\lambda \leq \sup\{|f(x)| : f \in V^o \cap E'_+\} .$$

Since λ was arbitrary, we conclude that

$$p_F(x) \leqslant \sup\{|f(x)| \ : \ f \in V^0 \cap E'_+\} \ .$$

Therefore the formula (3.1) holds.

(5.3.2) Lemma. Let $(E, E_+, \mathcal{P})$ and $(F, F_+, \mathcal{T})$ be locally o-convex spaces, and let $\mathcal{U}_E$ and $\mathcal{U}_F$ be respectively neighbourhood bases at 0 for $\mathcal{P}$ and $\mathcal{T}$ consisting of o-convex and circled sets. For any $V \in \mathcal{U}_E$ and $W \in \mathcal{U}_F$, let $p_V \otimes_{|\pi|} q_W$ be the gauge of $F_\pi(\Gamma(V \otimes W))$. Then

$$p_V \otimes_{|\pi|} q_W(z) = \sup\{|<z, \psi>| \ : \ \psi \in B_+(E, F), \ \psi \in (V \otimes W)^0\} \quad (z \in E \otimes F) \qquad (3.1)$$

and

$$p_V \otimes_\varepsilon q_W \leqslant p_V \otimes_{|\pi|} q_W \leqslant p_V \otimes_\pi q_W \ . \qquad (3.2)$$

Furthermore, if α is any strongly monotone seminorm on $(E \otimes F, C_\pi)$ such that

$$\alpha(x \otimes y) \leqslant p_V(x) q_W(y) \quad \text{for all} \ x \in E \ \text{and} \ y \in F, \qquad (3.3)$$

then $\alpha \leqslant p_V \otimes_{|\pi|} q_W$.

Proof. We identify $B(E, F)$ with the topological dual $(E \otimes_\pi F)'$ of $E \otimes_\pi F$, and $B_+(E, F)$ with the dual cone C'_π of C_π, formula (3.1) follows from (5.3.1).

As $\Gamma(V \otimes W) \subseteq F_\pi(\Gamma(V \otimes W))$ and $p \otimes_\pi q$ is the gauge of $\Gamma(V \otimes W)$, it follows that

$$p_V \otimes_{|\pi|} q_W \leqslant p_V \otimes_\pi q_W .$$

To prove that $p_V \otimes_\varepsilon q_W \leqslant p_V \otimes_{|\pi|} q_W$, let $z \in E \otimes F$, let $f \in V^0$ and $g \in W^0$.

Since $V^o = D(V^o)$ and $W^o = D(W^o)$, there exist $\lambda, \mu \in [0, 1]$, $f_1, f_2 \in V^o \cap E'_+$ and $g_1, g_2 \in W^o \cap F'_+$ such that

$$f = \lambda f_1 - (1 - \lambda) f_2 \quad \text{and} \quad g = \mu g_1 - (1 - \mu) g_2 .$$

Clearly

$$f_i \otimes g_j \in B_+(E, F) \quad \text{and} \quad f_i \otimes g_j \in (V \otimes W)^o \quad (i, j = 1, 2) .$$

By (3.1)

$$\begin{aligned} |\langle z, f \otimes g\rangle| &\leq \lambda\mu|\langle z, f_1 \otimes g_1\rangle| + \lambda(1 - \mu)|\langle z, f_1 \otimes g_2\rangle| \\ &\quad + \mu(1 - \lambda)|\langle z, g_1 \otimes f_2\rangle| + (1 - \lambda)(1 - \mu)|\langle z, f_2 \otimes g_2\rangle| \\ &\leq [\lambda\mu + \lambda(1 - \mu) + \mu(1 - \lambda)(1 - \lambda)(1 - \mu)] p_V \otimes_{|\pi|} q_W(z) \\ &= p_V \otimes_{|\pi|} q_W(z) . \end{aligned}$$

It then follows that

$$\begin{aligned} p_V \otimes_\varepsilon q_W(z) &= \sup\{|\langle z, f \otimes g\rangle| : f \in V^o, g \in W^o\} \\ &\leq p_V \otimes_{|\pi|} q_W(z) . \end{aligned}$$

Therefore inequalities (3.2) hold.

Finally, for any $z \in E \otimes F$ and any representation

$$z = \Sigma_{i=1}^{n} x_i \otimes y_i \quad \text{with} \quad x_i \in E \quad \text{and} \quad y_i \in F \quad (i = 1, \ldots, n) ,$$

we have

$$\alpha(z) \leq \Sigma_{i=1}^{n} \alpha(x_i \otimes y_i) \leq \Sigma_{i=1}^{n} p_V(x_i) q_W(y_i) ,$$

hence

$$\alpha(z) \leq p_V \otimes_\pi q(z) . \tag{3.4}$$

On the other hand, let

$$\Sigma_\alpha = \{z \in E \otimes F : \alpha(z) \leq 1\} .$$

Then (3.4) shows that $\Gamma(V \otimes W) \subseteq \Sigma_\alpha$, thus

$$F_\pi(\Gamma(V \otimes W)) \subseteq \Sigma_\alpha$$

since α is a strongly monotone seminorm on $(E \otimes F, C_\pi)$; consequently,

$$\alpha \leqslant p_V \otimes_{|\pi|} q_W .$$

(5.3.3) Theorem. Let $(E, E_+, \mathscr{P})$ and $(F, F_+, \mathscr{T})$ be locally o-convex spaces, let $\mathscr{U}_E$ and $\mathscr{U}_F$ be respectively neighbourhood bases at 0 for $\mathscr{P}$ and $\mathscr{T}$ consisting of o-convex and circled sets, and let $\mathscr{P} \otimes_{|\pi|} \mathscr{T}$ be the locally o-convex topology on $(E \otimes F, C_\pi)$ associated with the π-topology $\mathscr{P} \otimes_\pi \mathscr{T}$. Then $\mathscr{P} \otimes_{|\pi|} \mathscr{T}$ is determined by a family $\{p_V \otimes_{|\pi|} q_W : V \in \mathscr{U}_E,$ $W \in \mathscr{U}_F\}$ of strongly seminorms, and is the finest locally o-convex topology on $(E \otimes F, C_\pi)$ such that

$$\mathscr{P} \otimes_\varepsilon \mathscr{T} \leqslant \mathscr{P} \otimes_{|\pi|} \mathscr{T} \leqslant \mathscr{P} \otimes_\pi \mathscr{T} . \qquad (3.5)$$

Furthermore, $\mathscr{P} \otimes_{|\pi|} \mathscr{T}$ has the following universal property: for any locally o-convex space $(G, G_+, \mathcal{L})$ and any positive continuous bilinear map $\psi : E \times F \to G$, there exists a unique positive continuous linear map $T : (E \otimes F, C_\pi, \mathscr{P} \otimes_{|\pi|} \mathscr{T}) \to (G, G_+, \mathcal{L})$ such that

$$\psi = T \circ \omega ,$$

where $\omega : E \times F \to E \otimes F$ is the canonical bilinear map. $\mathscr{P} \otimes_{|\pi|} \mathscr{T}$ is the unique locally o-convex topology which has the mentioned property.

Proof. As $\{F_\pi(co(V \otimes W)) : V \in \mathscr{U}_E, W \in \mathscr{U}_F\}$ is a neighbourhood base at 0 for $\mathscr{P} \otimes_{|\pi|} \mathscr{T}$, it follows from (5.3.2) that $\mathscr{P} \otimes_{|\pi|} \mathscr{T}$ is the finest locally o-convex topology such that (3.5) holds, and is determined by

$\{p_V \otimes_{|\pi|} q_W : V \in \mathcal{U}_E, W \in \mathcal{U}_F\}$.

To verify that $\mathcal{P} \otimes_{|\pi|} \mathcal{T}$ has the universal property, we first notice from (4.4.1) that the map $T \to T \circ \omega$ is an algebraic isomorphism from $L(E \otimes_\pi F, G)$ onto $B(E \times F, G)$. As $C_\pi = co(E_+ \otimes F_+)$, it follows that $T(C_\pi) \subseteq G_+$ if and only if

$$T \circ \omega(u, v) \in G_+ \text{ for all } u \in E_+ \text{ and } v \in F_+ ;$$

in other words, T is positive if and only if $T \circ \omega$ is positive. Therefore, in order to show that $\mathcal{P} \otimes_{|\pi|} \mathcal{T}$ has the universal property, it suffices to verify that if $T \in L(E \otimes_\pi F, G)$ is positive (i.e., $T(C_\pi) \subseteq G_+$) , then T is continuous for $\mathcal{P} \otimes_{|\pi|} \mathcal{T}$ and $\mathcal{L}$.

In fact, let U be any o-convex, circled o-neighbourhood in G , and let $V \in \mathcal{U}_E$ and $W \in \mathcal{U}_F$ be such that $co(V \otimes W) \subseteq T^{-1}(U)$. As $T(C_\pi) \subseteq G_+$, it follows that $T^{-1}(W) = F_\pi(T^{-1}(W))$, and hence that

$$F_\pi(co(V \otimes W)) \subseteq T^{-1}(W) .$$

Therefore T is continuous for $\mathcal{P} \otimes_{|\pi|} \mathcal{T}$ and $\mathcal{L}$.

Finally, let $\mathcal{L}$ be any locally o-convex topology on $(E \otimes F, C_\pi)$ which has the universal property. As $(E \otimes F, C_\pi, \mathcal{P} \otimes_{|\pi|} \mathcal{T})$ is a locally o-convex space and ω is a positive continuous bilinear map from $(E \times F, E_+ \times F_+, \mathcal{P} \times \mathcal{T})$ into $(E \otimes F, C_\pi, \mathcal{P} \otimes_{|\pi|} \mathcal{T})$, it follows that the identity map on $E \otimes F$ is a continuous linear map from $(E \otimes F, C_\pi, \mathcal{L})$ onto $(E \otimes F, C_\pi, \mathcal{P} \otimes_{|\pi|} \mathcal{T})$, and hence that $\mathcal{P} \otimes_{|\pi|} \mathcal{T} \leq \mathcal{L}$. Interchanging $\mathcal{L}$ and $\mathcal{P} \otimes_{|\pi|} \mathcal{T}$, we obtain $\mathcal{L} = \mathcal{P} \otimes_{|\pi|} \mathcal{T}$, which proves the uniqueness.

Let $(E, E_+, \mathcal{P})$ and $(F, F_+, \mathcal{T})$ be locally o-convex spaces. The locally o-convex topology $\mathcal{P} \otimes_{|\pi|} \mathcal{T}$ on $(E \otimes F, C_\pi)$ associated with the π-topology $\mathcal{P} \otimes_\pi \mathcal{T}$ is referred to as the <u>ordered projective topology</u> of $\mathcal{P}$ and $\mathcal{T}$. We write $E \otimes_{|\pi|} F$ (or $(E \otimes_{|\pi|} F, C_\pi)$ when it is necessarily to emphasis the projective cone C_π) for the locally o-convex space $(E \otimes F, C_\pi, \mathcal{P} \otimes_{|\pi|} \mathcal{T})$.

As C_π' is identified with $B_+(E, F)$, it follows from a result of Namioka [1] (see Wong and Ng [1, (5.19)]) that the topological dual $(E \otimes_{|\pi|} F)'$ of $E \otimes_{|\pi|} F$ is $B_+(E, F) - B_+(E, F)$. Moreover, the $\mathcal{P} \otimes_{|\pi|} \mathcal{T}$ -closure of C_π coincide with the $\mathcal{P} \otimes_\pi \mathcal{T}$ -closure of C_π (see Wong [1, (2.15)]).

The ordered projective topology $\mathcal{P} \otimes_{|\pi|} \mathcal{T}$ of $\mathcal{P}$ and $\mathcal{T}$ had been considered by Popa (see Criteseu [1, p.224]), and parts of the preceding result were proved by Popa (see Criteseu [1, p.223]).

(5.3.4) Corollary. <u>Let</u> $(E, E_+, \mathcal{P})$ <u>and</u> $(F, F_+, \mathcal{T})$ <u>be locally solid spaces, let</u> $\mathcal{U}_E$ <u>and</u> $\mathcal{U}_F$ <u>be respectively neighbourhood bases at</u> 0 <u>for</u> $\mathcal{P}$ <u>and</u> $\mathcal{T}$ <u>consisting of solid, convex sets. Then the ordered projective topology</u> $\mathcal{P} \otimes_{|\pi|} \mathcal{T}$ <u>of</u> $\mathcal{P}$ <u>and</u> $\mathcal{T}$ <u>is a locally solid topology on</u> $(E \otimes F, C_\pi)$. <u>Furthermore, if we define for any</u> $V \in \mathcal{U}_E$ <u>and</u> $W \in \mathcal{U}_F$ <u>that</u>

$$(p_V \otimes_\pi q_W)_{S_\pi}(z) = \inf\{p_V \otimes_\pi q_W(t) : t \pm z \in C_\pi\} \quad (z \in E \otimes F), \tag{3.6}$$

<u>and</u>

$$(p_V \otimes_{|\pi|} q_W)_{S_\pi}(z) = \inf\{p_V \otimes_{|\pi|} q_W(t) : t \pm z \in C_\pi\} \quad (z \in E \otimes F), \tag{3.7}$$

<u>then</u> $\mathcal{P} \otimes_{|\pi|} \mathcal{T}$ <u>is determined by the family</u> $\{(p_V \otimes_\pi q_W)_{S_\pi} : V \in \mathcal{U}_E, W \in \mathcal{U}_F\}$ <u>of Riesz seminorms, as well as the family</u> $\{(p_V \otimes_{|\pi|} q_W)_{S_\pi} : V \in \mathcal{U}_E, W \in \mathcal{U}_F\}$ <u>of Riesz seminorms,</u>

$$p_V \otimes_\varepsilon q_W \leq p_V \otimes_{|\pi|} q_W \leq (p_V \otimes_{|\pi|} q_W)_{S_\pi} \leq (p_V \otimes_\pi q_W)_{S_\pi} \leq p_V \otimes_\pi q_W , \tag{3.8}$$

<u>and</u>

$$p_V \otimes_{|\pi|} q_W(t) = (p_V \otimes_{|\pi|} q_W)_{S_\pi}(t) \quad \underline{\text{for all}} \quad t \in C_\pi \tag{3.9}$$

<u>Proof</u>. It is known from (5.2.5)(a) that $(E \otimes F, C_\pi, \mathcal{P} \otimes_\pi \mathcal{J})$ is a locally decomposable space. As $\mathcal{P} \otimes_{|\pi|} \mathcal{J}$ is the locally o-convex topology associated with $\mathcal{P} \otimes_\pi \mathcal{J}$, it follows from Wong and Ng [1, (6.6)] that $\mathcal{P} \otimes_{|\pi|} \mathcal{J}$ coincides with the locally solid topology $(\mathcal{P} \otimes_\pi \mathcal{J})_{S_\pi}$ associated with $\mathcal{P} \otimes_\pi \mathcal{J}$, and hence that $\mathcal{P} \otimes_{|\pi|} \mathcal{J}$ is a locally solid topology on $(E \otimes F, C_\pi)$ such that the family

$$\{S_\pi(co(V \otimes W)) : V \in \mathcal{U}_E, W \in \mathcal{U}_F\} \tag{3.10}$$

is a neighbourhood base at 0 for $\mathcal{P} \otimes_{|\pi|} \mathcal{J}$ since (3.10) is a neighbourhood base at 0 for $(\mathcal{P} \otimes_\pi \mathcal{J})_{S_\pi}$. As $p_V \otimes_\pi q_W$ is the gauge of $co(V \otimes W)$, we have by Wong [1, (1.2.5)] that the functional $(p_V \otimes_\pi q_W)_{S_\pi}$, defined by (3.6), is the gauge of $S_\pi(co(V \otimes W))$, thus $\mathcal{P} \otimes_{|\pi|} \mathcal{J}$ is determined by the family $\{(p_V \otimes_\pi q_W)_{S_\pi} : V \in \mathcal{U}_E, W \in \mathcal{U}_F\}$ of Riesz seminorms.

On the other hand, it is known from Wong [1, (1.2.5)] that the functional $(p_V \otimes_{|\pi|} q_W)_{S_\pi}$, defined by (3.7), is the gauge of $S_\pi(F_\pi(co\, V \otimes W))$, hence the locally solid topology $(\mathcal{P} \otimes_{|\pi|} \mathcal{J})_{S_\pi}$ on $(E \otimes F, C_\pi)$ associated with $\mathcal{P} \otimes_{|\pi|} \mathcal{J}$ is determined by the family $\{(p_V \otimes_{|\pi|} q_W)_{S_\pi} : V \in \mathcal{U}_E, W \in \mathcal{U}_F\}$ of Riesz seminorms. As $\mathcal{P} \otimes_{|\pi|} \mathcal{J}$ is locally solid, it follows from Wong and Ng [1, (6.2)] that $\mathcal{P} \otimes_{|\pi|} \mathcal{J}$ coincides with $(\mathcal{P} \otimes_{|\pi|} \mathcal{J})_{S_\pi}$, and hence that $\mathcal{P} \otimes_{|\pi|} \mathcal{J}$ is determined by the family $\{(p_V \otimes_{|\pi|} q_W)_{S_\pi} : V \in \mathcal{U}_E, W \in \mathcal{U}_F\}$ of Riesz seminorms.

Of course, the convex hull of each absolutely dominated set is absolutely dominated, hence $co(V \otimes W)$ is absolutely dominated by (5.2.3), whenever $V \in \mathcal{U}_E$ and $W \in \mathcal{U}_F$. Also $F_\pi(co(V \otimes W))$ is convex and symmetric, hence

$$S_\pi(F_\pi(co(V \otimes W))) \subseteq F_\pi(F_\pi(co(V \otimes W))) = F_\pi(co(V \otimes W))$$

by Wong [1, (1.1.6)(f)], thus

$$co(V \otimes W) \subseteq S_\pi(co(V \otimes W)) \subseteq S_\pi(F_\pi(co(V \otimes W))) \subseteq F_\pi(co(V \otimes W)) .$$

which obtains (3.8) by making use of (3.2).

Finally, (3.7) shows that

$$(p_V \otimes_{|\pi|} q_W)_{S_\pi}(t) \leq p_V \otimes_{|\pi|} q_W(t) \quad \text{for all} \quad t \in C_\pi ,$$

thus (3.9) holds by (3.8).

(5.3.5) Corollary. _Let_ (X, X_+, p) _and_ (Y, Y_+, q) _be ordered normed spaces, let_ X_+ _be_ α-_normal in_ (X, p) _and let_ Y_+ _be_ β-_normal in_ (Y, q) . _Then_

$$q \otimes_{|\pi|} q(z) = \sup\{|\langle z, \phi\rangle| : \phi \in B_+(X, Y), \|\phi\| \leq 1\} \quad (z \in X \otimes Y) , \tag{3.11}$$

$p \otimes_{|\pi|} q$ _is a reasonable cross-norm on_ $X \otimes Y$, _and the projective cone_ C_π _of_ X_+ _and_ Y_+ _is_ 1-_normal in_ $(X \otimes Y, p \otimes_{|\pi|} q)$. _Furthermore, if in addition,_ p _and_ q _are Riesz norms, then the norm_ $\|\cdot\|_{|\pi|}$, _defined by_

$$\|z\|_{|\pi|} = \inf\{p \otimes_{|\pi|} q(t) : t \pm z \in C_\pi\} \quad (z \in X \otimes Y) , \tag{3.12}$$

is a Riesz norm on $(X \otimes Y, C_\pi)$ _which is a reasonable cross-norm, and_

$$p \otimes_{|\pi|} q(z) \leq \|z\|_{|\pi|} \leq \lambda p \otimes_{|\pi|} q(z) \quad \text{for all} \quad z \in X \otimes Y \tag{3.13}$$

for some $\lambda \geq 1$.

<u>Proof</u>. The ordered normed spaces (X, X_+, p) and (Y, Y_+, q) are locally o-convex spaces, and

$$(\Sigma_X \otimes \Sigma_Y)^\circ = \{\psi \in B(X, Y) : \|\psi\| \leq 1\} ,$$

where Σ_X and Σ_Y are respectively the unit balls in X and Y. In view of (5.3.2), the formula (3.11) holds, and $p \otimes_{|\pi|} q$ is a reasonable cross-norm.

To verify the 1-normality of C_π in $(X \otimes Y, p \otimes_{|\pi|} q)$, let

$$t_1 \leq_\pi z \leq_\pi t_2 \quad \text{with} \quad p \otimes_{|\pi|} q(t_i) \leq 1 \quad (i = 1, 2) .$$

For any $\psi \in B_+(X, Y)$ with $\|\psi\| \leq 1$, (3.11) shows that

$$|\langle z, \psi\rangle| \leq \max\{|\langle t_1, \psi\rangle| , |\langle t_2, \psi\rangle|\} \leq 1 ,$$

hence C_π is 1-normal in $(X \otimes Y, p \otimes_{|\pi|} q)$.

Finally, if p and q are Riesz norms, then

$$\|z\|_{|\pi|} = (p \otimes_{|\pi|} q)_{S_\pi} (z)$$

by (3.7) and (3.12), hence $\|\cdot\|_{|\pi|}$ is a Riesz norm on $(X \otimes Y, C_\pi)$ such that

$$p \otimes_\varepsilon q(z) \leq p \otimes_{|\pi|} q(z) \leq \|z\|_{|\pi|} \leq p \otimes_\pi q(z) ,$$

in other words, $\|\cdot\|_{|\pi|}$ is a reasonable cross-norm. Also, (5.3.4) ensures that the topology on $(X \otimes Y, C_\pi)$ induced by the norm $p \otimes_{|\pi|} q$ coincides with the topology induced by $\|\cdot\|_{|\pi|}$, hence (3.13) holds for some $\lambda \geq 1$.

Let (X, X_+, p) and (Y, Y_+, q) be ordered normed spaces such that X_+ is α-normal in (X, p) and Y_+ is β-normal in (Y, q). Then the reasonable cross-norm $p \otimes_{|\pi|} q$ on $(X \otimes Y, C_{\pi})$, defined by (3.11), is called the <u>ordered projective norm</u> of p and q.

If, in addition, p and q are Riesz norms, (3.11) and (3.12) show that $\|\cdot\|_{|\pi|}$ is the ordered projective norm in the sense of Wittstock [1]. Therefore, our terminology of ordered projective norm is different from that defined by Wittstock; in that it is required to be a Riesz norm. For Banach lattices setting, Fremlin [2] also studied this norm who call the positive-projective norm.

(5.3.6) Lemma. <u>Let</u> $(E, E_+, \mathcal{P})$ <u>and</u> $(F, F_+, \mathcal{J})$ <u>be locally decomposable spaces, and let</u> $\mathcal{U}_E$ <u>and</u> $\mathcal{U}_F$ <u>be respectively neighbourhood bases at</u> 0 <u>for</u> $\mathcal{P}$ <u>and</u> $\mathcal{J}$ <u>consisting of absolutely convex, decomposable sets. For any</u> $V \in \mathcal{U}_E$ <u>and</u> $W \in \mathcal{U}_F$, <u>let</u> $p_V \otimes_{|\varepsilon|} q_W$ <u>be the gauge of</u> $D_i((V^o \otimes W^o)^o)$. <u>Then</u>

$$p_V \otimes_{|\varepsilon|} q_W(z) = \inf\{p_V \otimes_\varepsilon q_W(t_1) + p_V \otimes_\varepsilon q_W(t_2) : t_1, t_2 \in C_i, t_1 - t_2 = z\}, \quad (3.14)$$

$$p_V \otimes_{|\varepsilon|} q_W(t) = p_V \otimes_\varepsilon q_W(t) \quad \text{for all} \quad t \in C_i, \quad (3.15)$$

<u>and</u>

$$p_V \otimes_\varepsilon q_W \leq p_V \otimes_{|\varepsilon|} q_W \leq p_V \otimes_\pi q_W. \quad (3.16)$$

<u>Proof</u>. As E_+ and F_+ are generating in E and F respectively, it follows from (5.2.1) that the biprojective cone C_i is generating in $E \otimes F$, and hence from Wong [1, (1.2.5)] that (3.14) and (3.15) hold.

It is known that $D_i((V^o \otimes W^o)^o) \subseteq (V^o \otimes W^o)^o$, and that $p_V \otimes_\varepsilon q$ is the gauge of $(V^o \otimes W^o)^o$, thus $p_V \otimes_\varepsilon q_W \leq p_V \otimes_{|\varepsilon|} q_W$. To prove that

$$p_V \otimes_{|\varepsilon|} q_W \leq p_V \otimes_\pi q_W ,$$

we first notice from (5.2.3) that each $co(V \otimes W)$ is decomposable in $(E \otimes F, C_\pi)$ As $C_\pi \subseteq C_i$ and $co(V \otimes W) \subseteq (V^o \otimes W^o)^o$, we have

$$co(V \otimes W) = D_\pi(co(V \otimes W)) \subseteq D_i(co(V \otimes W)) \subseteq D_i((V^o \otimes W^o)^o) ,$$

hence

$$p_V \otimes_{|\varepsilon|} q_W \leq p_V \otimes_\pi q_W .$$

Therefore (3.16) hold.

(5.3.7) Theorem. <u>Let</u> $(E, E_+, \mathcal{P})$ <u>and</u> $(F, F_+, \mathcal{J})$ <u>be locally decomposable spaces let</u> $\mathcal{U}_E$ <u>and</u> $\mathcal{U}_F$ <u>be respectively neighbourhood bases at</u> 0 <u>for</u> $\mathcal{P}$ <u>and</u> $\mathcal{J}$ <u>consisting of absolutely convex, decomposable sets, and let</u> $\mathcal{P} \otimes_{|\varepsilon|} \mathcal{J}$ <u>be the locally decomposable topology on</u> $(E \otimes F, C_i)$ <u>associated with the</u> ε<u>-topology</u> $\mathcal{P} \otimes_\varepsilon \mathcal{J}$. <u>Then</u> $\mathcal{P} \otimes_{|\varepsilon|} \mathcal{J}$ <u>is determined by a family</u> $\{p_V \otimes_{|\varepsilon|} q_W : V \in \mathcal{U}_E, W \in \mathcal{U}_F\}$ <u>of decomposable seminorms, and is the coarsest locally decomposable topology on</u> $(E \otimes F, C_i)$ <u>such that</u>

$$\mathcal{P} \otimes_\varepsilon \mathcal{J} \leq \mathcal{P} \otimes_{|\varepsilon|} \mathcal{J} \leq \mathcal{P} \otimes_\pi \mathcal{J} . \tag{3.17}$$

<u>Proof</u>. Since $\{(V^o \otimes W^o)^o : V \in \mathcal{U}_E , W \in \mathcal{U}_F\}$ is a neighbourhood base at 0 for $\mathcal{P} \otimes_\varepsilon \mathcal{J}$, it follows that $\{D_i((V^o \otimes W^o)^o) : V \in \mathcal{U}_E, W \in \mathcal{U}_F\}$ is a neighbourhood base at 0 for $\mathcal{P} \otimes_{|\varepsilon|} \mathcal{J}$, and hence from (5.3.6) that $\mathcal{P} \otimes_{|\varepsilon|} \mathcal{J}$ is determined by $\{p_V \otimes_{|\varepsilon|} q_W : V \in \mathcal{U}_E, W \in \mathcal{U}_F\}$. Also (3.16) of (5.3.6) shows that $\mathcal{P} \otimes_{|\varepsilon|} \mathcal{J}$ is the coarsest locally decomposable topology on $(E \otimes F, C_i)$ such that (3.17) holds.

Let $(E, E_+, \mathcal{P})$ and $(F, F_+, \mathcal{T})$ be locally decomposable spaces. The locally decomposable topology $\mathcal{P} \otimes_{|\varepsilon|} \mathcal{T}$ on $(E \otimes F, C_i)$ associated with the ε-topology $\mathcal{P} \otimes_\varepsilon \mathcal{T}$ is referred to as the ordered biprojective topology of $\mathcal{P}$ and $\mathcal{T}$. We write $E \otimes_{|\varepsilon|} F$ (or $(E \otimes_{|\varepsilon|} F, C_i)$ when it is necessarily to emphasis the biprojective cone C_i) for the locally decomposable space $(E \otimes F, C_i, \mathcal{P} \otimes_{|\varepsilon|} \mathcal{T})$.

(5.3.8) Corollary. Let $(E, E_+, \mathcal{P})$ and $(F, F_+, \mathcal{T})$ be locally solid spaces, let $\mathcal{U}_E$ and $\mathcal{U}_F$ be respectively neighbourhood bases at 0 for $\mathcal{P}$ and $\mathcal{T}$ consisting of solid, convex sets. Then the ordered biprojective topology $\mathcal{P} \otimes_{|\varepsilon|} \mathcal{T}$ of $\mathcal{P}$ and $\mathcal{T}$ is a locally solid topology on $(E \otimes F, C_i)$. Furthermore, if we define for any $V \in \mathcal{U}_E$ and $W \in \mathcal{U}_F$ that

$$(p_V \otimes_\varepsilon q_W)_{S_i}(z) = \inf\{p_V \otimes_\varepsilon q_W(t) : t \pm z \in C_i\} \quad (z \in E \otimes F), \tag{3.18}$$

then $\mathcal{P} \otimes_{|\varepsilon|} \mathcal{T}$ is determined by the family $\{(p_V \otimes_\varepsilon q_W)_{S_i} : V \in \mathcal{U}_E, W \in \mathcal{U}_F\}$ of Riesz seminorms on $(E \otimes F, C_i)$,

$$p_V \otimes_\varepsilon q_W \leq (p_V \otimes_\varepsilon q_W)_{S_i} \leq p_V \otimes_{|\varepsilon|} q_W, \quad \text{for all } V \in \mathcal{U}_E, W \in \mathcal{U}_F \tag{3.19}$$

and

$$(p_V \otimes_\varepsilon q_W)_{S_i} \leq (p_V \otimes_{|\pi|} q_W)_{S_\pi} \quad \text{for all } V \in \mathcal{U}_E, W \in \mathcal{U}_F, \tag{3.20}$$

in this case, we have

$$\mathcal{P} \otimes_{|\varepsilon|} \mathcal{T} \leq \mathcal{P} \otimes_{|\pi|} \mathcal{T} . \tag{3.21}$$

Proof. It is known from (5.2.5)(b) that $(E \otimes F, C_i, \mathcal{P} \otimes_\varepsilon \mathcal{T})$ is a locally o-convex space. As $\mathcal{P} \otimes_{|\varepsilon|} \mathcal{T}$ is the locally decomposable topology on $(E \otimes F, C_i)$ associated with $\mathcal{P} \otimes_\varepsilon \mathcal{T}$, it follows from Wong and Ng [1, (6.5)] that $\mathcal{P} \otimes_{|\varepsilon|} \mathcal{T}$ coincides with the locally solid topology on $(E \otimes F, C_i)$ associated with $\mathcal{P} \otimes_{\varepsilon} \mathcal{T}$, and hence that $\mathcal{P} \otimes_{|\varepsilon|} \mathcal{T}$ is a locally solid topology on $(E \otimes F, C_i)$ such that

$$\{S_i((V^o \otimes W^o)^o) : V \in \mathcal{U}_E, W \in \mathcal{U}_F\}$$

is a neighbourhood base at 0 for $\mathcal{P} \otimes_{|\varepsilon|} \mathcal{T}$. Since $p_V \otimes_{\varepsilon} q_W$ is the gauge of $(V^o \otimes W^o)^o$, the functional $(p_V \otimes_{\varepsilon} q_W)_{S_i}$, defined by (3.18), is the gauge of $S_i((V^o \otimes W^o)^o)$ (see Wong [1, (1.2.5)]), thus $\mathcal{P} \otimes_{|\varepsilon|} \mathcal{T}$ is determined by the family $\{(p_V \otimes_{\varepsilon} q_W)_{S_i} : V \in \mathcal{U}_E, W \in \mathcal{U}_F\}$ of Riesz seminorms on $(E \otimes F, C_i)$.

To verify (3.19), we first notice from (5.2.4) that each $(V^o \otimes W^o)^o$ is absolutely order-convex in $(E \otimes F, C_i)$, that is

$$S_i((V^o \otimes W^o)^o) \subseteq (V^o \otimes W^o)^o \quad . \tag{3.22}$$

On the other hand, since $C_\pi \subseteq C_i$ and $co(V \otimes W) \subseteq (V^o \otimes W^o)^o$, it follows that

$$p_V \otimes_{|\varepsilon|} q_W \leq p_V \otimes_{|\pi|} q_W \quad .$$

As $(V^o \otimes W^o)^o$ is convex, we have

$$D_i((V^o \otimes W^o)^o) \subseteq S_i((V^o \otimes W^o)^o) ,$$

thus (3.19) holds by (3.22) since $p_V \otimes_{|\varepsilon|} q_W$ is the gauge of $D_i((V^o \otimes W^o)^o)$ (see (5.3.6)).

Note that $C_\pi \subseteq C_i$ and that $p_V \otimes_{\varepsilon} q_W \leq p_V \otimes_{|\pi|} q_W$; it then follows from (3.7) and (3.18) that

$$(p_V \otimes_{|\pi|} q_W)_{S_\pi}(z) = \inf\{p_V \otimes_{|\pi|} q_W(t) : t \pm z \in C_\pi\}$$

$$\geq \inf\{p_V \otimes_{|\pi|} q_W(t) : t \pm z \in C_i\}$$

$$\geq \inf\{p_V \otimes_\varepsilon q_W(t) : t \pm z \in C_i\} = (p_V \otimes_\varepsilon q_W)_{S_i}$$

which obtains (3.20). Therefore (3.21) holds by (3.20) and (5.3.4) since $\mathcal{P} \otimes_{|\pi|} \mathcal{T}$ is determined by the family $\{(p_V \otimes_{|\pi|} q_W)_{S_\pi} : V \in \mathcal{U}_E, W \in \mathcal{U}_F\}$ of Riesz seminorms on $(E \otimes F, C_\pi)$.

(5.3.9) Corollary. <u>Let</u> (X, X_+, p) <u>and</u> (Y, Y_+, q) <u>be ordered normed spaces, let</u> X_+ <u>be</u> α-<u>generating in</u> (X, p) <u>and</u> Y_+ β-<u>generating in</u> (Y, q). <u>Then</u> $p \otimes_{|\varepsilon|} q$ <u>is a reasonable cross-norm on</u> $X \otimes Y$, <u>and the biprojective cone</u> C_i <u>of</u> X_+ <u>and</u> Y_+ <u>is</u> 1-<u>generating in</u> $(X \otimes Y, p \otimes_{|\varepsilon|} q)$. <u>Furthermore, if in addition,</u> p <u>and</u> q <u>are Riesz norms, then the norm</u> $\|\cdot\|_{|\varepsilon|}$, <u>defined by</u>

$$\|z\|_{|\varepsilon|} = \inf\{p \otimes_\varepsilon q(t) : t \pm z \in C_i\} \quad (z \in X \otimes Y), \tag{3.23}$$

<u>is a Riesz norm on</u> $(X \otimes Y, C_i)$ <u>which is also a reasonable cross-norm,</u>

$$\mu p \otimes_{|\varepsilon|} q(z) \leq \|z\|_{|\varepsilon|} \leq p \otimes_{|\varepsilon|} q(z) \quad (z \in X \otimes Y) \tag{3.24}$$

<u>for some</u> $\mu > 0$, <u>and</u>

$$\|z\|_{|\varepsilon|} \leq \|z\|_{|\pi|} \quad \text{for all} \quad z \in X \otimes Y. \tag{3.25}$$

<u>Proof.</u> The formula (3.16) shows that $p \otimes_{|\varepsilon|} q$ is a reasonable cross-norm, while (3.14) and (3.15) show that C_i is 1-generating in $(X \otimes Y, p \otimes_{|\varepsilon|} q)$.

If p and q are Riesz norms, then

$$\|z\|_{|\varepsilon|} = (p \otimes_\varepsilon q)_{S_i}(z) \quad \text{for all} \quad z \in X \otimes Y$$

by (3.18) and (3.23), hence the Riesz norm $\|\cdot\|_{|\varepsilon|}$ is a reasonable cross-norm by (3.19) and (3.16), and satisfies

$$\|z\|_{|\varepsilon|} \leq p \otimes_{|\varepsilon|} q(z) \quad \text{for all} \quad z \in X \otimes Y$$

by (3.19). Also (5.3.8) ensures that the topology on $(X \otimes Y, C_i)$ induced by $p \otimes_{|\varepsilon|} q$ coincides with the topology induced by $\|\cdot\|_{|\varepsilon|}$, thus (3.24) holds for some $\mu > 0$.

Observe that

$$\|z\|_{|\pi|} = (p \otimes_{|\pi|} q)_{S_\pi}(z) \quad \text{for all} \quad z \in X \otimes Y \; ;$$

the formula (3.25) then follows from (3.20).

Let (X, X_+, p) and (Y, Y_+, q) be ordered normed spaces such that X_+ is α-generating in (X, p) and Y_+ is β-generating in (Y, q). Then the reasonable cross-norm $p \otimes_{|\varepsilon|} q$ on $(X \otimes Y, C_i)$, defined by (3.14), is called the <u>ordered biprojective norm</u> of p and q.

If, in addition, p and q are Riesz norms, Wittstock [1] calls the Riesz norm $\|\cdot\|_{|\varepsilon|}$ defined by (3.23), the <u>ordered injective norm</u>.

By a <u>regular ordered normed space</u> is meant an ordered normed space (X, X_+, p) for which the norm p is a Riesz norm, that is

$$p(x) = \inf\{p(u) : u \pm x \in X_+\} \quad \text{for all} \quad x \in X .$$

Of course, regular ordered normed spaces are locally solid spaces. In view of a result of Jameson (see Wong [1, (1.1.9)]), we see that the Banach dual space of any regular ordered normed space is a regular ordered Banach space. Furthermore, if (X, X_+, p) and (Y, Y_+, q) are regular ordered normed spaces, then

$$\|\phi\| = \sup\{\phi(u, v) : u \in \Sigma_X \cap X_+,\ v \in \Sigma_Y \cap Y_+\} \quad \text{for all } \phi \in B_+(X, Y) \tag{3.26}$$

since the open unit ball in any regular ordered normed space is absolutely dominated, and thus

$$\|\psi\| \leq \|\phi\| \quad \text{whenever } \phi \pm \psi \in B_+(X, Y) , \tag{3.27}$$

namely, the bilinear norm on $B(X, Y)$ is absolutely monotone.

If (X, X_+, p) and (Y, Y_+, q) are regular ordered normed spaces, then (5.3.5) and (5.3.9) show that $(X \otimes Y, C_\pi, \|\cdot\|_{|\pi|})$ and $(X \otimes Y, C_i, \|\cdot\|_{|\varepsilon|})$ are regular ordered normed spaces, and

$$p \otimes_\varepsilon q \leq \|\ \|_{|\varepsilon|} \leq \|\cdot\|_{|\pi|} \leq p \otimes_\pi q \ ; \tag{3.28}$$

$(X \otimes Y, C_\pi, p \otimes_{|\pi|} q)$ and $(X \otimes Y, C_i, p \otimes_{|\varepsilon|} q)$ are, in general, not regular ordered normed spaces since $p \otimes_{|\pi|} q$ and $p \otimes_{|\varepsilon|} q$ are not necessarily Riesz norms, but $p \otimes_{|\pi|} q$ and $p \otimes_{|\varepsilon|} q$ are respectively equivalent to $\|\cdot\|_{|\pi|}$ and $\|\cdot\|_{|\varepsilon|}$ (see (3.13) and (3.24)). The questions arise naturally that under what conditions on X and Y, there are either $\|\cdot\|_{|\pi|} = p \otimes_\pi q$ or $\|\cdot\|_{|\varepsilon|} = p \otimes_\varepsilon q$. In order to answer these questions, we require the following terminology: We say that a regular ordered normed space (X, X_+, p) is

(a) a base normed space if p is additive on X_+, that is

$$p(u + v) = p(u) + p(v) \quad \text{for all } u, v \in X_+ ,$$

in this case, p is referred to as a base norm;

(b) an <u>order-unit normed space</u> if all elements in the open unit ball $\{x \in X : p(x) < 1\}$ are dominated by some $e \in X_+$ with $p(e) \leq 1$, in this case, e is called the order-unit, and p is called an <u>order-unit norm</u> (determined by the order-unit e) ;

(c) an <u>approximately order-unit normed space</u> if the open unit ball $\{x \in X : p(x) < 1\}$ is directed upwards, in this case, p is called an <u>approximately order-unit norm</u>.

Let (X, X_+, p) be an ordered normed space for which X_+ is complete in (X, p). Then (X, X_+, p) is a base normed space if and only if its ordered Banach dual space (X', X'_+, q^*) is an order-unit normed space as shown by Krein and Ellis (see Wong and Ng [1, (9.8)]), in this case there exists an order-unit e' in X'_+ such that

$$p(u) = \langle u, e' \rangle \quad \text{for all} \quad x \in X_+ ;$$

(X, X_+, p) is an approximately order-unit normed space if and only if (X', X'_+, p^*) is a base normed space as shown by Ng (see Wong and Ng [1, (9.9)]).

(5.3.10) Theorem. <u>Let</u> (X, X_+, p) <u>and</u> (Y, Y_+, q) <u>be base normed spaces for which</u> X_+ <u>and</u> Y_+ <u>are complete in</u> (X, p) <u>and</u> (Y, q) <u>respectively. Then</u>

(a) $(X \otimes Y, C_\pi, \|\cdot\|_{|\pi|})$ <u>is a base normed space and</u>

$$\|\cdot\|_{|\pi|} = p \otimes_\pi q \; ; \tag{3.29}$$

(b) $(X \otimes Y, C_\iota, \|\cdot\|_{|\varepsilon|})$ <u>is a base normed space</u>;

(c) <u>for any</u> $t \in C_\pi$, <u>we have</u>

$$\|t\|_{|\pi|} = \|t\|_{|\varepsilon|} \quad ;$$

<u>moreover, if in addition,</u> C_i <u>is the</u> $\|\cdot\|_{|\pi|}$<u>-closure of</u> C_π , <u>then</u>

$$\|z\|_{|\pi|} = \|z\|_{|\varepsilon|} \quad \underline{\text{for all}} \quad z \in X \otimes Y .$$

<u>Proof</u>. Let e' and d' be respectively the order-units in (X', X'_+) and (Y', Y'_+) for which

$$p(u) = \langle u, e'\rangle \quad (u \in X_+) \quad \text{and} \quad q(v) = \langle v, d'\rangle \quad (v \in Y_+) . \qquad (3.30)$$

As usual we identify $e' \otimes d'$ with a positive continuous bilinear form on $X \times Y$ obtained by the following equation

$$e' \otimes d'(x, y) = \langle x, e'\rangle\langle y, d'\rangle \quad \text{for all} \quad x \in X \quad \text{and} \quad y \in Y .$$

We claim that if $\psi \in B(X, Y)$ is such that $\|\psi\| \leq 1$, then

$$e' \otimes d' \pm \psi \in B_+(X, Y) , \qquad (3.31)$$

and that

$$\{\psi \in B_+(X, Y) : \|\psi\| \leq 1\} = \{\psi \in B_+(X, Y) : e' \otimes d' \pm \psi \in B_+(X, Y). \qquad (3.32)$$

In fact, for any $0 \neq u \in X_+$ and $0 \neq v \in Y_+$, we have from (3.23) that

$$\left|\psi\left(\frac{u}{\langle u, e'\rangle} , \frac{v}{\langle v, d'\rangle}\right)\right| \leq \|\psi\| \leq 1$$

or, equivalently,

$$|\psi(u, v)| \leq \langle u, e'\rangle\langle v, d'\rangle$$

which obtains (3.31). To prove (3.32), it has only to show, by (3.31), that if $\psi \in B_+(X, Y)$ is such that $e' \otimes d' \pm \psi \in B_+(X, Y)$, then $\|\psi\| \leq 1$. Indeed, since

$$|\psi(u, v)| \leq \langle u, e'\rangle\langle v, d'\rangle = p(u)q(v) \quad \text{for all} \quad u \in X_+ \quad \text{and} \quad v \in Y_+ ,$$

it follows from (3.26) that $\|\psi\| \leq 1$.

(a) It is clear that $\|e' \otimes d'\| \leq 1$. In view of the formulae (3.11) and (3.12) of (5.3.5), and (3.32), we obtain, for any $t \in C_\pi$, that

$$\|t\|_{|\pi|} = \sup\{<t, \psi> : \psi \in B_+(X, Y),\ e' \otimes d' \pm \psi \in B_+(X, Y)\}$$
$$\leq <t, e' \otimes d'> \leq \|t\|_{|\pi|} \qquad (3.33)$$

hence $\|\cdot\|_{|\pi|}$ is additive on C_π , and thus $(X \otimes Y, C_\pi, \|\cdot\|_{|\pi|})$ is a base normed space.

To verify (3.29), we notice that $(B(X, Y), \|\cdot\|) \equiv (X \otimes_\pi Y)'$, and from (3.31) that $B(X, Y)$ is the topological dual $(X \otimes Y, C_\pi, \|\cdot\|_{|\pi|})'$ of $(X \otimes Y, C_\pi, \|\cdot\|_{|\pi|})$. As $(X \otimes Y, C_\pi, \|\cdot\|_{|\pi|})$ is a base normed space, it follows from Krein-Ellis' theorem (see Wong and Ng [1, (9.8)]) that the ordered Banach dual space $(X \otimes Y, C_\pi, \|\cdot\|_{|\pi|})'$ is an order-unit normed space, and hence from (3.31) that

$$\{\phi \in B(X, Y) : \|\phi\| \leq 1\} \subseteq \{\phi \in B(X, Y) : \|\phi\|^*_{|\pi|} \leq 1\}$$

since $e' \otimes d'$ is an order-unit in the ordered Banach dual space $(X \otimes Y, C_\pi, \|\cdot\|_{|\pi|})$ where $\|\cdot\|^*_{|\pi|}$ is the dual norm of $\|\cdot\|_{|\pi|}$. Therefore

$$p \otimes_\pi q(z) = \sup\{|<z, \phi>| : \phi \in B(X, Y), \|\phi\| \leq 1\}$$
$$\leq \sup\{|<z, \psi>| : \psi \in B(X, Y), \|\psi\|_{|\pi|} \leq 1\} = \|z\|_{|\pi|} \quad (z \in X \otimes Y) ,$$

which obtains (3.29) since $\|\cdot\|_{|\pi|} \leq p \otimes_\pi q$.

(b) For any $t \in C_i$, (3.15) and (3.24) show that

$$\|t\|_{|\varepsilon|} = p \otimes_\varepsilon q(t) . \qquad (3.34)$$

Since the ordered Banach dual space of a base normed space is an order-unit normed space, it follows that

$$\{x' \in X' : p^*(x') \leq 1\} = \{x' \in X' : e' \pm x' \in X'_+\} \text{ and}$$
$$\{y' \in Y' : q^*(y') \leq 1\} = \{y' \in Y' : d' \pm y' \in Y'_+\} ,$$

and hence that

$$\langle t, e' \otimes d' \rangle \leq p \otimes_\varepsilon q(t) \leq \langle t, e' \otimes d' \rangle \quad \text{for all} \quad t \in C_i . \tag{3.35}$$

Therefore $\|\cdot\|_{|\varepsilon|}$ is additive on C_i by (3.34), and thus $(X \otimes Y, C_i, \|\cdot\|_{|\varepsilon|})$ is a base normed space.

(c) As $C_\pi \subseteq C_i$, it follows from (3.33), (3.34) and (3.35) that

$$\|t\|_{|\varepsilon|} = p \otimes_\varepsilon q(t) = \langle t, e' \otimes d' \rangle = \|t\|_{|\pi|} \quad \text{for all} \quad t \in C_\pi .$$

Moreover, if C_i is the $\|\cdot\|_{|\pi|}$-closure of C_π, as $\|\cdot\|_{|\pi|}$ and $\|\cdot\|_{|\varepsilon|}$ are Riesz norms, we have

$$\begin{aligned} \|z\|_{|\pi|} &= \inf\{\|t\|_{|\pi|} : t \pm z \in \overline{C}_\pi\} \\ &= \inf\{\|t\|_{|\varepsilon|} : t \pm z \in C_i\} = \|z\|_{|\varepsilon|} \quad \text{for all} \quad z \in X \otimes Y . \end{aligned}$$

Part (a) of (5.3.10) and part (a) of the following result are due to Ellis [1], while parts (b) and (c) of (5.3.10) and parts (b) and (c) of the following result are due to Wittstock [1].

(5.3.11) Theorem. Let (X, X_+, p) and (Y, Y_+, q) be order-unit normed spaces with order-units e and d in X and Y respectively. Then

(a) $(X \otimes Y, C_i, \|\cdot\|_{|\varepsilon|})$ is an order-unit normed space with the order-unit $e \otimes d$, and

$$\|\cdot\|_{|\varepsilon|} = p \otimes_\varepsilon q \tag{3.36}$$

(b) $(X \otimes Y, \overline{C}_\pi, \|\cdot\|_{|\pi|})$ _is an order-unit normed space with the order-unit_ $e \otimes d$, _where_ $\overline{C}_\pi$ _is the_ $\|\cdot\|_{|\pi|}$_-closure of_ C_π ;

(c) _if_ C_i _is the_ $\|\cdot\|_{|\pi|}$_-closure of_ C_π, _then_

$$\|z\|_{|\pi|} = \|z\|_{|\varepsilon|} \quad \text{for all} \quad z \in X \otimes Y .$$

Proof. We first notice that

$$p \otimes_\varepsilon q(e \otimes d) \leq \|e \otimes d\|_{|\varepsilon|} \leq \|e \otimes d\|_{|\pi|} \leq p \otimes_\pi q(e \otimes d) \leq p(e)q(d) \leq 1 , \tag{3.37}$$

$$p^*(u') = \sup\{<u, u'> : u \in \Sigma_X \cap X_+\} = <e, u'> \quad \text{for all} \quad u' \in X'_+ \tag{3.38}$$

and

$$q^*(v') = <d, v'> \quad \text{for all} \quad v' \in Y'_+ . \tag{3.39}$$

(a) We claim that

$$\{z \in X \otimes Y : p \otimes_\varepsilon q(z) \leq 1\} = \{z \in X \otimes Y : e \otimes d \pm z \in C_i\} . \tag{3.40}$$

In fact, since $\|\cdot\|_{|\varepsilon|}$ is a Riesz norm, it follows from (3.37) that

$$\{z \in X \otimes Y : e \otimes d \pm z \in C_i\} \subseteq \{z \in X \otimes Y : p \otimes_\varepsilon q(z) \leq 1\} .$$

If $z \in X \otimes Y$ is such that $p \otimes_\varepsilon q(z) \leq 1$, then

$$|<z, x' \otimes y'>| \leq p^*(x')q^*(y') \quad \text{for all} \quad x' \in X' \quad \text{and} \quad y' \in Y'$$

by the definition of the ε-norm. In particular, we obtain from (3.38) and (3.39) that

$$|<z, u' \otimes v'>| \leq p^*(u')q^*(v') = <e, u'><d, v'> \text{ for all } u' \in X'_+, v' \in Y'_+ ,$$

thus $e \otimes d \pm z \in C_i$.

In view of (3.40), we see that

$$\|z\|_{|\varepsilon|} \leq \|e \otimes d\|_{|\varepsilon|} \; p \otimes_\varepsilon q(z) \leq p \otimes_\varepsilon q(z) \quad \text{for all} \quad z \in X \otimes Y .$$

hence we obtain (3.36). It then follows from (3.40) that $(X \otimes Y, C_\iota, \|\cdot\|_{|\varepsilon|})$ is an order-unit normed space with the order-unit $e \otimes d$.

(b) As $B_+(X, Y) - B_+(X, Y)$ is the topological dual $(X \otimes Y, \overline{C}_\pi, \|\cdot\|_{|\pi|})$ of $(X \otimes Y, \overline{C}_\pi, \|\cdot\|_{|\pi|})$, and the open unit ball in $(X \otimes Y, \|\cdot\|_{|\pi|})$ is absolutely dominated, it has only to show that if $t \in C_\pi$ is such that $\|t\|_{|\pi|} \leq 1$, then

$$\phi(e, d) - \langle t, \phi\rangle \geq 0 \quad \text{for all} \quad \phi \in B_+(X, Y) .$$

In fact, for any $\phi \in B_+(X, Y)$, we have from (3.26) that

$$\begin{aligned} \|\phi\| &= \sup\{\phi(u, v) : u \in \Sigma_X \cap X_+ , v \in \Sigma_Y \cap Y_+\} \\ &= \phi(e, d) . \end{aligned}$$

In view of the formulae (3.11) and (3.12) of (5.3.5),

$$\|t\|_{|\pi|} = \sup\{\langle t, \phi\rangle : \phi \in B_+(X, Y), \|\phi\| \leq 1\} ,$$

thus

$$\langle t, \phi\rangle \leq \|\phi\| \, \|t\|_{|\pi|} \leq \|\phi\| = \phi(e, d)$$

which obtains our assertion.

(c) In view of parts (a) and (b), $e \otimes d$ is an order-unit in $(X \otimes Y, C_\iota)$ as well as in $(X \otimes Y, \overline{C}_\pi)$, thus

$$\begin{aligned} \|z\|_{|\varepsilon|} &= \inf\{\lambda > 0 : \lambda(e \otimes d) \pm z \in C_\iota\} \\ &= \inf\{\lambda > 0 : \lambda(e \otimes d) \pm z \in \overline{C}_\pi\} = \|z\|_{|\pi|} \end{aligned}$$

since $C_\iota = \overline{C}_\pi$ and $(X \otimes Y, C_\iota, \|\cdot\|_{|\varepsilon|})$, and $(X \otimes Y, \overline{C}_\pi, \|\cdot\|_{|\pi|})$ are order-unit normed spaces.

For approximately order-unit normed spaces we have the following result, due to Wittstock [1].

(5.3.12) Theorem. <u>Let</u> (X, X_+, p) <u>and</u> (Y, Y_+, q) <u>be regular ordered Banach spaces, let</u> $X \tilde{\otimes}_{|\pi|} Y$ <u>be the completion of</u> $(X \otimes Y, \|\cdot\|_{|\pi|})$ <u>and let</u> $\tilde{C}_\pi$ <u>be the closure of</u> C_π <u>in</u> $X \tilde{\otimes}_{|\pi|} Y$. <u>If</u> p <u>and</u> q <u>are approximately order-unit norms, then</u> $(X \tilde{\otimes}_{|\pi|} Y, \tilde{C}_\pi)$ <u>is an approximately order-unit normed space</u>.

<u>Proof</u>. As $B_+(X, Y) - B_+(X, Y)$ is the topological dual $(X \tilde{\otimes}_{|\pi|} Y, \tilde{C}_\pi)$ of $(X \tilde{\otimes}_{|\pi|} Y, \tilde{C}_\pi)$ and $B_+(X, Y)$ is the dual cone C'_π of C_π, it suffices to show, by a result of Ng (see Wong and Ng [1, (9.9)]) that the dual norm $\|\cdot\|^*_{|\pi|}$ on $B_+(X, Y) - B_+(X, Y)$ is a base norm.

In fact, let $\phi_\iota \in B_+(X, Y)$ ($\iota = 1, 2$). For any $\delta > 0$, (3.26) shows that there exist $u_\iota \in X_+$ and $v_\iota \in Y_+$ with $p(u_\iota) < 1$ and $q(v_\iota) < 1$ ($\iota = 1, 2$) such that

$$\|\phi_\iota\|^*_{|\pi|} < \phi_\iota(u_\iota, v_\iota) + \frac{\delta}{2} \quad (\iota = 1, 2). \tag{3.41}$$

Since p and q are approximately order-unit norms, there exist $u \in X_+$ and $v \in Y_+$ with $p(u) < 1$ and $q(v) < 1$ such that

$$u - u_\iota \in X_+ \quad \text{and} \quad v - v_\iota \in Y_+ \quad (\iota = 1, 2).$$

Since $\phi_\iota \in B_+(X, Y)$, it follows from (3.41) that

$$\|\phi_1\|^*_{|\pi|} + \|\phi_2\|^*_{|\pi|} < \phi_1(u_1, v_1) + \phi_2(u_2, v_2) + \delta$$
$$< \phi_1(u, v) + \phi_2(u, v) + \delta \leq \|\phi_1 + \phi_2\|^*_{|\pi|} + \delta ,$$

and hence that $\|\cdot\|^*_{|\pi|}$ is additive on $B_+(X, Y)$ since δ is arbitrary.

From (5.3.10)(c) and (5.3.11)(c), it is interesting to know under what condition on X, the $\|\cdot\|_{|\pi|}$-closure of C_π coincides with C_i. Wittstock [1] has given such a condition to ensuring that C_i is the $\|\cdot\|_{|\pi|}$-closure of C_π. Before proving this result, we require the following notation and results. Let (X, X_+, p) be a Banach lattice and $u \in X_+$. The vector subspace of X generated by the order-interval $[-u, u]$, denoted by X_u, is an ℓ-ideal in X. On X_u, we define

$$\|u\|_u = \inf\{\lambda > 0 : \lambda u \pm x \in X_+\} ,$$

then $\|\cdot\|_u$ is an order-unit norm on X_u such that

$$f(x) \leq \|x\|_u \quad \text{for all} \quad x \in X_u , \tag{3.42}$$

and $(X_u, (X_u)_+, \|\cdot\|_u)$ is a Banach lattice, where $(X_u)_+ = X_u \cap X_+$; in other words, $(X_u, (X_u)_+, \|\cdot\|_u)$ is an (AM)-space having the order-unit u. It then follows from the Kakutani representation theorem that there exists a compact Hausdorff space Ω such that the Banach lattice $C(\Omega)$ is isometrically order isomorphic to $(X_u, (X_u)_+, \|\cdot\|_u)$.

(5.3.13) Lemma. Let (X, X_+, p) be a Banach lattice, let $x_i \in X$ and $u \in X_+$ be such that $u \pm x_i \in X_+$ $(i = 1, 2, \ldots, n)$. For any $\delta > 0$ there exist $e_j \in (X_u)_+$ and $e'_j \in (X_u)'_+$ $(j = 1, \ldots, m)$ such that

$$\langle e_j , e'_k \rangle = \delta_j^{(k)} \quad (j, k = 1, \ldots, m)$$

and

$$\|x_i - \Sigma_{j=1}^{m} \langle x_i , e'_j \rangle e_j\|_u \leq \delta \quad (i = 1, \ldots, n) . \tag{3.43}$$

Proof. Let Ω be a compact Hausdorff space and let T be an isometrically order isomorphism from $(X_u, (X_u)_+, \|\cdot\|_u)$ onto $C(\Omega)$. Clearly, the map f , defined by

$$f(t) = (Tx_1(t), \ldots, Tx_n(t)) \quad \text{for all} \quad t \in \Omega ,$$

is a continuous map from Ω into $\mathbb{R}^n$, hence $K = f(\Omega)$ is a compact subset of $\mathbb{R}^n$. We now claim that there exist continuous functions $g_j : \mathbb{R} \to \mathbb{R}_+$ and $\zeta^{(j)} \in K$ $(j = 1, \ldots, m)$ with

$$\Sigma_{j=1}^{m} g_j = 1 \quad \text{and} \quad g_j(\zeta^{(k)}) = \delta_j^{(k)} \quad (j, k = 1, \ldots, m)$$

such that

$$\|a - \Sigma_{j=1}^{m} g_j(a)\, \zeta^{(j)}\|_{\mathbb{R}^n} = (\Sigma_{i=1}^{n} |a_i - \Sigma_{j=1}^{m} g_j(a)\, \zeta_i^{(j)}|^2)^{\frac{1}{2}} < \delta \quad \text{for all} \quad a = (a_1, \ldots, a_n) \in K \tag{3.44}$$

In fact, let $\zeta^{(1)} \in K$ and

$$S(\zeta^{(1)}, \delta) = \{b \in \mathbb{R}^n : \|b - \zeta^{(1)}\|_{\mathbb{R}^n} < \delta\} .$$

If $S(\zeta^{(1)}, \delta)$ does not cover K , we take $\zeta^{(2)} \in K \setminus S(\zeta^{(1)}, \delta)$. If K is not contained in $S(\zeta^{(1)}, \delta) \cup S(\zeta^{(2)}, \delta)$, we take $\zeta^{(3)} \in K \setminus (S(\zeta^{(1)}, \delta) \cup S(\zeta^{(2)}, \delta$ Since K is compact and $\|\zeta^{(i)} - \zeta^{(j)}\|_{\mathbb{R}^n} \geq \delta$ $(i \neq j)$, we obtain a finite sequence $\{\zeta^{(1)}, \ldots, \zeta^{(m)}\}$ in K with $\|\zeta^{(i)} - \zeta^{(j)}\|_{\mathbb{R}^n} \geq \delta$ such that

$$K \subset \bigcup_{j=1}^{m} S(\zeta^{(j)}, \delta) \quad . \tag{3.45}$$

Hence $\{S(\zeta^{(1)}, \delta), \ldots, S(\zeta^{(m)}, \delta), \mathbb{R}^n \setminus K\}$ is an open cover of $\mathbb{R}^n$. As $\mathbb{R}^n$

is a normal space, it follows from Dieudonne-Bochner's theorem concerning the continuous partition of unity that there exist continuous functions $g_j : \mathbb{R}^n \to \mathbb{R}_+$ $(j = 1, \ldots, m+1)$ with $\Sigma_{j=1}^{m+1} g_j = 1$ such that

$$g_j = 0 \text{ on } \mathbb{R}^n \setminus S(\zeta^{(j)}, \delta) \quad (j = 1, \ldots, m) \text{ and}$$

$$g_{m+1} = 0 \text{ on } K ;$$

in particular, $g_j(\zeta^{(k)}) = \delta_j^{(k)}$ $(j, k = 1, \ldots, m)$. For any $a \in K$, let $A = \{j : g_j(a) \neq 0\}$. Then $a \in S(\zeta^{(j)}, \delta)$ for all $j \in A$, hence

$$\begin{aligned} \|a - \Sigma_{j=1}^{m} g_j(a)\zeta^{(j)}\|_{\mathbb{R}^n} &= \|\Sigma_{j=1}^{m} g_j(a)a - \Sigma_{j=1}^{m} g_j(a)\zeta^{(j)}\|_{\mathbb{R}^n} \\ &\leq \Sigma_{j=1}^{m} g_j(a)\|a - \zeta^{(j)}\|_{\mathbb{R}^n} = \Sigma_{j\in A} g_j(a)\|a - \zeta^{(j)}\|_{\mathbb{R}^n} < \delta . \end{aligned}$$

Therefore we obtain our assertion.

Let $t_j \in \Omega$ be such that $\zeta^{(j)} = f(t_j)$ $(j = 1, \ldots, m)$. Then we have $g_j(f(t_k)) = \delta_j^{(k)}$ $(j, k = 1, \ldots, m)$, and from (3.44) that

$$\sup_{t\in\Omega} |Tx_i(t) - \Sigma_{j=1}^{m} g_j(f(t))Tx_i(t_j)| \leq \delta \quad \text{for all } i = 1, 2, \ldots, n . \tag{3.46}$$

Note that $0 \leq g_j \circ f \in C(\Omega)$ and that $\hat{t}_j$, defined by

$$\hat{t}_j(h) = h(t_j) \quad \text{for all } h \in C(\Omega) ,$$

is a positive continuous linear functional on $C(\Omega)$. There exist $e_j \in (X_u)_+$ and $e'_j \in (X_u)'_+$ such that

$$Te_j = g_j \circ f \quad \text{and} \quad T'(\hat{t}_j) = e'_j \quad (j = 1, \ldots, m)$$

where T' is the adjoint map of T. Since T is an isometry, we have from (3.46) that

$$\sup_{t\in\Omega} |Tx_i(t) - \Sigma_{j=1}^{m} Te_j(t)<Tx_i, \hat{t}_j>| = \sup_{t\in\Omega} |Tx_i(t) - \Sigma_{j=1}^{m} <x_i, e'_j>Te_j(t)|$$

$$= ||x_i - \Sigma_{j=1}^{m} <x_i, e'_j>e_j||_u \leq \delta \quad \text{for all } i = 1, 2, \ldots, n .$$

Furthermore,

$$<e_j, e'_k> = <e_j, T'(\hat{t}_k)> = <Te_j, \hat{t}_k> = <g_j \circ f, \hat{t}_k>$$

$$= g_j(t(t_k)) = \delta_j^{(k)} \quad (j, k = 1, \ldots, n).$$

This completes the proof.

(5.3.14) Lemma. Let (X, X_+, p) and (Y, Y_+, q) be regular ordered normed spaces for which X_+ and Y_+ are closed, let $C_{\pi''}$ and $C_{i''}$ be respectively the projective cone and the biprojective cone of X_+ and $(Y'')_+$, let

$$||z''||_{|\pi''|} = \inf\{p \otimes_{|\pi|} q^{**}(t) : t \pm z'' \in C_{\pi''}\} \text{ for any } z'' \in X \otimes Y'' \quad (3.47)$$

and

$$||z''||_{|\varepsilon''|} = \inf\{p \otimes_{\varepsilon} q^{**}(t) : t \pm z'' \in C_{i''}\} \text{ for any } z'' \in X \otimes Y'' . \quad (3.48)$$

Suppose further that I is the identity map on X and that $e_Y : Y \to Y''$ is the natural embedding. Then

(a) $I \otimes e_Y$ is an isometrically order embedding from $(X \otimes Y, \overline{C}_{\pi}, ||\cdot||_{|\pi|})$ into $(X \otimes Y'', \overline{C}_{\pi''}, ||\cdot||_{|\pi''|})$, and

(b) $I \otimes e_Y$ is an isometrically order embedding from $(X \otimes Y, C_i, ||\cdot||_{|\varepsilon|})$ into $(X \otimes Y'', C_{i''}, || \, ||_{|\varepsilon''|})$.

Proof. (a) For any $\phi \in B_+(X, Y)$ and $u \in X_+$, the functional f, defined by

$$f(y) = \phi(u, y) \quad \text{for all} \quad y \in Y ,$$

is a positive continuous linear functional on Y , hence the map $\tilde{\phi}$, defined by

$$\tilde{\phi}(u, y'') = \langle f, y''\rangle \quad \text{for all} \quad y'' \in Y'' ,$$

is a positive bilinear form on $X \times Y''$ such that

$$\tilde{\phi} = \phi \quad \text{on} \quad X \times Y \quad \text{and} \quad \|\tilde{\phi}\| = \|\phi\| .$$

As

$$\overline{C}_\pi = \{t \in X \otimes Y : \langle t, \phi\rangle \geq 0 \quad \text{for all} \quad \phi \in B_+(X, Y)\} ,$$

it follows that $t \in \overline{C}_\pi$ if and only if $I \otimes e_Y(t) \in \overline{C}_{\pi''}$, and hence that

$$\begin{aligned} p \otimes_{|\pi|} q(t) &= \sup\{\langle t, \phi\rangle : \phi \in B_+(X, Y), \|\phi\| \leq 1\} \\ &= \sup\{\langle I \otimes e_Y(t), \tilde{\phi}\rangle : \tilde{\phi} \in B_+(X, Y''), \|\tilde{\phi}\| \leq 1\} \\ &= p \otimes_{|\pi|} q^{**}(I \otimes e_Y(t)) \quad \text{for all} \quad t \in \overline{C}_\pi . \end{aligned}$$

Thus (a) holds by making use of (3.47) and (3.12).

(b) Y'_+ is $\sigma(Y''', Y')$-dense in $(Y''')_+$, hence

$$\langle t, u' \otimes v'\rangle \geq 0 \quad \text{for all} \quad u' \in X'_+, \ v' \in Y'_+$$

if and only if

$$\langle I \otimes e_Y(t), u' \otimes v'''\rangle \geq 0 \quad \text{for all} \quad u \in X'_+, \ v''' \in (Y''')_+ ;$$

in other words, $I \otimes e_Y$ is an order isomorphism from $(X \otimes Y, C_i)$ into $(X \otimes Y'', C_{i''})$. For any $t \in C_i$, we have from (1.12) of §4.1 that

$$p \otimes_\varepsilon q^{**}(I \otimes e_Y(t)) = p \otimes_\varepsilon q(t) ,$$

hence (b) holds by (3.48) and (3.23).

(5.3.15) Theorem. Let (X, X_+, p) be a regular ordered Banach space with the closed cone X_+. If X has the Riesz decomposition property, then for any regular ordered normed space (Y, Y_+, q) with closed cone Y_+, the biprojective cone C_i of X_+ and Y_+ is the $\|\cdot\|_{|\pi|}$-closure of the projective cone C_π of X_+ and Y_+.

Proof. We first assume that (X, X_+, p) is a Banach lattice. For any

$$t = \Sigma_{i=1}^{n} x_i \otimes y_i \in C_i ,$$

we choose an $u \in X_+$ such that $u \pm x_i \in X_+$ $(i = 1, 2, \ldots, n)$. For any $\delta > 0$, (5.3.13) shows that there exist $e_j \in (X_u)_+$ and $e'_j \in (X_u)'_+$ $(j = 1, \ldots, m)$ with $\langle e_j, e'_k\rangle = \delta_j^{(k)}$ $(j, k = 1, \ldots, m)$ such that

$$p(x_i - \Sigma_{j=1}^{m} \langle x_i, e'_j\rangle e_j) \leq \|x_i - \Sigma_{j=1}^{m} \langle x_i, e'_j\rangle e_j\|_u \leq \delta \quad (i = 1, \ldots, n) .$$

Let us define

$$d_j = \Sigma_{i=1}^{n} \langle x_i, e'_j\rangle y_i \quad (j = 1, \ldots, m) \quad \text{and} \quad s = \Sigma_{j=1}^{m} e_j \otimes d_j .$$

Note that $\{h|_{X_u} : h \in X'_+\}$ is $\sigma(X'_u, X_u)$-dense in $(X_u)'_+$ since $(X_u)_+ = X_u \cap X_+$ and the embedding map $J : X_u \to X$ is injective. As $t \in X_u \otimes Y$, it follows that

$$\langle t, e' \otimes v'\rangle \geq 0 \quad \text{for all} \quad e' \in (X_u)'_+ \quad \text{and} \quad v' \in Y'_+ ,$$

and hence from

$$\begin{aligned} \langle d_j, v'\rangle &= \Sigma_{i=1}^{n} \langle x_i, e'_j\rangle\langle y_i, v'\rangle = \langle \Sigma_{i=1}^{n} x_i \otimes y_i, e'_j \otimes v'\rangle \\ &= \langle t, e'_j \otimes v'\rangle \geq 0 \quad \text{for all } j = 1, \ldots, m \text{ and } v' \in Y'_+, \end{aligned}$$

that $d_j \in Y_+$ $(j = 1, \ldots, m)$ since Y_+ is closed. Therefore $s \in C_\pi$ and

$$\|t - s\|_{|\pi|} \leq p \otimes_\pi q(t - s) = p \otimes_\pi q(\Sigma_{i=1}^n x_i \otimes y_i - \Sigma_{i=1}^n \Sigma_{j=1}^m <x_i, e_j'>e_j \otimes y_i)$$

$$= p \otimes_\pi q(\Sigma_{i=1}^n (x_i - \Sigma_{j=1}^m <x_i, e_j'>e_j) \otimes y_i)$$

$$\leq \Sigma_{i=1}^n p(x_i - \Sigma_{j=1}^m <x_i, e_j'>e_j) q(y_i) \leq \delta \Sigma_{i=1}^n q(y_i)$$

which shows that $t \in \overline{C_\pi}$. Consequently $C_i = \overline{C_\pi}$ since $\overline{C_\pi} \subseteq C_i$ is always true.

Suppose now that X has the Riesz decomposition property. Then the ordered Banach dual of (X, X_+, p) is a Banach lattice, hence $(X'', (X'')_+, p^{**})$ is a Banach lattice, thus $\overline{C_{\pi''}} = C_{i''}$ by the first part, where $C_{\pi''}$ and $C_{i''}$ are respectively the projective cone and biprojective cone of $(X'')_+$ and Y_+. In view of (5.3.14),

$$\overline{C_\pi} = \overline{C_{\pi''}} \cap (X \otimes Y) = C_{i''} \cap (X \otimes Y) = C_i .$$

Therefore the proof is complete.

BIBLIOGRAPHY

Bellenot, S.F.

1. Factorable bounded operators and Schwartz spaces, Proc. Amer. Math. Soc. 42(1974), 551-554.

Brudovskii, B.S.

1. Associated nuclear topology, mappings of type s, and strong nuclear spaces, Dokl. Acad. Nauk SSSR (English translation) 178(1968), 271-273.
2. s-type mappings of locally convex spaces, Dokl. Acad. Nauk SSSR (English translation) 180(1968), 15-17.

Chaney, J.

1. Banach lattices of compact maps, Math. Z. 129(1972), 5-19.

Cohen, J.S.

1. Absolutely p-summing, p-nuclear operators and their conjugates, Math. Ann. 201(1973), 177-200.

Cristescu, R.

1. Topological vector spaces (Noordhoff (1977)).

Crofts, G. (see Dubinsky, Ed.,)

1. Concerning perfect Fréchet spaces and diagonal transformations, Math. Ann. 182(1969), 67-76.

Crone, L., Dubinsky, Ed. and W.B. Robinson,

1. Regular bases in products of power series spaces, J. Functional Analysis, 24(1977), 211-222.

Crone, L. and W.B. Robinson,

1. Diameters and diagonal maps in Köthe spaces.
2. Every nuclear Fréchet space with a regular basis has the quasi-equivalence property, Studia Math. T.L11(1975), 203-207.

Cross, J.,

1. The strong dual of a strongly nuclear space need not be nuclear, Proc. Amer. Math. Soc. 57(2)(1976), 250.

Day, M.M.

1. Normed linear spaces (Springer-Verlag, Berlin, 1962).

De Grande-De Kimpe, N.

1. Generalized sequence spaces, Bull. Soc. Math. Belg. 23(1971), 123-166.

2. Λ-mappings between locally convex spaces, Proc. Kon. Ned. Acad. v. Wet A74(1971), 261-274.

Diestal, J. and R.H. Lohman,

1. Applications of mapping theorem to Schwartz spaces and projections, Michigen Math. J. 20(1973), 39-44.

Djakow P. and B.S. Mitiagin,

1. Modified construction of nuclear (F)-spaces without basis, Preprint.

Dollinger, M.B.

1. Nuclear topologies consistent with a duality, Proc. Amer. Math. Soc. 23(1969), 565-568.

Douglas, R.G.

1. Banach algebra techniques in operator theory (Acad. Press, New York and London, 1972).

Dragilev, M.M.

1. On regular bases in nuclear spaces, Amer. Math. Soc. Transl. (2) 93(1970), 61-82.

Dubinsky, Ed. (see Crofts, G., Retherford, J.R. and Ramanujan, M.S.)

1. Infinite type power series subspaces of finite type power series spaces, Israel J. Math. 15(1973), 257-281.

2. Infinite type power series subspaces of infinite type power series spaces, <u>Israel J. Math</u>. 20(1975), 359-368.

3. Concrete subspaces of nuclear Frechet spaces, <u>Studia Math</u>. T.L11 (1975), 209-219.

Dubinsky, Ed. and G. Crofts,

1. Nuclear maps in sequence spaces, <u>Duke Math. J</u>. 36(1969), 207-214.

Dubinsky, Ed. and M.S. Ramanujan,

1. On λ-nuclearity, <u>Mem. Amer. Math. Soc</u>. No.128(1972).

2. Inclusion theorems for absolutely λ-summing maps, <u>Math. Ann</u>. 192 (1971), 177-190.

Dynin, A.S. and B.S. Mitiagin,

1. Criterion for nuclearity in terms of approximate dimension, <u>Bull. Acad. Polon. Sci</u>. 8(1960), 535-540.

Ellis, A.J.

1. Linear operators in partially ordered normed vector spaces, <u>J. London Math. Soc</u>. 41(1966), 323-332.

Figiel, T.

1. Factorization of compact operators and applications to the approximation problem, <u>Studia Math</u>. 45(1973), 191-210.

Fremlin, D.H.

1. Tensor products of Archimedean vector lattices, <u>Amer. J. Math</u>. 94 (1972), 777-798.

2. Tensor products of Banach lattices, <u>Math. Ann</u>. 211(1974), 87-106.

Gel'fand I.M. and N. Ya. Vilenkin,

1. <u>Generalized functions</u> vol.4 (Acad. Press, New York, London, 1964).

Greub, W.H.

1. <u>Linear Algebra</u>

Grothendieck, A.

1. Topological vector spaces (Gordon and Breach, New York, London, Paris, 1973).

2. Produits tensoriels topologiques et espaces nucleaires, Mem. Amer. Math. Soc. 16(1955).

Hasumi, M.

1. The extension property of complex Banach spaces, Tôhoku Math. J. 109(1958), 135-142.

Holub, J.R.,

1. Tensor product mappings, Math. Ann. 188(1970), 1-12.

2. Integral operators in Banach spaces, Proc. Amer. Math. Soc. 29(1) (1971), 75-80.

3. Hilbertian operators and reflexive tensor products, Pacific J. Math. 36(1)(1971), 185-194.

Horváth, J.,

1. Topological vector spaces and distributions (Addison Wesley Publishing Comp., London, Ontario, 1966).

2. Locally convex spaces (Summer School on topological vector spaces, Edited by L. Waelbroeck, Lecture Notes in Math. 331, Springer-Verlag, 1973), 41-83.

Johnson, W.B.,

1. Factorizing compact operators, Israel J. Math. 9(1971), 337-345.

Kelley, J.L.,

1. Banach spaces with the extension property. Trans. Amer. Math. Soc. 72(1952), 323-326.

Kōmura, T. and Y. Komura,

1. Über die Einbettung der nuklearen Raume in $(s)^A$, Math. Ann. 162(1966), 284-286.

Köthe, G.,

1. Topological vector spaces I (Springer-Verlag, Berlin, 1969).

2. Nuclear spaces (Lectures delivered at the University of California at Santa Barbara, 1969).

3. Über nukleare Folgenraume, Studia Math. T. XXX1(1968), 267-271.

4. Stark nukleare Folgenräume, J. Faculty of Science, Univ. of Tokyo Sec. I, vol. XVII(1970), 291-296.

5. Nukleare (F)-und (DF)-Folgenräume, Theory of sets and Topology (1972), 327-332.

Levin, V.L.

1. Tensor products and functors in categories of Banach spaces defined by KB-lineals, Trans. Moscow Math. Soc. 20(1969), 41-77.

Lindenstrauss, J. and H.P. Rasenthal,

1. The $\mathcal{L}_p$ Spaces, Israel J. Math. 7(1969), 325-349.

Lindenstrauss, J. and A. Pelczynski,

1. Absolutely summing operators in $\mathcal{L}_p$-spaces and their applications, Studia Math. 29(1968), 275-326.

Mitiagin, B.S. (see N. Zebin and A.S. Dynin),

1. Approximative dimension and bases in nuclear spaces, Usp. Mat. Nauk 16[4](1961), 73-132.

Mitiagin, B.S. and N. Zobin,

1. Contre exemple a l'existence d'un base dans un espace de Fréchet nucléaire, C.R. Acad. Sci. 279(1974), 255-256, 325-327.

Mori, Y.,

1. K̄omura's theorem in $\lambda(P)$-nuclear spaces, Math. Seminar Notes (Kobe Univ.) 5(1977), 41-46.

2. On bases in $\lambda(P)$-nuclear spaces, Math. Seminar Notes (Kobe Univ.) 5(1977), 49-58.

3. An example of a $\lambda(P)$-nuclear Fréchet space, Math. Seminar Notes (Kobe Univ.) 5(1977), 289-294.

Peressini, A.L. (see D.R. Sherbert),

1. Ordered topological vector spaces (Harper and Row, New York, 1967).

Peressini, A.L. and D.R. Sherbert,

1. Ordered topological tensor products, Proc. London Math. Soc. (3) 19(1969), 177-190.

Person, A.,

1. On some properties of p-nuclear and p-integral operators, Studia Math. T.XXXIII(1969), 213-222.

Pietsch. A.,

1. Nuclear locally convex spaces (Springer-Verlag, Berlin, 1972).

Ramanujan, M.S., (also see Dubinsky, Ed., and Terzioglu, T.,)

1. Power series spaces $\Lambda(\alpha)$ and associated $\Lambda(\alpha)$-nuclearity, Math. Ann. 169(1970), 161-168.

2. Absolutely λ-summing operators, λ a symmetric sequence space, Math. Z. 114(1970), 187-193.

3. Generalized nuclear maps in normed linear spaces, J. Reine Angew Math. 244(1970), 190-197.

Ramanujan, M.S., and T. Terzioğlu,

1. Diametral dimension of cartesian products, stability of smooth sequence spaces and applications, J. fur reine ang. Math.

Randtke, D.,

1. Characterization of precompact maps, Schwartz spaces and nuclear spaces, Trans. Amer. Math. Soc. 165(1972), 87-101.

2. A factorization theorem for compact operators, Proc. Amer. Math. Soc. 34(1)(1972), 201-202.

3. Representation theorem for compact operators, Proc. Amer. Math. Soc. 37(2)(1973), 481-485.

4. A simple example of a universal Schwartz space, Proc. Amer. Math. Soc. 37(1)(1973), 185-188.

5. A structure theorem for Schwartz spaces, Math. Ann. 201(1973), 171-176.

6. On the embedding of Schwartz spaces into product spaces, Proc. Amer. Math. Soc. 55(1)(1976), 87-92.

Robertson, A.P. and W.J. Robertson,

1. Topological vector spaces (Cambridge University Press, 1964).

Robinson, W., (see also Crone, L.),

1. On $\Lambda_1(\alpha)$-nuclearity, Duke Math. J. 40(1973), 541-546.

2. Relationships between λ-nuclearity and Pseudo-μ-nuclearity, Trans. Amer. Math. Soc. 201(1975), 291-303.

Rosenberger, B.,

1. Universal generators for rarieties of nuclear spaces, Trans. Amer. Math. Soc. 184(1973), 273-290.

Rosier, R.C.,

1. Dual spaces of certain vector sequence spaces, Pacific J. 46(2)(1973), 487-501.

Saxon, S.A.,

1. Embedding nuclear spaces in products of an arbitrary Banach space, Proc. Amer. Math. Soc. 34(1)(1972), 138-140.

Schaefer, H.H.,

1. Topological vector spaces (Springer-Verlag, Berlin, 1971, 3rd print).

2. Banach lattices and positive operators (Springer-Verlag, Berlin, 1974).

Schatten, R.,

1. A theory of cross-spaces (Princeton, 1950).

2. Norm ideals of completely continuous operators (Springer-Verlag, Berlin, 1960).

Stegall, C.P. and J.R. Retherford,

1. Fully nuclear and completely nuclear operators with applications to $\mathcal{L}$ and $\mathcal{L}_\infty$ spaces, Trans. Amer. Math. Soc. 163(1972), 457-492.

Swartz, C.,

1. Absolutely summing and dominated operators on spaces of vector-valued continuous functions, Trans. Amer. Math. Soc. 179(1973), 123-131.

Terzioğlu, T.,

1. Approximation property of co-nuclear spaces, Math. Ann. 191(1971), 35-37.

2. A characterization of compact linear mappings, Arch. Math. 22(1971), 76-78.

3. On Schwartz spaces, Math. Ann. 182(1969), 236-242.

4. Nuclear and strongly nuclear sequence spaces, Rev. Fac. Sci. Univ. Istanbul. Sec. A 34(1969), 1-5.

5. Smooth sequence spaces and associated nuclearity, Proc. Amer. Math. Soc. 37(2)(1973), 497-504.

6. Smooth sequence spaces, Proc. Symposium on Functional Analysis, Istanbul (1973), 31-41.

7. Stability of smooth sequence spaces, J. für die reine und angewandte Math. (1975), 184-189.

8. Remarks on compact and infinite-nuclear mappings, Math. Bulkanica 2(1972), 251-255.

9. Linear operators on smooth sequence spaces, Rev. Fac. Sci. Univ. Istanbul.

10. On the diametral dimension of the projective tensor products, Rev. Fac. Sci. Univ. Istanbul.

11. Symmetric bases of nuclear spaces, J. für die reine und ang. Math. 252(1972), 200-204.

Treves, F.,

1. Topological vector spaces, distributions and kernel (Acad. Press, New York and London, 1967).

Wickstead, A.W.

1. Spaces of linear operators between partially ordered Banach spaces, Proc. London Math. Soc. 28(1974), 141-158.

Wittstock, G.,

1. Ordered normed tensor products (Lecture Notes in Physics 29 Springer-Verlag, Berlin (1974), 67-84).

Wojtyński, W.,

1. On conditional bases in non-nuclear Fréchet spaces, Studia Math. 35(1970), 77-96.

Wong, Yau-Chuen,

1. The topology of uniform convergence on order-bounded sets (Lecture Notes in Math. 531, Springer-Verlag, Berlin, 1976).
2. Locally o-convex spaces, Proc. London Math. Soc. 19(1969) 289-309.

Wong, Yau-Chuen and Kung-Fu Ng,

1. Partially ordered topological vector spaces (Oxford Math. Monographs, Clarendon Press, Oxford, 1973).

INDEX AND SYMBOLS